CHEMICAL DYNAMICS AT LOW TEMPERATURES

ADVANCES IN CHEMICAL PHYSICS

VOLUME LXXXVIII

EDITORIAL BOARD

CHEMICAL DYNAMICS AT LOW TEMPERATURES

VICTOR A. BENDERSKII

Institute for Chemical Physics at Chernogolovka
Chernogolovka, Russia

DMITRII E. MAKAROV

School of Chemical Sciences
University of Illinois at Urbana-Champaign
Urbana, Illinois

CHARLES A. WIGHT

Department of Chemistry
University of Utah
Salt Lake City, Utah

ADVANCES IN CHEMICAL PHYSICS
VOLUME LXXXVIII

Series Editors

Ilya Prigogine

University of Brussels
Brussels, Belgium
and
University of Texas
Austin, Texas

Stuart A. Rice

Department of Chemistry
and
The James Franck Institute
University of Chicago
Chicago, Illinois

AN INTERSCIENCE® PUBLICATION
JOHN WILEY & SONS, INC.
NEW YORK • CHICHESTER • BRISBANE • TORONTO • SINGAPORE

Library of Congress Cataloging Number **58-9935**
ISBN 0-471-58585-8

Printed in the United States of America

10 9 8 7 6 5 4 3 2

INTRODUCTION

Few of us can any longer keep up with the flood of scientific literature, even in specialized subfields. Any attempt to do more and be broadly educated with respect to a large domain of science has the appearance of tilting at windmills. Yet the synthesis of ideas drawn from different subjects into new, powerful, general concepts is as valuable as ever, and the desire to remain educated persists in all scientists. This series, *Advances in Chemical Physics*, is devoted to helping the reader obtain general information about a wide variety of topics in chemical physics, a field which we interpret very broadly. Our intent is to have experts present comprehensive analyses of subjects of interest and to encourage the expression of individual points of view. We hope that this approach to the presentation of an overview of a subject will both stimulate new research and serve as a personalized learning text for beginners in a field.

ILYA PRIGOGINE
STUART A. RICE

PREFACE

Chemical dynamics at low temperatures is connected with elementary reactions that surmount potential energy barriers separating reactants from products in the absence of thermal activation. The first experimental evidence of this type of reactions was obtained in the early 1970s in studies of solid-state conversion of free radicals. These investigations clearly demonstrated that there is a sufficiently sharp transition from Arrhenius-like exponential temperature dependence, characteristic of thermal activation, to much weaker power-like temperature dependence down to the low-temperature limit of the rate constant.

In principle, the explanation of this phenomenon was known to be associated with tunneling through a barrier. Only after a substantial body of experimental data was accumulated were adequate models developed that elucidated the multidimensional character of tunneling and the effects of nontunneling intra- and intermolecular vibrational modes. Similar ideas have been considered independently in the quantum transition-state theory, which has been applied mainly to gas-phase reactions proceeding in the region below the energetic threshold. The supersonic jet cooling technique, in combination with high-resolution molecular spectroscopy, has revealed numerous examples of multidimensional tunneling in isolated molecules and dimers.

There are a number of specialized reviews covering the advances in each of these separate areas of research, but there is now a need for a survey of the entire field. The joint consideration of multidimensional tunneling and its manifestation in the various branches of chemical physics can provide theoreticians with a guide to a huge set of yet unsolved problems to which modern quantum mechanical methods can be applied. Experimentalists need information about the deep analogies between tunneling phenomena taking place in a variety of fields that, at first sight, might seem unrelated. Our goal is to address both of these needs, which dictates the structure of the current volume and the choice of materials.

In the first chapter, the history of the development of chemical dynamics at low temperatures is surveyed. The formulation of general problems and the main approximations used to solve them are given. The second chapter considers specific features of tunneling chemical dynamics. The results are presented without derivation. The third chapter contains a consistent description of one-dimensional tunneling in the path integral formalism. The more traditional consideration of the same problem in the WKB approximation is given in Appendix A. This chapter is designed for newcomers to the study of the quantum theory of chemical reactions. The fourth and fifth chapters are devoted to special problems of two- and multidimensional tunneling. Readers who are not interested in theoretical aspects can skip these two Chapters, because Chapter 2 contains the basic information necessary for understanding Chapters 6–9, in which pertinent experimental results are presented.

We are grateful to V. I. Goldanskii, who was the initiator of the investigation in this field, W. H. Miller., N. Makri, and D. Truhlar for useful advice, which helped us in selection of materials, and our collaborators, P. Grinevich, E. Ya. Misochko, and T. J. Tague, Jr., for support and discussions. One of us (V.A.B.) is grateful to the National Science Foundation for support and to the University of Utah for hospitality while this volume was being written. Finally, we are indebted to Anna Tapia and Alexander Benderskii for their skillful assistance in the preparation of the manuscript.

Chernogolovka, Russia   Victor A. Benderskii
Urbana, Illinois   Dmitrii E. Makarov
Salt Lake City, Utah   Charles A. Wight
January 1994

CONTENTS

CHEMICAL DYNAMICS AT LOW TEMPERATURES

ADVANCES IN CHEMICAL PHYSICS

VOLUME LXXXVIII

INTRODUCTION

CONTENTS

1.1. HISTORICAL BACKGROUND

Every chemical reaction involves the rearrangement of chemical bonds. Under ordinary circumstances, the reactants and products are stable on the time scale of their vibrational frequencies. In this respect they represent quasistationary states, and the reaction itself is associated with surmounting an energy barrier that divides them. The stability of the reactants under thermal conditions implies that the population distribution among their energy levels is close to equilibrium. For this to take place, the barrier height should be greater than both the thermal energy and the energy level spacing; that is

$$V_0 \gg \hbar\omega_0 \qquad \beta V_0 \gg 1 \qquad \beta = (k_B T)^{-1} \qquad (1.1)$$

Under these conditions the rate constant is determined by the statistically averaged reactive flux from the initial to the final state.

The quintessential expression of classical chemical kinetics is the Arrhenius law:

$$k(T) = k_0 \exp(-\beta V_0) \qquad (1.2)$$

The rate constant in this expression can be interpreted loosely as some characteristic attempt frequency multiplied by a Boltzmann factor, which represents the probability of occupying the initial states that lie just above the top of the barrier. The Arrhenius law predicts that even for the lowest barrier still satisfying Eq. (1.1) the rate constant vanishes at sufficiently low temperature. For instance, even for a very fast reaction with $k_0 = 10^{13} \, \text{s}^{-1}$, $V_0 = 1.2 \, \text{kcal/mol}$, $k = 10^{12} \, \text{s}^{-1}$ at 300 K, the rate constant decreases to $\leq 10^{-9} \, \text{s}^{-1}$ at $T = 10 \, \text{K}$. Such a low value of k completely precludes the possibly of measuring any conversion on a laboratory time scale.

1

Quantum mechanics brought new essential features to chemical reaction theory. First, the nature of the chemical bond itself has been established, and the concept of a potential energy surface (PES) on which the chemical reaction occurs has been introduced. Second, the partition functions for quantized states of the reactants and products were considered in the development of classical transition state theory (CLTST) [Eyring, 1953; Eyring et al., 1983; Glasstone et al., 1941]. This theory, developed originally by Eyring, treats only transitions that surmount the barrier with $E > V_0$. The history of CLTST has been written up by Laidler [1969]. A significant refinement of the theory was to postulate the existence of a transition state, or dividing surface, that defines a "point of no return" such that once a classical trajectory of free motion along the reaction coordinate crosses the transition state, it never returns to the reactant region [Miller, 1976; Wigner, 1938]. Third, quantum mechanics allows for tunneling, i.e., the penetration of particles through classically forbidden regions. This possibility was originally considered within CLTST only as giving rise to small corrections in the rate constant. Wigner [1932, 1938] was the first to calculate this correction for a parabolic barrier. He discovered that the apparent activation energy

$$E_a = k_B T^2 \frac{\partial \ln K}{\partial T} \tag{1.3}$$

becomes less than the barrier height and decreases with decreasing temperature, viz.,

$$E_a = V_0 - \frac{\beta}{24}(\hbar \omega_{\#})^2 \tag{1.4}$$

where $\omega_{\#}$ is the characteristic frequency of the upside-down parabolic barrier

$$V(Q) = V_0 - \tfrac{1}{2} m \omega_{\#}^2 Q^2 \tag{1.5}$$

and the reaction coordinate Q is separated from the total set of degrees of freedom. It was noted earlier [Hund, 1927; Roginsky and Rozenkevitsch, 1930] that the tunneling corrections should be noticeable in the case of high and narrow barriers, and they should be taken into account for light particle transfer reactions (e.g., for hydrogen atoms). Bell's solution [Bell, 1933, 1935, 1937] to Wigner's problem for low temperatures ($\beta \hbar \omega_{\#}/2\pi > 1$) demonstrated that at $T \to 0$ the rate coefficient is no longer subject to the Arrhenius law. However, this result was virtually unnoticed, partly because of the exotic nature of the experimental conditions (for that time), and partly because of the apparently inappropriate form of the parabolic barrier (1.5) at $E \ll V_0$, where the potential should approach a minimum.

The next three decades were a period of overwhelming domination of CLTST, which became a universal basis for microscopic descriptions of gas- and liquid-phase chemical reactions. Tunneling was drawing little attention in the context of chemical kinetics during those years. Only Bell's long-term studies, summarized in his well-known books [Bell, 1973, 1980], ascertained the two main consequences of proton tunneling in liquid-phase reactions: the lowering of the apparent activation energy and growth of the H/D kinetic isotope effect with decreasing temperature. The same rules were found by the early 1960s in gas phase reactions of H atoms [Johnston, 1960]. In both cases tunneling was considered to occur from thermally activated reactant states just below the barrier top and to give rise only to small corrections to the Arrhenius law.

In 1959 Goldanskii showed that at sufficiently low temperatures (i.e., when the population of thermally activated energy levels vanishes), the sole contribution to the reactive flux consists of tunneling from the ground state, and the rate constant approaches its low-temperature quantum limit k_c, becoming temperature-independent [Goldanskii, 1959, 1979]. The transition from the Arrhenius region to the low-temperature plateau occurs over a relatively narrow temperature range determined by a characteristic temperature T_c, which has later been called the cross over temperature. It depends not only on V_0, but also on the barrier width d and tunneling mass m, viz.,

$$\beta_c^{-1} = k_B T_c = a\left(\frac{\hbar^2 V_0}{2md^2}\right)^{1/2} \tag{1.6}$$

In this expression d is the barrier width corresponding to the zero-point energy in the initial state, and the factor a is of the order unity depending on the barrier shape. It equals $1/2$, $2/\pi$, and $3/4$ for rectangular, parabolic, and triangular barriers, respectively, and is unity for a barrier constructed from two shifted parabolas. For the parabolic barrier (1.5), Eq. (1.6) assumes the form

$$k_B T_c = \frac{\hbar\omega_\#}{2\pi} \tag{1.7}$$

The low-temperature limiting rate constant k_c is related to V_0 and T_c through the approximate formula

$$k_c = A k_0 \exp(-\beta_c V_0) \tag{1.8}$$

where the factor $A \sim (\beta_c V_0)^{-1}$ accounts for the decrease in $k(T)$ in the intermediate region between the Arrhenius dependence and low-temperature plateau. As follows from (1.6)–(1.8), the rate constant of the tunneling reaction, unlike the classical case, depends not only on the barrier height, but on the additional parameter $\hbar^2/2md^2$ (or $\hbar\omega_\#$), which does not appear in CLTST. The dependence of k_c on this parameter is as strong as on V_0. For example, k_c can vary by more than 10–12 orders of magnitude using typical ranges of d and m ($d = 0.5$–2.5 Å, $m/m_H = 1$–20) at a fixed barrier height. The low-temperature rate constant may thus extend over the entire region of experimentally accessible values. Several one-dimensional models of tunneling are presented in more detail in Chapter 3 of this volume.

Experimental studies of a large number of low-temperature solid-phase reactions undertaken by many groups in the 1970s and 1980s have confirmed the two basic predictions of Goldanskii's model for $k(T)$: the existence of a low-temperature limit k_c and a characteristic crossover temperature T_c. The aforementioned difference between quantum chemical reactions and the classical ones has also been found, namely the values of k_c for different systems range over many orders of magnitude, even for reactions with similar values of V_0 and thus with similar Arrhenius behaviors. For illustration, a number of typical experimental $k(T)$ dependencies is represented in Figure 1.1.

At $T < T_c$ tunneling manifests itself not only in irreversible chemical reactions, but also in spectroscopic splittings. Tunneling breaks the degeneracy of symmetric potential wells and gives rise to tunnelling multiplets that can be detected by a variety of spectroscopic techniques, from inelastic neutron scattering to optical and microwave spectroscopy.

The most illustrative examples of this sort are the tunneling inversion of NH_3 and rotations of methyl groups in organic molecules. In these cases it is impossible to prepare an initial nonstationary state; that is, one cannot talk about reactants and products. Rather, the coherent tunneling motion through the barriers introduces tunneling splittings Δ in the spectra that typically range from 10^{-5} to 10^2 cm^{-1} ($3 \times 10^5 - 3 \times 10^{12}$ s^{-1}). Surprisingly, they can be measured at temperatures where the thermal energy is several orders greater than Δ. At $T > T_c$ tunneling splittings are not detected because over-barrier transitions from thermally activated energy levels at $E > V_0$ effectively compete with tunneling, and thereby destroy the coherence of the motion between potential wells.

In nearly all cases, the values of k_c and T_c derived from the experimental curves $k(T)$ are not in agreement with one-dimensional tunneling calculations that utilize crystallographic and spectroscopic data to define the tunneling distance d. Furthermore, in stark contrast to the

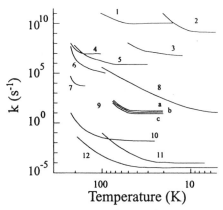

Figure 1.1 Examples of temperature dependences of rate constants for the reactions in which the low-temperature rate constant limit has been observed; 1, hydrogen transfer in excited singlet state of molecule (6.14); 2, molecular reorientation in methane crystal; 3, internal rotation of CH_3 group in radical (7.42); 4, inversion of oxyranyl radical (8.18); 5, hydrogen transfer in the excited triplet state of molecule (6.20); 6, isomerization in the excited triplet state of molecule (6.22); 7, tautomerization in the ground state of 7-azoindole dimer (6.15); 8, polymerization of formaldehyde; 9, limiting stage of chain (a) hydrobromination, (b) chlorination, and (c) bromination of ethylene; 10, isomerization of sterically hindered aryl radical (6.44); 11, abstraction of a hydrogen atom by methyl radical from a methanol matrix in reaction (6.41); 12, radical pair isomerization in dimethylglyoxime crystal (Figure 6.25).

prediction of Eq. (1.6), the crossover temperature T_c was found to have only a weak dependence on the mass of the tunneling particle. The key to understanding this behavior is in realizing that the PES of even the simplest chemical reaction is multidimensional (the dimensionality $N \geq 2$), and the potential barrier $V(Q)$ defines a saddle point that represents a maximum along Q but a minimum on the $(N-1)$-dimensional surface dividing the reactant and product valleys. Reduction of the N-dimensional reaction surface to a level that is mathematically and/or computationally tractable remains a significant challenge for theorists. A discussion of general strategies for reducing the dimensionality of the reactive surface is presented in the second section of this chapter.

In general, all the N degrees of freedom participate in the classical motion of the system, which involves, in particular, low-frequency modes such as intermolecular phonon and libron modes. These modes may be especially important when an atom or fragment is transferred between molecules in neighboring lattice sites. When the frequencies of these vibrations ω are less than $\omega_{\#}$, the Arrhenius dependence extends into the region where $T < T_c$ (albeit with smaller E_a) until these vibrations freeze out at $k_B T < \hbar\omega/2$, or until the low-frequency vibrations cease to participate in the transition.

In the first case the crossover temperature is given by

$$k_B T'_c = \frac{\hbar\omega}{2} < k_B T_c \qquad (1.9)$$

and is not connected with $\omega_\#$ at all. In the second case where the low-frequency modes decouple from the reaction, the tunneling trajectories should differ from the classical ones in that they do not pass through the saddle point. Consequently, the height of the barrier through which the particle tunnels is actually greater than that for the classical transition. Both effects show up in tunneling reactions and this explains in part why k_c and T_c differ from the values calculated within one-dimensional static-barrier models.

The recognition of these shortcomings of the one-dimensional models has spurred the development of new multidimensional models in which the reactant motion creates a dynamic barrier that has qualitatively different characteristics compared with the static barrier. A simple dynamical model of low-temperature tunneling reactions, referred to as the *fluctuational barrier preparation* model (or *vibration-assisted tunneling*) was introduced in the early 1980s [Benderskii et al., 1980; Ovchinnikova, 1979; Trakhtenberg et al., 1982]. Only more recently has the close relationship of this model to the contemporary quantum transition state theory (QTST) been recognized. The continuing development of QTST since the 1970s [Babamov and Marcus, 1981; Miller, 1974; 1983; Truhlar and Kuppermann, 1971] has stimulated both theoretical and experimental studies of quantum chemical reactions. Because these reactions, unlike the classical ones, correspond to small transition probabilities and are observed in the absence of thermal activation, many of the early results of low-temperature reactions that went almost unnoticed have at last been thrust into the mainstream of development of fundamental concepts in chemical kinetics. Advances in quantum chemical dynamics during the past decade now permit unified descriptions of such seemingly disparate phenomena as spectroscopic tunneling splitting (with transition frequencies $10^{10}-10^{11}\,s^{-1}$) and slow low-temperature chemical conversions (with rate constants as slow as $10^{-5}\,s^{-1}$).

Computations have shown that in the quantum temperature regime, different most probable transition paths (ranging from the classical minimum energy path (MEP) to the straightforward one-dimensional tunneling of early models) are possible, depending on the geometry of the potential energy surface (PES). Several extensions of the one-dimensional models to two and higher dimensions are discussed in Chapters 4 and 5.

Quantum effects in low-temperature chemistry are discussed in the recent book by Goldanskii et al. [1989a] and in reviews by Jortner and Pullman [1986], Goldanskii et al. [1989b], and Benderskii and Goldanskii [1992]. The analyses of a limited number of solid-state reactions given in the book and reviews are based on the theory of radiationless transitions [Kubo and Toyazawa, 1955], which was developed for studies of optical properties of impurity centers in crystals. These do not reflect the sweeping progress in the general quantum theory of chemical reactions. Most of the early applications of QTST were in the area of gas-phase chemical reactions, though quantum effects are not usually dominant here because of the relatively high interaction energies required by the experimental techniques. However, modern quantum theory holds great promise for examining the dynamics of low-temperature reactions, which in turn provide fertile ground for testing this theory.

Much of the experimental results in the area of low-temperature chemical reactions are collected in the review by Benderskii and Goldanskii [1992], but the spectroscopic manifestations of tunneling are not considered in detail. Nevertheless, high-resolution spectroscopy, in combination with supersonic cooling, provides unique opportunities for studying tunneling in isolated molecules and dimers at extremely low temperatures. The simplicity of these gas-phase systems in comparison with solids permits the description of tunneling effects at a higher quantitative level, thereby expanding our understanding in this field. In this volume we consider contributions of both gas-phase and condensed-phase experiments to contemporary low-temperature dynamics in a unified approach. The results of spectroscopic and kinetic studies are compared with theoretical calculations using several different types of models and levels of sophistication. In each case, the objective is to determine the true nature of the transition dynamics as $T \to 0$.

1.2. THE ROUTES OF SIMPLIFYING THE PROBLEM

The central problem of elementary chemical reaction dynamics in condensed phases is to find the rate constant of transition in the reaction complex interacting with its environment. This problem is closely related to the more general problem of statistical mechanics of irreversible processes (see, e.g., Blum [1981] and Kubo et al. [1985] for a treatment of relaxation of an initially nonequilibrium state of a particle in the presence of a heat reservoir). If the particle is coupled to the reservoir weakly enough, then the properties of the latter are fully determined by the spectral characteristics of its susceptibility coefficients. Moreover, in this linear response (weak coupling) limit any reservoir may be thought of as an infinite number of oscillators $\{q_j\}$ with an appropriately chosen spectral density, each coupled linearly in q_j to the particle coordinates. The coordinates q_j may not have a direct physical sense; they may be just unobservable variables whose role is to provide the correct response properties of the reservoir.

In a chemical reaction the role of a particle is played by the reaction complex, which itself includes many degrees of freedom. Therefore, the separation of reservoir and particle does not suffice for making the problem tractable, and the successive reduction of internal degrees of freedom in the reaction complex is required. Several possible ways to accomplish such a reduction are summarized in Table 1.1.

By convention, we divide the total system characterized by nuclear coordinates and composed of reactant complex and environment into the PES coordinates and the heat bath, by using the weak coupling criterion. Thus, the set of PES coordinates will not in general be identical to the reactant complex because it contains only those degrees of freedom that cannot be separated and put into the harmonic heat bath because of their strong coupling to the reaction coordinate. Because the calculation of any multidimensional PES is a difficult and time-consuming task, in most cases the choice of the internal PES coordinates is based on models that take into account information about the structure, barrier height, and characteristic frequencies of the reaction complex and its environment.

In most cases the consideration of low-temperature chemical reactions is, for practical purposes, restricted to a single PES. However, in general, a chemical reaction is a passage between two (or several) electronic states, each characterized by its own PES (reactant and product *diabatic terms*). Interaction between these states (*diabatic coupling*) often causes the states to avoid each other and creates two new adiabatic PES's, one of which (the lower) connects the reactant and product states in a continuous way. The upper and lower surfaces are separated by the adiabatic splitting, which is a measure of the strength of interaction between the zero-order diabatic terms. For low-temperature chemical reactions, the usual case is that the adiabatic splitting is large. Tunneling occurs through a barrier on a single adiabatic surface, and the role of nonadiabaticity is to modify the preexponential factor.

Once an adiabatic potential energy surface has been defined, the next step is to determine the minimum energy path (MEP). Within the

TABLE 1.1
Reduction of the General Quantum Dynamics Problem

Reduction steps		Problems

Initial electron–nuclear system

Born-Oppenheimer approximation

Adiabatic vs nonadiabatic transition

Initial system (nuclear coordinates)

Reaction complex	Environment

Separation of harmonic linearly coupled bath

PES coordinates	*Linear coupling*	Harmonic bath

Dissipative tunneling (quantum Kramers problem)

Separation of reaction coordinate; vibrationally adiatatic approximation

Reaction coordinate in vibrationally adiabatic potential	*Linear coupling*	Harmonic bath
		$\omega > \omega_c$	$\omega < \omega_c$

Renormalization of tunneling splitting

Two-level system	*Linear coupling*	Low-frequency harmonic bath

Reaction coordinate of fast system	*Averaging over slow subsystem configurations*	Slow subsystem coordinates

Vibration-assisted tunneling

Sudden approximation

Two-level system

Multidimensional nonseparable PES

Multidimensional tunneling

vibrationally adiabatic approximation, high-frequency modes will, in general, be effectively decoupled from the reaction so that motion along these coordinates preserves the respective quantum numbers (usually $n = 0$). Specifically, by separating degrees of freedom having frequencies

ω_j that are greater than ω_0 (and $\omega_\#$), a vibrationally adiabatic potential [Miller, 1983] can be introduced, viz.,

$$V_{\text{vad}}(Q) = V(Q) + \sum_j \frac{\hbar\omega_j(Q)}{2} \qquad (1.10)$$

If all the PES coordinates are split off in this way, the original multi-dimensional problem reduces to that of one-dimensional tunneling in the effective barrier (1.10) of a particle that is coupled to the heat bath. This problem is known as the dissipative tunneling problem, which has been a topic of intense study for the past 15 years primarily in connection with tunneling phenomena in solid state physics [Caldeira and Leggett, 1983]. Interaction with the heat bath leads to a friction force that acts on the particle moving in the one-dimensional potential (1.10). As a consequence, the barrier frequency $\omega_\#$ is replaced by the Kramers' frequency $\lambda_\#$ [Kramers, 1940] defined by

$$\lambda_\# = \omega_\# \left\{ \left[1 + \left(\frac{\eta}{2\omega_\#} \right)^2 \right]^{1/2} - \frac{\eta}{2\omega_\#} \right\} \qquad (1.11)$$

where η is the friction coefficient. According to Hanggi [1986] and Hanggi et al. [1990], the crossover temperature decreases to

$$k_B T_c = \frac{\hbar\lambda_\#}{2\pi} \qquad (1.12)$$

For CLTST to be valid, the friction should maintain thermal equilibrium in the initial state without significantly affecting the transition rate for over-barrier trajectories. Therefore, in addition to the conditions in Eq. (1.1), we impose the additional requirement

$$(\beta V_0)^{-1} < \frac{\eta}{2\omega_\#} < 1 \qquad (1.13)$$

At low temperatures ($\beta\hbar\omega_0 \gg 1$) only the lowest initial eigenstate actually participates in the transition, and the problem permits a further reduction to a two-level system (TLS) [Leggett et al., 1987; Suarez and Silbey, 1991b]. In doing so, one separates the high-frequency part of the bath spectrum (vibrations with frequencies $\omega > \omega_c$ where ω_c is a characteristic cutoff frequency). The tunneling is considered instantaneous on the time scale of the low-frequency bath vibrations ($\omega < \omega_c$) and thus is described by a tunneling matrix element Δ, which is renormalized by the high-frequency vibrations [Leggett et al., 1987]. In essence, the rate theory version based on the model of a TLS coupled to a low-frequency heat bath is a golden rule approach that utilizes the basic formula

$$k = \frac{2\pi}{\hbar} \left(\frac{\Delta}{2}\right)^2 \rho_f FC \qquad (1.14)$$

where $\Delta/2$ is the tunneling matrix element (half the tunneling splitting in the isolated TLS), ρ_f is the density of energy levels in the final state, and FC is the Franck-Condon factor (i.e., the square of the overlap integral of oscillator wave functions for the initial and final states).

With this model it is possible to relate the frequency spectrum of the bath to the temperature dependence of the rate constant. For a bath of classical oscillators ($\beta\hbar\omega_c \ll 1$) the Franck-Condon factor is proportional to $\exp(-\beta E_r/4)$, where the reorganization energy E_r is given by

$$E_r = \frac{1}{2} \sum m_j \omega_j^2 \Delta q_j^2 \qquad (1.15)$$

and Δq_j is the shift of the jth oscillator associated with transition. The heat bath reorganization therefore contributes an amount $V_0 = E_r/4$ to the barrier.

At low temperatures ($\beta\hbar\omega_c \gg 1$) the contributions from one-phonon, two-phonon processes, etc., can be systematically extracted from the general expression for the rate constant, and the type of the dominant process is determined by the bath spectrum and temperature. The results of Leggett et al. [1987] show that quantum dynamics of a TLS crucially depends on the spectrum of the bath. For sufficiently strong coupling, the bath may dramatically slow the tunneling rate, even to the point of localizing the particle in one of the wells. This strong dependence on the bath spectrum is inherent to the quantum dynamics and does not show up in classical transitions.

Whereas the reorganization energy of the bath modes may slow the tunneling rate by raising the effective barrier, there are other modes that may be efficient promoters of the tunneling by lowering the barrier and/or by reducing the tunneling distance. The dissipative tunneling model, as it is formulated in Leggett et al. [1987], does not properly account for these "promoting" vibrational modes, which are strongly coupled to the particle and do not belong to the heat bath in the above weak coupling sense. Consider, for example, tunneling of a hydrogen atom in an $OH \cdots O$ fragment, as illustrated in Figure 1.2. The height and width of the barrier for the H atom depends on the O–O distance. On the other hand, the O–O distance is the same in the initial and final states, i.e., the reorganization energy corresponding with the O–O vibration equals zero. If the promoting vibration has a high frequency, it

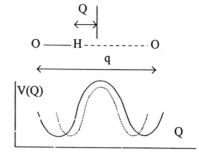

Figure 1.2. Transfer of the hydrogen atom in the linear $OH \cdots O$ fragment promoted by stretching O–O vibration. The solid curve corresponds to the equilibrium O–O distance, $q = q_0$; the dotted curve corresponds to $q < q_0$.

can be incorporated by using the vibrationally adiabatic potential along the MEP (1.10). On the other hand, if the promoting mode has a very low frequency, then tunneling occurs instantaneously on the time scale of promoting vibrations of the heavy fragments, and its rate results from averaging over the slow subsystem configurations [Benderskii et al., 1980; Ovchinnikova, 1979; Trakhtenberg et al., 1982]. The most probable tunneling path in this case differs from the MEP. This idea, later christened the sudden approximation, has recently been developed in detail by Levine et al. [1989] in connection with dissipative tunneling within the framework of quantum transition state theory. Following Siebrand et al. [1984], Suarez and Silbey [1991b], and Borgis et al. [1989], the promoting vibrations (q_p) are incorporated into the TLS model by introducing the tunneling matrix element dependence

$$\Delta = \Delta_0 \exp(-\gamma q_p) \tag{1.16}$$

where γ^{-1} is small compared to the tunneling distance. For $OH \cdots O$ fragments the value of γ^{-1} is in the range 0.03–0.04 Å [Borgis et al., 1989].

In realistic systems the separation of the modes by their frequencies and subsequent reduction to one dimension with the methods described above is often not possible. In this case an accurate multidimensional analysis is needed. Another case in which a multidimensional study is required and which obviously cannot be accounted for within the dissipative tunneling model is that of a complex PES with several saddle points and therefore several MEPs and tunneling paths. Whereas the goal of the previous models is to carry out analytical calculations and gain insight into the physical picture, the multidimensional calculations are expected to give a quantitative description of concrete chemical systems. However, at present we are just at the beginning of this process, and only a few examples of numerical multidimensional computations, mostly on rather idealized PES's, have been performed so far. Nonetheless, these primary studies have established a number of novel features of tunneling reactions, which do not show up in the effectively one-dimensional models.

The most general problem should be of the particle in a nonseparable potential, linearly coupled to an oscillator heat bath, in which the motion of the particle in the classically accessible region is subjected to frictional forces of the bath. However, this multidimensional quantum Kramers' problem has not been explored as yet.

Before concluding this section, it is worth highlighting one more basic peculiarity of quantum dynamics that distinguishes it from the more conventional classical dynamics. Because trajectories with energies $E < V_0$ are forbidden in classical mechanics, a tunneling trajectory occupies a region of phase space in which the coordinates are real and conjugate momenta imaginary. This is most conveniently done by introducing the concept of imaginary time for motion in classically inaccessible regions. An appropriate language for trajectory studies of this type is the Feynman path integral formulation of quantum mechanics [Feynman, 1972; Feynman and Hibbs, 1965] where the probability amplitude of transition equals the path integral over all possible ways of connecting the initial and final states. Much of the theoretical framework adopted for describing tunneling processes in this volume is presented within the context of this path integral approach.

The theoretical part of this volume largely follows the reduction scheme described above. In Chapter 2 we discuss the main features of tunneling reactions that distinguish them from thermally activated ones. The most salient theoretical conclusions are given without derivation. The next three chapters pursue the pedagogical task and systematically deal with the theoretical imaginary time path integral approach. In these chapters we follow mainly the review by Benderskii et al. [1993]. Chapters 6–9 then describe several different aspects of experimentally observed quantum effects ranging from hydrogen transfer to chemical reactions involving transfer of relatively heavy particles. To appreciate the subtle relationships that connect these apparently unrelated phenomena, it will help to understand the general features that characterize this relatively new field of low-temperature chemical dynamics. The conclusions that are necessary for understanding the last four chapters (6–9) are cited in Chapter 2.

FROM THERMAL ACTIVATION TO TUNNELING

CONTENTS

2.1. CROSSOVER TEMPERATURE

Although the inadequacies of the one-dimensional model [Goldanskii, 1959, 1979] are well understood by now, we begin our discussion with the simplest one-dimensional version of theory, because it will permit us to elucidate some important features of tunneling reactions inherent in more realistic approaches. The one-dimensional model relies on the following assumptions:

1. The reaction coordinate is selected from the total set of PES coordinates.

2. The energy spectra of reactants and products are continuous.

3. Thermal equilibrium is maintained in the course of transition.

4. Tunneling and over-barrier transitions proceed along the same coordinate.

Assumptions 1–3, according to Wigner [1938], form the basis of CLTST, and one of the advantages of the one-dimensional model is that it extends CLTST to the subbarrier energy region in a straightforward manner. The additional assumption needed is to satisfy the condition of Eq. (1.13).

The rate constant is a statistical average of the reactive flux from the initial to the final state:

$$k = Z_0^{-1} \int_0^\infty \rho(E) w(E) \exp(-\beta E)\, dE \qquad (2.1)$$

15

where $\rho(E)$ is the density of energy levels in the initial state, $w(E)$ is the barrier transparency at energy E, and Z_0 is the partition function in the initial state included in (2.1) to normalize the flux. In this formula the reader will readily recognize the traditional CLTST expression (see, e.g., Eyring et al. [1983]), in which the transformation from phase space variables (P, Q)

$$k_{cl} = Z_0^{-1} \int \frac{P}{m} \exp[-\beta E(P, Q)]\theta(E - V_0)\delta(Q)\, dP\, dQ \qquad (2.2)$$

to new variables (E, t) has been carried out. Here, t is the time of motion along the classical trajectory with the energy E. Equation (2.2) defines the statistically averaged flux of particles with energy $E = p^2/2m + V(Q)$ and $p > 0$ across the dividing surface at $Q^{\#} = 0$. The step function $\theta(E - V_0)$ is introduced because the classical passage is possible only at $E > V_0$. In classically forbidden regions $E < V_0$ the transmission coefficient of the barrier is exponentially small and is given by the well known WKB expression (see, e.g., Landau and Lifshitz [1981])

$$\rho(E)w(E) = (2\pi\hbar)^{-1} \exp\left[-\frac{2S(E)}{\hbar}\right] \qquad (2.3)$$

where $S(E)$ is the quasiclassical action in the barrier

$$S(E) = \int_{Q_1(E)}^{Q_2(E)} dQ\, \{2m[V(Q) - E]\}^{1/2} \qquad (2.4)$$

and $Q_1(E)$ and $Q_2(E)$ are the turning points defined by $V(Q_1) = V(Q_2) = E$ (see Figure 2.1). After substituting (2.3) and (2.4) for (2.1), the integral can be carried out using the method of steepest descent. The stationary point $E = E_a$ is given by the equation derived by Miller and

Figure 2.1. One-dimensional barrier along the coordinate of an exoergic reaction. $Q_1(E)$, $Q_1'(E)$, $Q_2(E)$, and $Q_2'(E)$ are the turning points; ω_0 and ω^* are the initial well and upside-down barrier frequencies; V_0 is the barrier height; and $-\Delta E$ is the reaction heat. Classically accessible regions are 1 and 3; the tunneling region is 2.

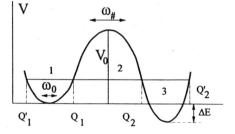

George [1972]:

$$\hbar\beta = -2\frac{\partial S(E)}{\partial E} = 2\int_{Q_1(E_a)}^{Q_2(E_a)} dQ\left\{\frac{m}{2[V(Q) - E_a]}\right\}^{1/2} \equiv \tau(E_a) \quad (2.5)$$

where $\tau(E_a)$ is the vibrational period for a classical particle at energy E_a in the upside-down barrier. Finally (2.1) reduces to

$$k = k_0 \exp\left[-\frac{2S(E_a)}{\hbar}\right] \exp(-\beta E_a) \quad (2.6)$$

where the prefactor equals

$$k_0 = Z_0^{-1}(2\pi\hbar)^{-1}\left[\frac{\pi}{(\partial^2 S/\partial E^2)_{E=E_a}}\right]^{1/2} \quad (2.7)$$

and

$$2\frac{\partial^2 S}{\partial E^2} = \frac{\partial \tau}{\partial E} = \left\{\frac{\partial^2[2S(E) + \beta\hbar E]}{\partial(\hbar\beta)^2}\right\}^{-1}$$

According to (2.6) both the apparent activation energy E_a and the apparent prefactor $k_a = k_0 \exp(-2S(E_a)/\hbar)$ decrease with decreasing temperature. As $T \to 0$, the activation energy $E_a \to 0$ and $\tau(E_a) \to \infty$, but the rate constant k given by (2.6) approaches a finite value k_c. As temperature increases, the vibrational period τ decreases. It is evident, though, that it cannot be smaller than $2\pi/\omega_\#$, where

$$\omega_\# = \left[m^{-1}\frac{d^2V(Q)}{dQ^2}\right]^{1/2} \quad (2.8)$$

This means that there is a crossover temperature defined by (1.7) (or $\hbar\beta_c = 2\pi/\omega_\#$), above which tunneling "switches off," because the quasi-classical trajectories that give the extremum to the integrand in (2.1) cease to exist. This change in the character of semiclassical motion is universal for barriers of sufficiently general shape.

The relative contribution of over-barrier $(E > V_0)$ transitions and tunneling $(E < V_0)$ to integral (2.1) is governed by the dimensionless parameter

$$\xi = \frac{\delta_T}{\delta_0} = \left(\frac{2k_B T}{\hbar\omega_\#}\right)^{1/2} \quad (2.9)$$

where δ_T is the amplitude of thermal vibrations $\delta_T^2 = k_B T / m\omega_\#^2$, and δ_0 is the corresponding zero-point amplitude $\delta_0^2 = \hbar/2m\omega_\#$ for the upside-down barrier. When $\xi \gg 1$, the amplitude of thermal vibrations is large compared with δ_0. The thermal fluctuations give rise to Arrhenius temperature dependence of the rate constant, where the activation energy is approximately equal to the barrier height. When $\xi \ll 1$, the thermal fluctuations are small, and $k(T)$ approaches the low-temperature limit k_c with $E_a = 0$. At intermediate values of $\xi \approx 1$ the major contribution to the integral in Eq. (2.1) comes from energies in the range $0 < E_a < V_0$. That is, tunneling occurs primarily from thermally activated energy levels.

Of special interest is the case where the barrier is parabolic, as in Eq. (1.5). Here, it is possible to examine the crossover between the classical and quantum regimes in detail. Note that the above derivation does not hold in this case because the integrand in (2.1) has no stationary points. Using the exact formula for the transmission coefficient of the parabolic barrier [Landau and Lifshitz, 1981]

$$w(E) = \left\{ 1 + \exp\left[\frac{2\pi(V_0 - E)}{\hbar\omega_\#} \right] \right\}^{-1} \qquad (2.10)$$

which holds above the barrier as well as below it, and taking the integral (2.1) with infinite limits, one finds

$$k = Z_0^{-1}(2\pi\hbar\beta)^{-1} \frac{\hbar\beta\omega_\#/2}{\sin(\hbar\beta\omega_\#/2)} \exp(-\beta V_0) \qquad (2.11)$$

When $\hbar\beta\omega_\#/2 \ll 1$, we recover the basic CLTST relation

$$k(T) = Z_0^{-1} \frac{k_B T}{2\pi\hbar} \exp(-\beta V_0) \qquad (2.12)$$

Using the harmonic approximation for the initial state

$$Z_0^{-1} = 2\sinh\left(\frac{\hbar\beta\omega_0}{2}\right) \qquad (2.13)$$

we find that at sufficiently high temperatures, $\hbar\beta\omega_0/2 \ll 1$, we arrive at the classical limit of the rate constant

$$k_L(T) = \frac{\omega_0}{2\pi} \exp(-\beta V_0) \qquad (2.14)$$

Equation (2.11) was derived by Wigner [1932], who pointed out that the factor

$$f = \frac{\hbar\beta\omega_\# / 2}{\sin(\hbar\beta\omega_\# / 2)} \tag{2.15}$$

constitutes the quantum corrections to CLTST. His formula (1.4) follows from (2.11) when the sine function is expanded to third order in $\hbar\beta\omega_\# / 2$. The prefactor in (2.11) diverges at the crossover temperature (1.7). This artifact occurs because the parabolic approximation is acceptable only near the top of the barrier, and at $T \ll T_c$, when the particle explores the bottom of the well, this approximation does not suffice. The role of zero-point vibrations and anharmonicities at $T \ll T_c$ was recognized by Goldanskii [1959], who used the Eckart barrier rather than the parabolic one for this reason.

Because the harmonic oscillator period is independent of energy, $\partial\tau / \partial E = 0$ for the parabolic barrier, and Eq. (2.5) therefore has no solution. Thus, at $T < T_c$ the apparent activation energy falls abruptly until it reaches the region where the potential is no longer parabolic. Figure 2.2 demonstrates that to describe variationally a realistic barrier shape (Eckart potential) by an effective parabolic one, the frequency of the latter, ω_{eff}, should drop with decreasing temperature. At high temperatures, $T > T_c$, transitions near the barrier top dominate, and the parabolic approximation with $\omega_{eff} = \omega_\#$ is accurate.

Let us now turn to the case $T \to 0$. First, it is somewhat suspicious that a continuous integral (2.1) should be combined with a discrete partition function Z_0 (2.13), the latter being customary in CLTST. This serious deficiency cannot be circumvented within the framework of this CLTST-based formalism, and a more rigorous reasoning is needed to describe the quantum situation. Adequate methods for this will be introduced within the framework of path integral formalism. Here, we simply note that formulae (2.6) and (2.7) combined with (2.13) still turn out to be correct in the quasiclassical sense [Waxman and Leggett, 1985, Hanggi and

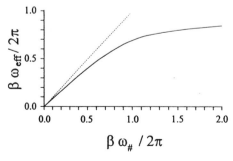

Figure 2.2. Variationally determined effective parabolic barrier frequency $\omega_{eff}^\#$ for the Eckart barrier in units $2\pi/(\hbar\beta)$. The dotted line is the high-temperature limit $\omega_{eff}^\# = \omega^\#$. (From Voth et al. [1989b].)

Hontscha, 1988; 1991]. For the prefactor in (2.6) to be finite, the exponential increase in Z_0^{-1} should be compensated by a decrease in $|\partial \tau / \partial E|^{-1/2}$. As shown in Appendix B, for β satisfying (2.5) at $\beta \to \infty$ the following equation holds:

$$-\hbar \frac{\partial \beta}{\partial E} = -\frac{\partial \tau}{\partial E} = \frac{1}{\omega_0 E} \qquad (2.16)$$

The solution of this equation is

$$E = E_0 \exp(-\beta \hbar \omega_0) \qquad (2.17)$$

where E_0 is a characteristic energy that depends on the barrier shape and is of the order of V_0. When substituting (2.16) and (2.17) for (2.6), the exponents $\exp(\pm \beta \omega_0 / 2)$ cancel out in the prefactor, and k_c takes the form

$$k_c = \frac{\omega_0}{2\pi} \left(\frac{2\pi E_0}{\hbar \omega_0}\right)^{1/2} \exp\left(-\frac{2S_0}{\hbar}\right) \qquad (2.18)$$

where S_0 is the action $S(E)$ (2.4) at $E = 0$. The square root in (2.18) is responsible for the zero-point vibrations in the initial well and is of the order of $(S_0/\hbar)^{1/2}$. Exploration of the region $0 < T < T_c$ requires numerical calculations using Eqs. (2.5) and (2.6). Since the change in prefactor is small compared to that in the leading exponential term (compare (2.14) and (2.18), the Arrhenius plot $k(\beta)$ is often drawn simply by setting $k_0 = \omega_0/2\pi$ (as in Figure 2.3). Typical dependences of the apparent prefactor k_a and the activation energy E_a on temperature are illustrated in Figure 2.4. The narrow intermediate region between the Arrhenius

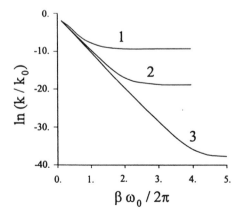

Figure 2.3. Arrhenius plot of $k(T)$ for one-dimensional barrier with $\omega^\#/\omega_0 = 1$, 0.5, and 0.25 for curves 1–3, respectively. $2\pi V_0/\hbar\omega_0 = 10$. The prefactor is constant.

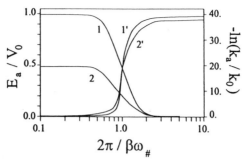

Figure 2.4. Apparent activation energy $(1, 2)$ and logarithm of apparent prefactor $\ln k_{\mathrm{a}}$ $(1', 2')$ versus temperature. The value of $2\pi V_0/\hbar\omega_0$ is 40 and 20 for curves 1, 1' and 2, 2', respectively.

behavior and the low-temperature limit has a width

$$\frac{\Delta T}{T_{\mathrm{c}}} \cong \frac{k_{\mathrm{B}} T_{\mathrm{c}}}{V_0} \tag{2.19}$$

The above discussion concerning formula (2.1) implied that tunneling transitions were incoherent and characterized by a rate constant. This is a direct consequence of assumption 2, cited at the beginning of this section. On the other hand, the study of spectroscopic manifestations of tunneling requires giving up this assumption. Consider, for example, the states in a symmetric double-well potential. Tunneling through the central barrier results in the splitting of each bound energy level by an amount given semiclassically by

$$\hbar\Delta = \frac{\hbar\omega(E)}{\pi} \exp\left(-\frac{S(E)}{\hbar}\right) \tag{2.20}$$

where E is the (nonsplit) eigenenergy in the isolated well. This splitting shows up as a doublet of spectral lines, for example, in neutron scattering spectroscopy or high-resolution optical spectroscopy.

As the temperature of such a system increases, the lines broaden and eventually coalesce to one central line. At high temperatures the width of this central line is determined by incoherent thermally activated hopping between the wells. Therefore, there should be a characteristic temperature that separates the two different regimes of incoherent hopping and coherent oscillations between the wells. The former is characterized by the rate constant k with Arrhenius behavior (2.14), while the latter is governed by the oscillation frequency Δ. Although strictly speaking, these two quantities cannot be compared directly, we suggest the following spectroscopic criterion. The doublet disappears when the thermal hopping fully "smears out" two spectral lines, i.e., when $\Delta \cong k$. The temperature T^* at which this happens comes directly from $\exp(-S/\hbar) = \exp(-V_0/k_B T^*)$. If we compare T^* with T_c given by (1.6), we shall see that T^* is twice as large. This reflects the difference between the incoherent tunneling transition, whose rate is proportional to $\exp(-2S/\hbar)$, and coherent oscillations, which have a frequency proportional to $\exp(-S/\hbar)$. In the one-dimensional model at hand the mechanism by which coherence is destroyed is associated with the interference of amplitudes in the continuous energy spectrum above the barrier. However, interaction with a reservoir of weakly coupled degrees of freedom (oscillators) provides a more efficient mechanism of phase destruction. This mechanism will be considered in Section 2.3.

The one-dimensional model for incoherent tunneling considered thus far applies to irreversible exoergic chemical reactions (i.e., those with an energy difference $E_0^f - E_0^i < 0$ between the zero-point levels of the final and initial states). For endoergic reactions the lower bound on the integral (2.1) should be replaced by $E_0^f - E_0^i$, since tunneling is possible only from initial states with $E \geq E_0^f$. When $T < T_c$, the apparent activation energy of endoergic reactions approaches its low-temperature limit equal to $E_0^f - E_0^i \leq V_0$. The Arrhenius plot therefore consists of two nearly straight lines corresponding to activation energies V_0 and $E_0^f - E_0^i$ at $T > T_c$ and $T < T_c$, respectively.

2.2.   TUNNELING AND DISSIPATION

The classical motion of a particle interacting with its environment can be phenomenologically described by the Langevin equation:

$$m\frac{d^2Q}{dt^2} + \eta\frac{dQ(t)}{dt} + \frac{dV(Q)}{dQ} = f(t) \tag{2.21}$$

where $f(t)$ is a Gaussian random force. The most remarkable advance achieved since the pioneering Kramers' paper [1940] in analyzing chemical transitions governed by (2.21) is associated with the idea that the effect of friction is totally equivalent to that of linear coupling to a bath of harmonic oscillators (see, for example, Calderia and Leggett [1983], Hanggi et al. [1990], and Dekker [1991]). A convenient quantity to characterize the overall bath effect is its spectral density:

$$J(\omega) = \frac{\pi}{2}\sum_j m_j^{-1}\omega_j^{-1}C_j^2\delta(\omega - \omega_j) \tag{2.22}$$

where C_j is a parameter of coupling to the oscillator with frequency ω_j. No information is actually available about the coupling constants in molecular crystals, and introduction of the $J(\omega)$ is a phenomenological way to account for environment effects. To obtain (2.21) one has to choose

$$J(\omega) = \eta\omega \qquad (2.23)$$

Other spectral densities correspond to memory effects in the generalized Langevin equation, which will be considered in Chapter 5. It is the equivalence of the friction force to the influence of the oscillator bath that allows one to extend (2.21) to the quantum region. Here, the friction coefficient η and $f(t)$ are related by the fluctuation–dissipation theorem (FDT):

$$\int_{-\infty}^{\infty} dt \, \langle f(t)f(0) \rangle \exp(i\omega t) = \eta\hbar\omega \coth\left(\frac{\beta\hbar\omega}{2}\right) \qquad (2.24)$$

where $\langle f(t)f(0) \rangle$ is the symmetrized correlation function (see, e.g., Chandler [1987]). In the classical limit $f(t)$ is δ-correlated:

$$\langle f(t)f(t') \rangle = 2\beta^{-1}\eta\delta(t - t') \qquad (2.25)$$

As first shown by Caldeira and Leggett [1981], friction reduces the transition probability in the tunneling regime by a factor

$$\exp\left[-\frac{A\eta(\Delta Q)^2}{\hbar}\right] \qquad (2.26)$$

where ΔQ is the tunneling distance and A is a factor of the order unity that depends on the shape of barrier. The way in which friction affects $k(T)$ can be illustrated by a simple example of a cusp-shaped parabolic term depicted in Figure 2.5. This potential is an appropriate model for a

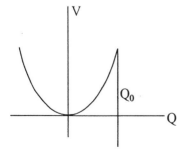

Figure 2.5. Cusp-shaped potential, made up of a parabola and a vertical wall.

strongly exoergic chemical reaction with a steep descent to the product valley. The rate constant is proportional to the probability of reaching the barrier top $Q = Q_\#$, which in turn is described by a Gaussian:

$$k \propto \exp\left(-\frac{Q_\#^2}{2\delta^2}\right) \qquad (2.27)$$

where the damped harmonic oscillator spread

$$\delta^2(\eta, \beta) = \frac{\hbar}{\pi m} \int_0^\infty d\omega \, \frac{\eta\omega}{(\omega_0^2 - \omega^2)^2 + (\eta/m)^2\omega^2} \coth\left(\frac{\beta\hbar\omega}{2}\right) \quad (2.28)$$

may be found from the FDT. This relation may be interpreted as the mean-square amplitude of a quantum harmonic oscillator $\delta^2(\omega) = \hbar/2m\omega \coth(\beta\hbar\omega/2)$ averaged over a Lorentzian distribution of the system's normal modes. In the absence of friction, (2.27) describes thermally activated as well as tunneling processes when $\beta\hbar\omega_0/2 < 1$ and > 1, respectively. At first glance it may seem surprising that the same formula holds true for the damped and undamped oscillator. This statement has been borne out by Grabert et al. [1984a], and the explanation for it lies in the aforementioned averaging. The asymptotic formulae for δ^2 are

$$\delta^2 = \begin{cases} \dfrac{k_B T}{m\omega_0^2} & k_B T \gg \hbar\omega_0 \\[3mm] \dfrac{\hbar}{2m\omega_0}\left(1 - \dfrac{\eta}{\pi m\omega_0}\right) & T = 0 \quad \dfrac{\eta}{m} \ll \omega_0 \\[3mm] \dfrac{2\hbar}{\pi\eta}\ln\left(\dfrac{\eta}{m\omega_0}\right) & T = 0 \quad \dfrac{\eta}{m} \gg \omega_0 \end{cases} \qquad (2.29)$$

According to (2.29) the effect of dissipation is to reduce the space sampled by the harmonic oscillator, making it smaller than the quantum uncertainty of position for an undamped oscillator (de Broglie wavelength). With exponential accuracy, (2.27) agrees with the Caldeira-Leggett formula (2.26), and similar expressions may be obtained for more realistic potentials.

A few comments on (2.27), (2.29), and (1.12) are appropriate at this point. The activation energy in the Arrhenius region is independent of η, since friction changes only the velocity at which a classical particle crosses the barrier and thus affects only the preexponential factor. However, friction reduces both k_c and T_c and thereby widens the Arrhenius region. Dissipation has a noticeable effect on the temperature dependence of

$k(T)$ when η is strong enough, namely when η/m is comparable to or greater than $\omega_{\#}$. Dependences $k(T)$ at various friction coefficients, calculated from the data by Grabert et al. [1987], are presented in Figure 2.6.

Friction also changes the way $k(T)$ approaches its low-temperature limit and widens the intermediate region between the two asymptotes of $k(\beta)$. At temperatures far below the crossover point, $k(T)$ behaves as

$$k(T) = k(0) \exp[A(T)] \qquad A(T) \propto T^n \qquad (2.30)$$

where n depends on the spectrum of the bath. For example, when the friction is frequency-independent, as in Eq. (2.23), the exponent is $n = 2$. For the deformation potential model (e.g., defects in a 3D crystal lattice that has a Debye phonon spectrum), $J(\omega) \propto \omega^3$, and the resulting exponent is $n = 4$. Finally, in the absence of friction, $A(T) \propto \exp(-\hbar\omega_0/k_B T)$.

Thus far we have discussed the direct mechanism of dissipation, when the reaction coordinate is coupled directly to a bath having a continuous spectrum. For chemical reactions this situation is rather rare, since low-frequency acoustic phonon modes have wavelengths much greater than the size of the reaction complex. This means that the relative displacement of any two neighboring reactant molecules is generally small. However, the direct mechanism may play an important role in long-range electron transfer reactions in dielectric media, where the reorganization energy is associated with displacement of molecules from their equilibrium positions by low-frequency polarization phonons. Another cause of friction may be anharmonicity of solids, which leads to

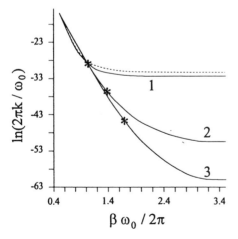

Figure 2.6. Arrhenius plot of the dissipative tunneling rate in a cubic potential with $V_0 = 5\hbar\omega_0$ and $\eta/2\omega^{\#} = 0$, 0.25, and 0.5 for curves 1–3, respectively. The crossover temperatures are indicated by asterisks. The dashed line shows $k(T)$ for parabolic barrier with the same $\omega^{\#}$ and V_0. (From Grabert et al. [1987].)

multiphonon processes. In particular, Raman processes may provide small energy losses when high-frequency motion along the reaction coordinate scatters from the lattice, exciting low-frequency phonon modes.

A discussion of the indirect dissipation mechanism is more pertinent to the present topic of incoherent tunneling reactions. In this mechanism, the reaction coordinate is coupled to one or several active modes that characterize the reaction complex. These modes are damped because of coupling to a continuous bath. The overall effect of active oscillators and bath may be represented by an effective spectral density $J_{\mathrm{eff}}(\omega)$. For instance, in the case of one harmonic active oscillator with frequency ω_1, mass m_1, and friction coefficient η, $J_{\mathrm{eff}}(\omega)$ is proportional to the imaginary part of its susceptibility and equals [Garg et al., 1985]

$$J_{\mathrm{eff}}(\omega) = m_1 \omega_1^4 \chi''(\omega) = \frac{\eta \omega \omega_1^4}{(\omega_1^2 - \omega^2)^2 + (\eta/m)^2 \omega^2} \qquad (2.31)$$

Since the susceptibilities can be extracted from optical spectra of these active modes, it is possible to develop a quantitative description of reaction dynamics based on dissipative tunneling. Such a description should consist of a detailed analysis of motion on the PES of the reaction complex while accounting for dissipation of the active modes. The advantage of this approach is that it would allow one to restrict the number of degrees of freedom in the PES to relevant modes, while incorporating effects of the environment in a phenomenological manner.

There are alternative purely phenomenological approaches to the problem of dissipation in quantum mechanics, based on modifying the Schrödinger equation so as to account for the energy losses. These approaches have been reviewed in the context of tunneling by Razavy and Pimpale [1988]. We refer an interested reader to this review for more details and mention only two basic ideas: Gisin's equation [Gisin, 1981, 1982] and the complex potential model [Razavy, 1978] (see also the review by Razavy and Pimpale [1988]). In the former, the damping coefficient is introduced in the time-dependent Schrödinger equation. Its solution yields the wave function of a damped quantum system in terms of the eigenfunctions of the same system in the absence of dissipation. The nonlinearity of Gisin's equation allows one to construct a wave packet corresponding to the initial nonstationary state and to study its decay. In the complex potential method, extra care should be taken about preserving the norm of the wave function, because the imaginary part of the potential entails a decaying wave function that cannot be properly normalized. To preserve the norm, an extra term is added to the Schrödinger equation:

$$i\hbar \frac{\partial \Psi}{\partial t} = \left[-\frac{\hbar^2}{2m} \nabla^2 + V(x) + iV_1(x) \right]\Psi - i\langle \Psi|V_1|\Psi\rangle \Psi \qquad (2.32)$$

The imaginary part of the nth energy eigenvalue obtained by solving this equation is equal to

$$\Gamma_n = -\frac{\langle \Psi_n|V_1|\Psi_n\rangle}{\langle \Psi_n|\Psi_n\rangle} \qquad (2.33)$$

2.3. COHERENT VERSUS INCOHERENT TUNNELING

Aside from the obvious computational difficulties of calculating the rate constant as $T \to 0$, the existence of a low-temperature limit for k poses a conceptual problem. This difficulty can be illustrated by comparing the cases of coherent and incoherent tunneling. If the potential has a symmetric double-well shape, for example, then quantum mechanics predicts that the probability density will experience coherent oscillations between the wells. These oscillations are associated with tunneling splitting measured spectroscopically, so the relative energy (splitting) of the stationary states provide direct information about the frequency of the coherent oscillations. Therefore, the rate constant for a simple one-dimensional system (i.e., one with no bath states) does not exist at $T = 0$, unless it is characterized by an unbound Hamiltonian. In practice, however, there are exchange chemical reactions, characterized by nearly symmetric double-well potentials, for which rate coefficients have been measured in the low-temperature limit. To account for this, one must postulate the existence of some external mechanism whose role is to destroy the phase coherence. It is here that the need to introduce a heat bath comes about.

In this section we consider in some detail the mechanism of coherence breakdown due to the bath, in order to clarify the physical assumptions that underlie the concept of rate constant at low temperatures. The particular tunneling model we choose is the two-level system (TLS) having a Hamiltonian

$$\hat{H}_0 = \frac{\hbar}{2}(\Delta_0 \sigma_x + \epsilon \sigma_z) \qquad (2.34)$$

where σ_x and σ_z are the Pauli matrices, $\Delta_0/2$ is the *tunneling matrix element* (expressed in angular frequency units), and $\hbar\epsilon$ is the energy *bias* between the two wells. The Hamiltonian (2.34) includes only the lowest energy doublet of the actual double well. Given a double well, one is able to approximate it by a TLS (2.34) when $\epsilon \ll \omega_0$, $\Delta_0 \ll \omega_0$, and, moreover, the temperature is so low that the higher energy levels are not populated, $k_B T \ll \hbar\omega_0$. The most complicated case is of no asymmetry, i.e., $\epsilon = 0$, and it is this problem with which we are most concerned. When $\epsilon = 0$, the system described by $\hat{H}_0$ has two energy levels, $E_\pm = \mp \hbar\Delta_0/2$. If the particle is initially put into the left well, the quantum mechanical amplitudes of being in the left and right wells oscillate, respectively, as

$$c_L(t) = \cos\left(\frac{\Delta_0 t}{2}\right)$$

$$c_R(t) = i \sin\left(\frac{\Delta_0 t}{2}\right) \tag{2.35}$$

and the probability that the particle can be found in the left well at time t is simply

$$P_L(t) = |c_L|^2 = \tfrac{1}{2}[1 + \cos(\Delta_0 t)] \tag{2.36}$$

The simplest scheme that accounts for the destruction of phase coherence is the so-called *stochastic interruption* model [Nikitin and Korst, 1965; Simonius, 1978; Silbey and Harris, 1989]. Suppose the process of free tunneling is interrupted by a sequence of *collisions* separated by time periods $v_0^{-1} = t_0 \ll \Delta_0^{-1}$. After each collision the system "forgets" its initial phase, i.e., the off-diagonal matrix elements of the density matrix go to zero:

$$\rho = \begin{bmatrix} |c_L|^2 & c_L c_R^* \\ c_R c_L^* & |c_R|^2 \end{bmatrix} \rightarrow \rho' = \begin{bmatrix} |c_L|^2 & 0 \\ 0 & |c_R|^2 \end{bmatrix} \tag{2.37}$$

The density matrix ρ describes a pure state, as seen from the equality $\rho^2 = \rho$, while ρ' does not. The transition described by (2.37) corresponds to the *strong collision* limit in which the particle is completely localized. In the more general case, the off-diagonal elements may not completely vanish. This, however, does not affect the qualitative picture.

After each strong collision, the system, having been localized in the left or right well, resumes free tunneling from the diagonal state. Thus, after N collisions the probability of surviving in the left well is

$$P_L(t) = \left[\cos^2\left(\frac{\Delta_0 t_0}{2}\right)\right]^N \cong \left[1 - \frac{1}{8}(\Delta_0 t_0)^2\right]^{2N} \cong \exp\left(-\frac{\Delta_0^2 t_0 t}{4}\right) \tag{2.38}$$

where $t = Nt_0$. The first-order rate coefficient for this population decay is therefore

$$k = \frac{\Delta_0^2}{4\nu_0} \qquad (2.39)$$

This simple gas-phase model asserts that the rate constant is proportional to the square of tunneling matrix element divided by some characteristic bath frequency.

Now, to put this model on a firmer foundation and make it more realistic, we specify the total Hamiltonian (including the TLS and bath) to be

$$\hat{H} = \hat{H}_0 + \hat{H}_b + \hat{H}_{int} \qquad \hat{H}_{int} = \hat{f}\sigma_z \qquad (2.40)$$

where $\hat{H}_b$ is the free bath Hamiltonian and the operator $\hat{f}$ acts on the bath variables only. The environment, as it were, "observes" whether the particle is on the right or on the left, through the interaction $\hat{f}\sigma_z$, and thereby disrupts the interference between the two eigenstates of the matrix σ_z. Of particular interest is the model as a set of harmonic oscillators q_j with frequencies ω_j that are linearly coupled to the tunneling coordinate:

$$\hat{H}_b = \sum \frac{p_j^2}{2m_j} + \frac{m_j\omega_j^2 q_j^2}{2} \qquad \hat{f} = \sum C_j Q_0 q_j \qquad (2.41)$$

where $2Q_0$ is the interwell distance. The quantity $Q = Q_0\sigma_z$ is simply the coordinate of the particle, which, in the present approximation, takes two values, $\pm Q_0$. The bath is characterized by its spectral density $J(\omega)$ given by Eq. (2.22), which is proportional to the mean square of force acting on a particle from oscillators with frequency ω. In the solid state, the spectral density is related to the phonon density $\rho(\omega)$ by

$$J(\omega) = \frac{\pi}{2} \frac{C^2(\omega)\rho(\omega)}{m\omega} \qquad (2.42)$$

This model, called the spin-boson Hamiltonian, is probably the only problem (except maybe for some very artificial ones) whose full solution can be obtained without any additional approximations. The equation of motion for the expectation value $\langle \sigma_z \rangle$ in the weak coupling limit has a

Langevin-like form:

$$\frac{d^2\langle\sigma_z\rangle}{dt^2} + \eta_{TLS}\frac{d\langle\sigma_z\rangle}{dt} + \Delta_0^2\langle\sigma_z\rangle = 0 \tag{2.43}$$

where the damping coefficient η_{TLS} is determined by the spectral density at $\omega = \Delta_0$ [Silbey and Harris, 1983] according to

$$\eta_{TLS} = \frac{\pi}{\hbar}Q_0^2\coth\left(\frac{\beta\hbar\Delta_0}{2}\right)\sum\frac{C_j^2\delta(\omega_j-\Delta_0)}{m_j\omega_j}$$

$$= \frac{2}{\hbar}Q_0^2 J(\Delta_0)\coth\left(\frac{\beta\hbar\Delta_0}{2}\right) \tag{2.44}$$

This damping coefficient is simply the rate constant of transitions between the energy levels of the doublet, and it may be represented as

$$\eta_{TLS} = k_\uparrow + k_\downarrow \tag{2.45}$$

where $k_\uparrow$ is the probability (per unit time) to escape from the lower energy level to the upper one, and $k_\downarrow$ is the probability of reverse transition. The explicit form for $k_\uparrow$ and $k_\downarrow$ is the golden rule:

$$k_\uparrow = \frac{2\pi}{\hbar}\sum_j\sum_{n=1}^{\infty}|\langle n-1|C_jQ_0q_j|n\rangle|^2\delta(\Delta_0-\omega_j)Z_j^{-1}\exp\left[-\beta\hbar\omega_j\left(n+\frac{1}{2}\right)\right]$$

$$= \frac{2}{\hbar}Q_0^2 J(\Delta_0)[\exp(\beta\hbar\Delta_0)-1]^{-1} \tag{2.46}$$

where Z_j is the partition function of jth oscillator. Using the principle of microscopic reversibility and detailed balance we may write

$$k_\downarrow = k_\uparrow\exp(\beta\hbar\Delta_0) \tag{2.47}$$

Conservation of energy demands that each intradoublet transition is accompanied by emission or absorption of a phonon with energy $\hbar\Delta_0$. This requirement is taken into account by the δ function in Eq. (2.46).

Equation (2.43) describes a system that decays either by damped oscillations (when $\eta_{TLS} < 2\Delta_0$) or by exponential relaxation ($\eta_{TLS} > 2\Delta_0$). Since η_{TLS} grows with increasing temperature, there may be a crossover between these two regimes at β^* such that $2\hbar^{-1}Q_0^2 J(\Delta_0)\coth(\beta^*\hbar\Delta_0/$

2) $= 2\Delta_0$. In the case where the friction coefficient is large,[1] i.e., $\eta \gg 2\Delta_0$, then the long time behavior of the solution to (2.43) is determined by the exponent $\exp(-\Delta_0^2 t/\eta_{TLS})$. The rate constant is then

$$k = \frac{\Delta_0^2}{\eta_{TLS}} \qquad (2.48)$$

Expression (2.48) has the same form as (2.39) if we define the collision frequency as $\nu_0 = \eta_{TLS}/4$. Both of these formulae can be expressed in the golden rule form (cf. (1.14))

$$k = 2\pi\hbar\left(\frac{\Delta_0}{2}\right)^2 \rho_f \qquad (2.49)$$

if the density of final states ρ_f takes the form

$$\rho_f = \frac{2}{\pi\hbar\eta_{TLS}} = \frac{1}{2\pi\hbar\nu_0} \qquad (2.50)$$

The identities in (2.50) are very illuminating. The first one illustrates that the coupling to bath modes broadens the energy levels of the TLS, such that the density of final states is inversely proportional to the damping coefficient. The second shows that the mean level spacing of the system is formally equal to $2\pi\hbar\nu_0$, where ν_0 is the characteristic bath frequency. Finally, the same results (2.48)–(2.50) are obtained if one supposes that the energy level in the right well (in the original basis prior to diagonalization of the Hamiltonian) has an imaginary part $i\Gamma/2 = i\eta_{TLS}/2$ [Rom et al., 1991]. This imaginary part may be interpreted as being caused by a process that removes products from the final state. If the width of the energy level associated with this process exceeds the tunneling splitting Δ, then the tunneling becomes irreversible.

The solution of the spin-boson problem with arbitrary coupling has been discussed in detail by Leggett et al., [1987]. The displacement of the bath oscillators from their equilibrium positions during the transition results in an effective renormalization of the tunneling matrix element by the bath overlap integral:

$$\Delta_{eff} = \Delta_0 \exp\left(-\frac{\Phi_0}{2}\right)$$

---

[1] Our conclusions about the case for large η_{TLS} are speculative in nature, and are meant to be merely illustrative, because (2.43) and (2.44) are obtained only in the weak coupling limit.

where

$$\Phi_0 = 2 \sum Q_0^2 C_j^2 \hbar^{-1} m_j^{-1} \omega_j^{-3} = 4Q_0^2 (\pi\hbar)^{-1} \int_0^\infty d\omega \, \omega^{-2} J(\omega) \qquad (2.51)$$

where the quantity $\exp(-\Phi_0)$ is nothing but the Franck-Condon factor for the bath. This relationship expresses the fact that the bath oscillators actually *cause* a dynamical asymmetry of the initial and final states, and tunneling occurs only when a bath fluctuation symmetrizes the potential. Most of the possible situations are given by Leggett et al. [1987] and we simply quote the main results here. The spectral density is assumed to have the form

$$J(\omega) \propto \omega^n \xi\left(\frac{\omega}{\omega_c}\right) \qquad (2.52)$$

where ξ is a cutoff function that is unity at $\omega \ll \omega_c$ and vanishes at $\omega \gg \omega_c$ (for example, $\xi = \exp(-\omega/\omega_c)$). The cutoff frequency is assumed to be greater than both the bare tunneling splitting and the thermal energy, $\Delta_0 \ll \omega_c$, $k_B T \ll \hbar\omega_c$. The case $n = 1$ (cf. (2.23)) is referred to as ohmic dissipation, whereas $n < 1$ and $n > 1$ are called subohmic and superohmic, respectively. The distinction between these cases lies in the different lobes of low-frequency vibrations in $J(\omega)$; this is evident from (2.51) since Φ_0 diverges for $n \leq 1$. Coupling to these low-frequency modes (at $n < 1$) results in localization of the particle in one of the wells (symmetry breaking) at $T = 0$. The system exhibits an exponential relaxation characterized by a rate constant $\ln k \propto -(\hbar\omega_c/k_B T)^{1-n}$. This subohmic case, requiring special treatment, is of little importance for chemical systems.

In the superohmic case at $T = 0$, the system exhibits weakly damped coherent oscillations characterized by the damping coefficient η_{TLS} from (2.44), but with Δ_0 replaced by Δ_{eff}. In the region $1 < n < 2$, there is a crossover from oscillations to exponential decay, in accordance with our weak-coupling predictions.

The ohmic case is most complex. A particular result is that the system is localized in one of the wells at $T = 0$, for sufficiently strong friction, viz., $\eta > \pi\hbar/2Q_0^2$. At higher temperatures there is an exponential relaxation with the rate $\ln k \propto (4\eta Q_0^2/\pi\hbar - 1)\ln T$. Of special interest is the particular case $\eta = \pi\hbar/4Q_0^2$. It turns out that under these conditions the system exhibits an exponential decay characterized by a rate constant $k = \pi\Delta_0^2/2\omega_c$ that is completely independent of temperature. Comparing this with (2.39) one sees that the *collision frequency* turns out to be precisely equal to the cutoff vibration frequency $\nu_0 = \omega_c/2\pi$.

When the potential is sufficiently asymmetric that the energy bias ϵ is large compared with the renormalized tunneling splitting, $\epsilon > \Delta_{eff}$, the coherent oscillations are completely suppressed. It therefore may be concluded that a transition from coherent to incoherent behavior can be induced by an increase in temperature or asymmetry or both. The concept of a low-temperature limit of the rate constant k_c is therefore valid for most systems having a biased potential. In fact, the required degree of asymmetry is actually very small (e.g., smaller than the level spacing $\hbar\omega_0$), and it is this circumstance that permits one to study exchange reactions, like strongly exoergic reactions, with the golden rule approach.

We note in passing that an often-used model in the theory of diffusion of impurities in 3D Debye crystals is the so-called deformational potential approximation. This model has the characteristics $C(\omega) \propto \omega$, $\rho(\omega) \propto \omega^2$, and $J(\omega) \propto \omega^3$, which, for a strictly symmetric potential, displays weakly damped oscillations and does not have a well-defined rate constant at $T = 0$. If the system permits definition of the rate constant at $T = 0$, it is proportional to the square of the tunneling matrix element times the Franck-Condon factor. Accurate determination of the prefactor requires specifying the particular spectrum of the bath.

The case of coherent tunneling is invariably studied experimentally by spectroscopic methods. For example, the neutron scattering structure factor determining the spectral line shape is equal to

$$S(\mathbf{k}, \omega) = \cos^2(\mathbf{kQ_0})\delta(\omega) + \frac{\sin^2(\mathbf{kQ_0})}{\pi[1 + \exp(\beta\hbar\omega)]} \int_{-\infty}^{\infty} dt \, \exp(i\omega t)C(t) \quad (2.53)$$

where $\mathbf{k}$ is the wave vector, $\hbar\omega$ is the absorbed energy, $\mathbf{Q} = \mathbf{Q_0}\sigma_z$ is the displacement of the scatterer, and $C(t)$ is the symmetrized correlator

$$C(t) = \tfrac{1}{2}\langle \sigma_z(0)\sigma_z(t) + \sigma_z(t)\sigma_z(0)\rangle \quad (2.54)$$

The δ part in (2.53) is responsible for elastic scattering, whereas the second term, which is proportional to the Fourier transform of $C(t)$, leads to the gain and loss spectral lines. When the system undergoes undamped oscillations with frequency Δ_0, this leads to two delta peaks in the structure factor, placed at $\omega = \pm\Delta_0$. Damping results in broadening of the spectral line. The spectral theory clearly requires knowing an object different from $\langle \sigma_z(t)\rangle$, the correlation function [Dattaggupta et al., 1989].

Nevertheless, Leggett et al. [1987] have argued, with some provisos,[2] that $\langle \sigma_z(t) \rangle$ (with the initial condition $\sigma_z(0) = 1$) and $C(t)$ may, for all practical purposes, be taken to be equal. If $C(t)$ then obeys the damped oscillator equation (2.43), then the inelastic part of the structure factor has a Lorentzian form with the peaks at $\omega = \pm(\Delta_0^2 - \eta_{TLS}^2/4)^{1/2}$ and linewidths $\Gamma = \eta_{TLS}/2$. As the temperature increases, the peaks broaden and approach each other, ultimately joining together in the incoherent tunneling limit. An example of dependence of the structure factor on ω is illustrated in Figure 2.7 for the special exactly soluble case of ohmic friction with $\eta = \pi\hbar/4Q_0^2$ [Sasetti and Weiss, 1990]. In accordance with our predictions, the peaks become indistinguishable, and thus the system fully loses coherence above the particular temperature $T^* \cong \hbar\Delta_0^2/4k_B\omega_c$.

At temperatures such that $k_B T \cong \hbar\omega_0$, the two-state approximation breaks down and the interdoublet dynamics start to play an essential role in broadening the tunneling splitting spectral line [Parris and Silbey, 1985; Dekker, 1991]. The energy level scheme for a symmetric double well is represented in Figure 2.8. Each well has a vibrational ladder with spacing between the levels $\Delta E \cong \hbar\omega_0$, and each level is split into a doublet with tunneling splitting $\hbar\Delta_n \ll \Delta E$. The contribution of interdoublet transitions to the linewidth may be found using the golden rule and is given by an expression similar to (2.46). In the specific case of linear coupling, transitions from the ground state $|0\rangle$ to the lth excited doublet $|l\rangle$ with

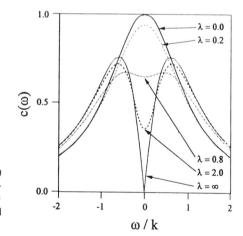

Figure 2.7. Spectral line shape $C(\omega)$ for a tunneling doublet at different values of scaled inverse temperature $\lambda = \hbar\beta k/2\pi$, $k = \pi\Delta^2/2\omega_0$. (From Sasetti and Weis [1990].)

---

[2] The long time behavior of $\langle \sigma_z(t) \rangle$ may essentially differ from that of $C(t)$, but this affects mostly the form of the spectral line at $\omega \cong 0$, and seemingly this is immaterial for determining the tunneling splitting [Sasetti and Weiss, 1990].

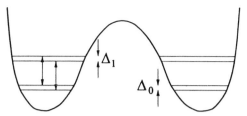

Figure 2.8. Energy levels in a symmetric double well. Arrows indicate interdoublet transitions induced by vibrations.

absorption of one phonon are characterized by the overall rate constant:

$$\bar{k}_\uparrow = \sum_{l \geq 1} k_\uparrow(0, l) = 2\hbar^{-1} \sum_{l \geq 1} |0|Q|l\rangle|^2 J(E_l - E_0)\bar{n}(E_l - E_0) \quad (2.55)$$

where the mean number of phonons $\bar{n}$ of energy $E_l - E_0$ is given by

$$\bar{n}(E) = [\exp(\beta E) - 1]^{-1} \quad (2.56)$$

The corresponding level broadening equals half of $\bar{k}_\uparrow$. In fact, $\bar{k}_\uparrow$ is the diagonal kinetic coefficient characterizing the rate of phonon-assisted escape from the ground state [Ambegaokar, 1987]. In the harmonic approximation for the well, the only nonzero matrix element is that with $l = 1$, $|\langle 0|Q|1\rangle|^2 = \delta_0^2$, where δ_0 is the zero-point spread of the harmonic oscillator. For an anharmonic potential, other matrix elements contribute to (2.55). From comparison of (2.46) and (2.55) it can be seen that at $T = 0$, only the intradoublet broadening mechanism works. At higher temperatures, $\omega_0^{-1} < \hbar\beta < \Delta_0^{-1}$, the interdoublet contribution causes an exponential growth of the linewidths proportional to $\exp(-\beta\hbar\omega_0)$. When $\hbar\beta \ll \omega_0^{-1}$, the relative contribution of the two mechanisms depends on the spectral density,

$$\frac{\bar{k}_\uparrow}{k_\uparrow} = \left(\frac{\hbar\Delta_0}{2m\omega_0^2 Q_2^2}\right)\frac{J(\omega_0)}{J(\Delta_0)} \quad (2.57)$$

The interdoublet transitions may prevail over intradoublet ones if the spectral density $J(\omega)$ grows with ω faster than ω^2.

The situation changes drastically when the coupling to the oscillators is symmetrical, $C_j f(Q)q_j$, where $f(Q)$ is an even function, or, in the two-state model, when the coupling is proportional to σ_x. In this case the intradoublet matrix elements are identically zero due to symmetry, and therefore the intradoublet broadening may appear only in higher orders of perturbation theory (i.e., for multiphonon transitions). Note that the vibrations coupled in this way have the property that they do not contribute to the Franck-Condon factor because their equilibrium positions are the same in the initial and final states. As we shall see in Section 2.5, it is these vibrations that modulate the barrier, thereby enhancing tunneling. When the intradoublet transitions are forbidden, the tunneling splitting may be observed even at temperatures close to $k_B T_c \cong \hbar \omega_0$. A relevant example of this is found in tunneling rotations, where the symmetry of the potential is not broken by phonons (see Section 7.3).

2.4. VIBRONIC RELAXATION AND ELECTRON TRANSFER

Let us return to the case where reaction takes place via a nonadiabatic transition. This situation typically occurs when the PES is constructed from a Hamiltonian in which one or more terms have been neglected. These terms then couple the initial and final states, thereby providing a mechanism for reaction to take place. The neglected terms may include, for example,

1. Coupling of electronic states due to symmetry breaking
2. Terms neglected in the Born-Oppenheimer approximation
3. Spin–orbit coupling

The first type of interaction is often associated with the overlap of wave functions localized at different sites in the initial and final states. The strength of the interaction therefore determines the rate of electron transfer between the sites. The second and third types are especially important when considering vibronic relaxation of excited electronic states.

Within the framework of first-order perturbation theory, the rate constant is given by the statistically averaged Fermi golden rule formula:

$$k = \frac{2\pi}{\hbar} \sum_{i,f} |V_{if}|^2 \exp(-\beta E_i) \delta(E_i - E_f) \qquad (2.58)$$

where the matrix element V_{if} equals

$$V_{if} = \langle \psi_i(Q)\psi_i^e(r, Q)|V|\psi_f^e(r, Q)\psi_f(Q) \rangle \qquad (2.59)$$

and the wave functions are expressed as products of electronic (ψ^e) and nuclear functions (ψ).

Processes associated with electron transfer and/or vibronic relaxation are ubiquitous in chemistry, and many review papers have discussed them in detail (see, e.g., Ovchinnikov and Ovchinnikova [1982] and Ulstrup [1979]). We are concerned with them only insofar as tunneling of the nuclei is involved.

The fundamental problem can be formulated as follows. There are initial and final electronic surfaces that cross at some seam that lies in a classically forbidden region of phase space. According to the Born-Oppenheimer principle, the electronic transition is a vertical process; i.e., it occurs at fixed nuclear configuration. It is therefore tunneling of the nuclei into the classically forbidden region that creates such a configuration. This situation, as shown by Robinson and Frosh [1963], is typical of vibronic relaxation in polyatomic molecules. A diagram of energy levels is shown in Figure 2.9a. An optical transition to the excited electronic state A_1 is followed by rapid vibrational relaxation [Hill and Dlott, 1988], such that only the lowest vibrational sublevels are populated. These sublevels lie below the point at which the A_1 surface crosses that of the final state A_2. The transition is exoergic in the sense that minimum E_2^0, is situated much lower than E_1^0:

$$\Delta E = E_1^0 - E_2^0 \gg \hbar\omega \tag{2.60}$$

Thus, the transition $A_1 \rightarrow A_2$ entails the creation of many vibrational quanta. Because of the dense spectrum of highest vibrational sublevels and their rapid vibrational relaxation in the A_2 state, this radiationless transition (RLT) is irreversible and thus it may be characterized by a rate constant k. The irreversibility condition formulated by Bixon and Jortner [1968] is

$$\tau_R = 2\pi\hbar\rho_f \gg k^{-1} \tag{2.61}$$

where τ_R is the recurrence time, i.e., the time it takes the system with the density of the energy levels ρ_f to return to the initial state. This inequality

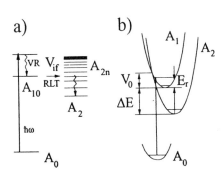

Figure 2.9. (a) Diagram of energy levels for a polyatomic molecule. Optical transition occurs from the ground state A_0 to the excited electronic state A_1. A_{2n} are the vibrational sublevels of optically forbidden electronic state A_2. Wavy arrows indicate vibrational relaxation (VR) in the states A_1 and A_2, and radiationless intersystem transition (RLT). (b) Crossing of the terms A_1 and A_2. The reorganization energy E_r, barrier height V_0, and reaction heat $-\Delta E$ are indicated.

defines the statistical limit of the RLT rate constant and is a quantitative embodiment of the second key assumption of TST (see Section 2.1). When the characteristic time of vibrational relaxation τ_V is much shorter than τ_R, then the rate constant is independent of τ_V. For small molecules consisting of only a few atoms, the inequality (2.61) is usually not satisfied. Moreover, τ_V may even become larger than τ_R. This situation is beyond our present consideration.

The total set of resonant sublevels participating in the RLT typically consists of a small number of active *acceptor* modes with nonzero matrix elements (2.59) and many inactive modes with $V_{if} = 0$. The latter play the role of the reservoir and ensure the resonance condition $E_i = E_f$. For aromatic hydrocarbon molecules, for example, the main acceptor modes are strongly anharmonic C–H vibrations that accept most of the electronic energy in singlet–triplet transitions. The inactive modes in this case are the stretching and bending vibrations of the carbon skeleton. The value of ρ_f afforded by these intramolecular vibrations is often so large that they behave as an essentially continuous bath even in the absence of intermolecular vibrations. This statement is supported by the observation that RLT rates for many molecules embedded in crystal lattices are similar to those of the isolated gas-phase molecules.

Use of the Condon approximation for the active and inactive modes causes the matrix element (2.59) to break up into a product of overlap integrals (for the inactive modes) and a constant factor V responsible for interaction of the potential energy terms (for the active modes). In this approximation the time dependence of the survival probability of A_1 is given by

$$P_i(t) = \exp(-kt) \cos^2\left(\frac{2\pi t}{\tau_R}\right) \tag{2.62}$$

where the RLT rate constant is

$$k = \frac{2\pi}{\hbar} V^2 \rho_f \mathrm{FC} \tag{2.63}$$

and FC is the statistically weighted Franck-Condon factor:

$$\mathrm{FC} = \sum_{n_k} \left[\prod_k |\langle \psi_i(Q)|\psi_{f,n_k}(Q)\rangle|^2 \right] \tag{2.64}$$

The sum in (2.64) is over all sets of energy-conserving vibrational quantum numbers n_k such that $\sum n_k \hbar \omega_k = E_1^0 - E_2^0$.

The reorganization of nuclear configuration in exoergic electron

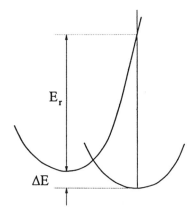

Figure 2.10. Marcus model of two harmonic terms in the limit of strong coupling. Reorganization energy E_r is shown.

transfer reactions is usually considered in the same framework. A typical diagram of terms is depicted in Figure 2.10. Comparison of (1.14), (2.49), (2.63) reveals the universality of the golden rule in describing both nonadiabatic and adiabatic chemical reactions. However, the matrix elements entering into the golden rule formula have a quite different nature. In the case of adiabatic reactions it comes from tunneling along the reaction coordinate, whereas for nonadiabatic reactions it originates from the electronic interaction discussed above. The constant matrix element approximation usually used in RLT theory reduces the problem of transition in a strongly asymmetric potential (e.g., Figures 2.9 and 2.10) to essentially that of a two-level system. For the nonadiabatic problem, the reduction to a TLS is a direct result of treating the electronic interaction as a perturbation. In the case of adiabatic tunneling, there is no good small parameter, and this reduction therefore requires invoking some rather delicate reasoning [Leggett et al., 1987].

Although the general descriptions of RLT and electron transfer are similar, very different types of vibration are involved in each case. In the former case, the accepting modes are high-frequency intramolecular vibrations, whereas in the second case the major role is played by a continuous spectrum of polarization phonons in condensed media [Dogonadze and Kuznetsov, 1975]. The localization effects associated with the low-frequency part of the phonon spectrum (mentioned in the previous section) still do not show up in electron transfer reactions due to asymmetry of the potential.

Another conventional simplification is replacing the whole vibrational spectrum by a single harmonic vibration with an effective frequency $\bar{\omega}$. In doing so, one eliminates consideration of the reversibility of the process. It is again the model of an active oscillator mentioned in Section 2.2 and, in fact, it is friction in the active mode that renders the transition irreversible. Such an approach leads to the well-known Kubo-Toyazawa problem [Kubo and Toyazawa, 1955], in which the Franck-Condon factor FC depends on two parameters, the order of multiphonon process N and the coupling parameter $\bar{S}$:

$$N = \frac{\Delta E}{\hbar \bar{\omega}}$$

$$\bar{S} = \frac{E_r}{\hbar \bar{\omega}} = \frac{m \bar{\omega} (\Delta R_{if})^2}{2\hbar} \tag{2.65}$$

where the reorganization energy E_r indicated in Figures 2.9b and 2.10 is determined by the displacement of the oscillators from their equilibrium positions. In this model there is a quantitative difference between RLT and electron transfer stemming from the aforementioned difference in phonon spectra. RLT is the weak coupling case $\bar{S} \leq 1$, whereas for electron transfer in polar media the strong coupling limit is reached, i.e., $\bar{S} \gg 1$. For the case of intersystem crossing in aromatic hydrocarbons, the coupling parameter is typically $\bar{S} = 0.5$–1.0.

In the strong coupling limit at high temperatures the electron transfer rate constant is given by the Marcus formula [Marcus, 1964]:

$$k = \frac{V^2}{\hbar} \left(\frac{\pi \beta}{E_r} \right)^{1/2} \exp \left[\frac{-\beta (E_r - \Delta E)^2}{4 E_r} \right] \tag{2.66}$$

The transition described by (2.66) is classical and is characterized by an activation energy equal to the potential at the crossing point. The prefactor is the attempt frequency $\bar{\omega}/2\pi$ times the Landau-Zener transmission coefficient B for the nonadiabatic transition [Landau and Lifshitz, 1981]:

$$B = 4\pi \delta(\bar{v}) \qquad \text{where} \quad \delta = \frac{V^2}{\hbar \bar{v} |\Delta F|} \tag{2.67}$$

In this expression δ is the Massey parameter, which depends on the mean thermal velocity $\bar{v}$, and ΔF is the difference in slopes of the initial and final terms at the crossing point. Of course, (2.66) is of typical TST form. It can be generalized to arbitrarily large electronic interactions in a straightforward manner by replacing B with the Zener transmission factor

$$B = 1 - \exp[-2\pi \delta(\bar{v})] \tag{2.68}$$

Adiabatic reactions, i.e., those occurring on a single PES, correspond to $B = 1$, and the adiabatic barrier height takes the role of E_a. The low-temperature limit of the nonadiabatic reaction rate constant is

$$k_c = \frac{2\pi}{\hbar^2 \bar{\omega}} V^2 \exp\left[\frac{\Delta E}{\hbar \omega} \left(1 - \ln \frac{\Delta E}{E_r} \right) - \frac{E_r}{\hbar \omega} \right] \qquad (2.69)$$

This formula, aside from the prefactor, is simply a one-dimensional Gamov factor for tunneling in the barrier shown in Figure 2.10. The temperature dependence of k, being Arrhenius at high temperatures, levels off to k_c near the crossover temperature, which, for $\Delta E = 0$, is equal to $k_B T_c \cong \hbar \bar{\omega}/4$.

As an illustration of these considerations, the Arrhenius plot of the electron transfer rate constant, observed by De Vault and Chance [1966], is shown in Figure 2.11. Note that only E_r, which actually is the sum of reorganization energies for all degrees of freedom, enters into the high-temperature rate constant formula (2.66). At low temperature, however, to preserve E_r, one has to fit an additional parameter $\bar{\omega}$, which has no direct physical significance for a real multiphonon problem.

In contrast to the Marcus formula, the barrier height in the RLT case increases with increasing ΔE. As seen from Figure 2.9b, the classically available regions for both terms lie on the same side of the crossing point. The tunneling behavior at $\hbar \bar{\omega} \beta \gg 1$ is due to the large disparity of imaginary momenta in the initial and final states. The low-temperature limit for RLT is given by

$$k_c = \frac{2\pi}{\hbar} V^2 \left(\frac{2\pi}{\Delta E \hbar \omega} \right)^{1/2} \exp\left(- \frac{\Delta E}{\hbar \omega} \ln \frac{\Delta E}{\hbar \omega} - \frac{E_r}{\hbar \omega} \right) \qquad (2.70)$$

The exponent in this formula is readily obtained by calculating the difference of quasiclassical actions between the turning and crossing points for each term. The most remarkable difference between (2.69) and (2.70) is that the electron transfer rate constant grows with increased ΔE, while the RLT rate constant decreases. This exponential dependence

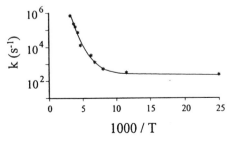

Figure 2.11. Arrhenius plot of $k(T)$ for electron transfer from cytochrome c to the special pair of bacteriochlorophylls. (From de-Vault and Chance [1966].)

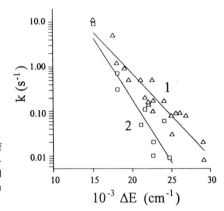

Figure 2.12. Energy gap dependence of the rate constant of intersystem ST conversion for aromatic hydrocarbons (1) and their perdeuterated analogues (2). (From Siebrand [1967].)

$k_c(\Delta E)$ [Siebrand, 1967], known as the energy gap law, is illustrated in Figure 2.12 for ST conversion.

According to CLTST, the activation energy E_a depends on ΔE according to empirical Broensted-Polanyi-Semenov (BPS) rule (see, e.g., Glasstone et al. [1941]):

$$E_a(\Delta E) = E_a(0) - \alpha_{cl}\Delta E \qquad 0 < \alpha_{cl} < 1 \qquad (2.71)$$

The symmetry coefficient $\alpha_{cl} = -\beta^{-1}\partial \ln k/\partial\Delta E$ is usually close to $1/2$, in agreement with the Marcus formula. Turning to the quantum limit, one observes that the barrier transparency increases with increased ΔE as a result of both thinning and lowering of the barrier. Therefore, k_c grows faster than the Arrhenius rate constant. At $T = 0$

$$\ln k_c(\Delta E) \cong \ln k_c(0)\left[1 + \frac{\Delta E}{E_r}\left(\ln\left|\frac{\Delta E}{E_r}\right| - 1\right)\right] \qquad (2.72)$$

This relationship is the analogue of the BPS rule for tunneling reactions. The quantum symmetry coefficient $\alpha_q = -\beta_c^{-1}\partial \ln k_c/\partial\Delta E$ is greater than α_{cl} and it may exceed 1.

The origin of the kinetic isotope effect lies in the dependence of ω_0 and $\omega_{\#}$ on the mass of the reacting particle. Classically, this dependence comes about only via the prefactor ω_0 (see (2.14)), and the ratio of rate constants of transfer of isotopes with masses m_1 and m_2 is temperature-independent and equal to

$$\frac{k_L(m_1)}{k_L(m_2)} = \frac{\omega_0(m_1)}{\omega_0(m_2)} = \left(\frac{m_2}{m_1}\right)^{1/2} \qquad (2.73)$$

In CLTST the kinetic isotope effect arises partly from the difference in partition functions in the initial state (see (2.12)), and at $\beta\hbar\omega_0/2 > 1$

$$\frac{k(m_1)}{k(m_2)} = \exp\frac{\beta\hbar\omega_0[(m_2/m_1)^{1/2} - 1]}{2} \tag{2.74}$$

That is, the exponential increase of the isotope effect with β is determined by the difference of the zero-point energies. The crossover temperature (1.7) depends on mass as

$$\frac{T_c(m_1)}{T_c(m_2)} = \frac{\omega_\#(m_1)}{\omega_\#(m_2)} = \left(\frac{m_2}{m_1}\right)^{1/2} \tag{2.75}$$

In the H/D isotope effect case, $m_2/m_1 = 2$. The interval of temperatures between $T_c(H)$ and $T_c(D)$ is wider than ΔT predicted by (2.19), and in this interval the H atom tunnels while the D atom may classically overcome the barrier. For this reason the isotope effect becomes several orders larger than that described by (2.74). At $T < T_c(m_1)$ the tunneling isotope effect becomes temperature-independent:

$$\frac{k_c(m_1)}{k_c(m_2)} = \exp\left\{\beta_c V_0\left[\left(\frac{m_2}{m_1}\right)^{1/2} - 1\right]\right\} \tag{2.76}$$

The dependence of k_H/k_D for H and D transfer on temperature is presented in Figure 2.13. Qualitatively, the conclusions about the kinetic isotope effect drawn here on the basis of the one-dimensional model remain correct for higher dimensions. However, an important difference is that in higher dimensions the mass dependence of k_c is weaker than $-\ln k_c \propto m^{1/2}$. This enables the observation of tunneling by much heavier masses ($m \le 20m_H$).

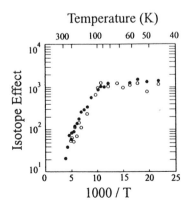

Figure 2.13. Temperature dependence of $k_H(T)/k_D(T)$ for H and D transfer in reaction (6.20). (From Al-Soufi et al. [1991].)

The aforementioned description of vibronic relaxation is based on the statistical Rice-Rampsberger-Kassel-Marcus (RRKM) theory for unimolecular reactions, assuming that an initial excitation spreads rapidly and uniformly over all states within the energy shell before a conversion such as intersystem crossing, dissociation, or isomerization takes place (see, for example, Wardlaw and Marcus [1988]). In many cases, the predictions of the statistical theory closely match the observed behavior. The intramolecular vibrational energy redistribution and relaxation depend on the vibrational state density, which sharply increases with energy and number of modes, $\rho(E, N) \propto (E/\hbar\bar{\omega})^{N-1}$. The coupling terms in the vibrational Hamiltonian induce an increasing number of Fermi resonances as the density of states becomes larger. Overlapping of Fermi resonances yields the irregular (chaotic) behavior of highly excited vibrational states, so that in the high-energy region vibrational quantum numbers can no longer be intrinsically assigned in contrast with vibrational levels at low energy [Marcus, 1985].

Dynamical manifestation of vibrational energy redistribution in jet-cooled isolated molecules were obtained by Felker and Zewail [1985, 1988]. In the low-energy regime $E \sim \hbar\bar{\omega}$, the optical transition prepares only one vibrational eigenstate, which undergoes no further vibrational evolution prior to the fluorescence transition. For somewhat higher energies, the vibrational coupling becomes appreciable and results in the beat modulation of fluorescence from a single vibrational state. When $\rho(E) \gtrsim 10^2$ cm, the vibrational relaxation dominates, and the fluorescence decay of pure vibrational states is an indicator of quasiperiodic or chaotic vibrational dynamics, corresponding to validity of RRKM theory. This is the reason why for large polyatomics where the density of highly excited vibrational states reaches 10^{10} cm, the Bixon-Jortner condition (2.61) for RLT seems to be applicable. Heller [1990] proposed another mode mixing scenario, in which classically chaotic dynamics arises because of hopping from one Born-Oppenheimer surface to another. Even if the motion on each surface is regular, i.e., there are good conserved action variables (or "modes"), they may be destroyed as a result of energy transfer between the modes corresponding to different surfaces. Heller's scheme, based on the classical surface hopping model of Tully and Preston [1971], generally supports the use of RRKM theory. However, the question how tunneling changes this picture of mode mixing still awaits its solution.

2.5. VIBRATION-ASSISTED TUNNELING

In early papers devoted to the analysis of low-temperature experimental data, the barrier height V_0 was assumed to be the same as in gas-phase reactions. The barrier width d was then found by fitting the experimental data (usually the temperature-dependent rate constant) in various one-dimensional tunneling models, e.g., to Eq. (2.1). The problem with this approach is that the calculated barrier widths were much smaller than the values calculated from known bond lengths in the crystals based on spectroscopic and crystallographic data. On the other hand, use of the experimentally known interreactant distances with the known barrier heights in these one-dimensional models yielded predicted tunneling rates that are immeasurably small. Therefore, V_0 and d could not be reconciled within the one-dimensional model of Section 2.1 because the van der Waals distances between reactants in a low-temperature lattice are usually much longer than in gas-phase reaction complexes.

To circumvent this difficulty, one must take into account that the reactants themselves take part in intermolecular vibrations, which may bring them to distances sufficiently short that tunneling, as well as classical over-barrier transition, is greatly facilitated. Of course, there is an energy cost associated with shortening the interreactant distance. However, the intermolecular modes are generally much softer than the intramolecular ones, so this energy cost is much smaller if taken mostly along a low-frequency intermolecular mode rather than the high-frequency mode associated with transfer of a light particle (e.g., a proton or H atom).

To illustrate the point, let us consider the collinear reaction AB + C → A + BC. It is known (c.f., Baer [1982]) that motion of the system in the center-of-mass frame is equivalent to motion of a single particle of mass

$$m = \left(\frac{m_A m_B m_C}{m_A + m_B + m_C}\right)^{1/2} \qquad (2.77)$$

moving on a two-dimensional PES $U(r, R)$ obtained from $U(R_{AB}, R_{BC})$ by scaling the distances R_{AB} and R_{AC} and reducing the angle between the axes from $\pi/2$ to the skew angle

$$\beta = \tan^{-1}\left(\frac{m_B}{m}\right) \qquad (2.78)$$

as illustrated in Figure 2.14. For the usual case of light-particle tunneling between two heavier atoms, $m_B \ll m_A, m_C$ and the skew angle β between the product and reactant valleys is small. Longitudinal motion along the reactant valley corresponds to relative motion of the heavy molecule AB relative to the heavy atom C, whereas transverse motion in this valley corresponds to high-frequency AB vibrations. In view of the light mass of B, tunneling along the transverse coordinate is a fast process on the time

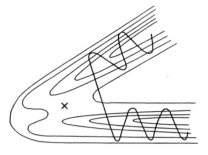

Figure 2.14. PES of collinear exchange reaction $AB + A \rightarrow A + BA$ in the case $m_A \gg m_B$. The cross indicates the saddle point. One possible corner-cutting trajectory is shown (schematic). The unbound initial state corresponds to a gas-phase reaction. (From Ovchinnikova [1979].)

scale of the slow A–C motion. Therefore, the tunneling transition may be assumed to occur at fixed A–C distances. This is represented in Figure 2.14 by a line that cuts straight across the angle between the reactant and product valleys.

The overall rate constant then comes from averaging the tunneling transmission factor over the probability distribution of R_{AC}. This reasoning was set forth by Johnston and Rapp [1961] and developed by Ovchinnikova [1979] , Miller [1975b], Truhlar and Kupperman [1971], Babamov and Marcus [1981], and Babamov et al. [1983] for gas-phase hydrogen transfer reactions. A similar model has been put forth to explain the transfer of light impurities in metals [Flynn and Stoneham, 1970; Kagan and Klinger, 1974]. Simple analytical expressions have been found for an illustrative model [Benderskii et al., 1980] in which the A–B and B–C bonds have been assumed to be represented by parabolic terms. In the case of strongly exoergic reactions, the final term turns into an absorbing wall and the transition is completed whenever the distance AB reaches a certain value and the A–B bond is broken. This simplified model potential is shown in Figure 2.5. The intra- and intermolecular coordinates Q and q are harmonic and have frequencies ω_0 and ω_1, and reduced masses m_0 and m_1. At fixed intermolecular displacement the tunneling probability equals

$$w(Q_0, q) = w_0 \exp\left[-\frac{(Q_0 - q)^2}{2\delta_0^2} \right] \qquad (2.79)$$

where Q_0 is the total distance particle B should overcome, and δ_0 is the rms amplitude for the intramolecular (transverse) vibration. As we have argued above, this probability is to be averaged over the equilibrium distribution for the q oscillator:

$$P(q) = \delta_1^{-1}(2\pi)^{-1/2} \exp\left(-\frac{q^2}{2\delta_1^2} \right) \qquad (2.80)$$

The result is

$$k = w_0 \frac{\delta_0}{(\delta_0^2 + \delta_1^2)^{1/2}} \exp\left[-\frac{Q_0^2}{2(\delta_0^2 + \delta_1^2)} \right] \qquad (2.81)$$

where

$$\delta_i^2 = \frac{\hbar}{2m_i\omega_i} \coth(\beta\hbar\omega_i/2) \qquad (2.82)$$

A typical situation is when $\delta_1 > \delta_0$, so that the tunneling distance d is shortened mostly at the expense of the low-frequency intermolecular vibration q. Although the probability distribution of this motion $P(q)$ has a small range, the net effect of shortening the tunneling distance $(Q_0 - q)$ may be very large. For example, in the case of H atom tunneling in an asymmetric O_1–H$\cdots O_2$ fragment, the O_1–O_2 vibrations reduce the tunneling distance from 0.8–1.2 Å to 0.4–0.7 Å, but this causes the tunneling probability to increase by several orders of magnitude. The expression (2.81) is equally valid for the displacement of a harmonic oscillator and for an arbitrary Gaussian random value q. In a solid the intermolecular displacement may have contributions for various lattice motions, and the above two-mode model may not work. However, Eq. (2.81) is valid as long as $P(q)$ is Gaussian, no matter how complex the motion is.

The two-mode model has two characteristic crossover temperatures that correspond to the "freezing" of each vibration. At temperatures above $T_{c0} = \hbar\omega_0/2k_B$, the rate constant $k(T)$ exhibits its ordinary Arrhenius form, in which the activation energy is determined by the effective barrier height

$$V^{\#} = \frac{1}{2} m_0\omega_0^2 Q_0^2 \frac{m_1\omega_1^2}{m_0\omega_0^2 + m_1\omega_1^2} \qquad (2.83)$$

which is lower than the one-dimensional potential barrier $V_0 = \frac{1}{2}m_0\omega_0^2 Q_0^2$. In this regime, the reaction rate is dominated by classical (over-barrier) transitions. At temperatures between the two crossover points $T_{c0} > T > T_{c1} = \hbar\omega_1/2k_B$, the tunneling transition along Q is modulated by the classical low-frequency q vibration, and the apparent activation energy is smaller than $V^{\#}$. The rate constant levels off to its low-temperature limit k_c only at $T < T_{c1}$, where the overall reaction rate is dominated by tunneling from the ground state of the initial parabolic term. The effective barrier in this case is neither $V^{\#}$ nor V_0, but

$$V_{\text{eff}} \cong V^{\#}\left[1 + \frac{\omega_0\delta_1^2}{\omega_1(\delta_0^2 + \delta_1^2)}\right] \tag{2.84}$$

We note that it is the lower crossover temperature T_{c1} that is usually measured in experiments. The above simple analysis shows that this temperature is determined by the intermolecular vibrational frequencies rather than by the properties of the reaction complex or by the static barrier. It is not surprising then, that in most solid-state reactions the observed value of T_{c1} is of order of the Debye temperature of the crystal. Although the result (2.81) was obtained using the approximation $\omega_1 \ll \omega_0$, Benderskii et al. [1991a,b] have shown that the leading exponential term turns out to be exact for arbitrary values of the two frequencies. It is instructive to compare (2.81) with (2.27) and see that friction slows tunneling down, while the q mode promotes it.

Let us now turn to the influence of vibrations on exchange chemical reactions, such as transfer of a proton between two O atoms in Figure 1.2. The potential is symmetric and, depending on the coupling symmetry, there are two possible types of contour plot, schematically drawn in Figure 2.15. The O atoms participate in different intra- and intermolecular vibrations. Those normal skeletal vibrations that change the O–O distance keeping the position of O–O center constant are symmetrically coupled to the proton coordinate, while the other modes that displace the O–O fragment as a whole (with respect to the H atom) have antisymmetric coupling. The second case is associated with the Franck-Condon factor introduced in Sections 2.3 and 2.4, while the vibrations of the first type do not contribute to the reorganization energy because the equilibrium O–O distance remains unchanged. For this reason these vibrations influence tunneling in an entirely different way. For a model in which the reactant

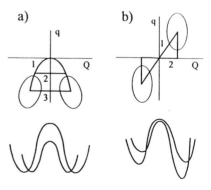

Figure 2.15. Schematic contour plots for q vibration coupled (a) symmetrically and (b) antisymmetrically to the reaction coordinate Q. Saddle points are situated at the origin. Lines 1, 2, and 3 correspond to the MEP, the sudden trajectory, and the path in the static barrier, respectively. Below, a sketch of the potential along the tunneling coordinate Q is represented at different q.

and product valleys are represented by paraboloids with frequencies ω_0 and ω_1, the transition probability has been found to be [Benderskii et al., 1991a,b,c, 1992b]

$$k_s \propto \exp\left(-\frac{Q_0^2}{\delta_0^2 \cos^2\varphi + \delta_1^2 \sin^2\varphi}\right)$$

$$k_a \propto \exp\left(-\frac{Q_0^2}{\delta_0^2} - \frac{q_0^2}{\delta_1^2}\right) \tag{2.85}$$

for symmetric and antisymmetric cases, respectively. Here, 2φ is the angle between the reactant and product valleys, q_0 is the displacement from the equilibrium position, and the thermally averaged amplitudes δ_i are those from (2.82) but taken at $\beta/2$.

Let us discuss in some detail the reaction path in both cases. At $T > T_{c0}$ since only the barrier height enters into the Arrhenius factor, the relevant trajectories generally cross the dividing surface near the saddle point where the barrier is lowest. In tunneling regime, the barrier transparency depends on both its height and width, so the optimum reaction path is a compromise of these competing factors. Consider first the symmetric case. From what we have said above, it is clear that the reaction path (at $T = 0$) should lie between the two extreme lines, the MEP, minimizing the barrier height, and the static barrier, minimizing the path length. For transfer of light particles, when the angle between the valleys is small and the MEP is strongly curved, the length factor prevails over that of height, and the reaction path cuts straight across the corner between the valleys. The probability of reaching the saddle point is roughly $\ln w_s \propto -\omega_1 q_1^2$, where q_1 is the distance from the minimum to the saddle point. On the other hand, the probability of cutting the corner is $\ln w_c \propto -\omega_0 Q_0^2$. Therefore, this corner-cutting path is preferred when $w_c \gg w_s$, i.e.,

$$1 > \frac{\omega_1}{\omega_0} \gg \sin^2\varphi \tag{2.86}$$

where $\sin\varphi = Q_0/q_1$. Physically, the corner-cutting trajectory implies that the particle crosses the barrier suddenly on the timescale of the slow q vibrational period. In the literature this approximation is usually called the "sudden," "frozen bath," or "fast flip" approximation, or the large curvature case. In the opposite case of small curvature (called also adiabatic or "slow flip" approximation), $\omega_1/\omega_0 \ll \sin^2\varphi$, which is relevant for transfer of fairly heavy masses, and the reaction path follows the MEP to a good approximation.

In the antisymmetric case the possible reaction path ranges from the MEP (when ω_1 and ω_0 are comparable) to the sudden path ($\omega_1 \ll \omega_0$), when the system "waits" until the q vibration "symmetrizes" the potential (the segment of path with $Q = Q_0$) and then instantaneously tunnels in the symmetric potential along the line $q = 0$. All of these types of paths are depicted in Figure 2.15.

When the mass of the tunneling particle is extremely small, it tunnels through a one-dimensional static barrier. With increasing mass, the contribution from the intermolecular vibrations also increases, and this leads to a weaker mass dependence of k_c than that predicted by the one-dimensional theory. That is why the strong isotope H/D effect is observed along with weak dependence $k_c(m)$ for heavy transferred particles, as illustrated in Figure 2.16. It is this circumstance that makes the transfer of heavy reactants (with masses $m \leq 20$–30) possible.

The different roles of antisymmetric and symmetric vibrational modes are clearly manifested in the vibrational selectivity of tunneling [Redington et al., 1988; Fuke and Kaya, 1989; Sekiya et al., 1990a,b]. The tunneling splitting Δ in progressions of vibrational levels corresponding to modes symmetrically and antisymmetrically coupled to the tunneling coordinate behaves in entirely different ways as a function of the vibrational quantum number n. In the former, the splitting increases severalfold with increasing n from $n = 0$ to $n = 1$–2, while in the latter a decrease of Δ is observed with increasing n. This effect can be explained with the aid of Figure 2.15. If the coupling is strong enough that the q vibration experiences a displacement q^* greater than its zero-point amplitude, when moving along the two-dimensional tunneling trajectory, exciting the vibrations of both types changes Δ in a similar way, namely

$$\ln[\Delta(n)] = \ln[\Delta(0)] + (2n + 1) \ln\left(\frac{q^*}{\delta_{1n}}\right) - \ln\left(\frac{q^*}{\delta_{10}}\right) \qquad (2.87)$$

Figure 2.16. Low-temperature limit k_c calculated from (2.85) versus $(m/m_H)^{1/2}$. The hydrogen transfer rate is assumed to be equal to $10^4\,\text{s}^{-1}$, the effective symmetric vibration mass 125 amu. The ratio of force constants corresponding to the intra- (K_0) and intermolecular (K_1) vibrations is $(K_1/K_0)^{1/2} = 2.5 \times 10^{-2}$, 5×10^{-2}, and 10^{-1} for curves 1–3, respectively. (From Benderskii et al. [1991a].)

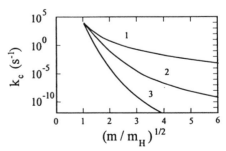

where

$$\delta^2_{1n} = \frac{(n + \frac{1}{2})\hbar}{m_1 \omega_1} \tag{2.88}$$

This increase in Δ with increasing n is simply due to shortening of the tunneling distance with increasing vibration amplitude δ_{1n}, and it is equivalent to the effect of increasing temperature for the incoherent tunneling rate [Benderskii et al., 1992b].

For the symmetric case, the qualitative picture does not change in the weak coupling limit, because all the vibration can do is to modulate the barrier, thus facilitating tunneling. The situation is more subtle for antisymmetrically coupled mode. As shown in Figure 2.15 this vibration, in contrast to the symmetric mode, asymmetrizes the potential such that at any particular instant, the initial and final states are not in resonance. This effect leads to a reduction in the splitting. Consider the problem perturbatively. If the vibration and the potential $V(Q)$ were uncoupled, each tunneling doublet E_0, E_1 (we consider only the lowest one) of the uncoupled potential $V(Q)$ would give rise to progression of vibrational levels with energies

$$E_{0,1}(n) = \hbar\omega_1(n + \tfrac{1}{2}) + E_{0,1}(0) \tag{2.89}$$

The nonzero matrix elements of the antisymmetric coupling $V_{\text{int}} = CQq$ are

$$\langle n, 0|CqQ|n + 1, 1\rangle = C\delta_{10}(n + 1)^{1/2}Q_0$$
$$\langle n, 0|CqQ|n - 1, 1\rangle = C\delta_{10}n^{1/2}Q_0 \tag{2.90}$$

Second-order perturbation theory gives the shift of the lower level as

$$\Delta E_0(n) = C^2\delta^2_{10}Q_0^2\left(\frac{n}{\hbar\omega_1 - \Delta_0} - \frac{n + 1}{\hbar\omega_1 + \Delta_0}\right) \tag{2.91}$$

where $\Delta_0 = E_1 - E_0$. The shift of the upper level, $\Delta E_1(n)$ is obtained from (2.91) simply by making the change $\Delta_0 \to -\Delta_0$. Since $\Delta_0 \ll \omega_1$, we finally obtain the renormalized splitting as a function of n:

$$\Delta(n) = \Delta_0\left[1 - \frac{2C^2(n + \frac{1}{2})Q_0^2}{\hbar m_1 \omega_1^3}\right] \tag{2.92}$$

Therefore, the tunneling splitting decreases with increasing n, in accordance with the experiment. The weak coupling formula holds for $C^2Q_0^2/$

$m_1 \omega_1^3 \ll 1$. Both the weak and strong coupling results (2.87) and (2.92) could be formally obtained from multiplying Δ_0 by the overlap integral (square root of the Franck-Condon factor) for the harmonic q oscillator:

$$\mathrm{FC}^{1/2} = \int dq \, \psi_n(q + q^*)\psi_n(q - q^*) = \exp\left(-\frac{q^{*2}}{2\delta_{10}^2}\right)L_n\left(\frac{q^{*2}}{\delta_{10}^2}\right) \quad (2.93)$$

where $q^* = CQ_0/m_1\omega_1^2$, and L_n is the Laguerre polynomial. When $n \neq 0$, the harmonic oscillator wave functions have nodes, and for $q^* \leq \delta_{10}$, overlapping of the functions with opposite signs reduced the integral [Huller, 1980].

2.6. IS THERE AN ALTERNATIVE TO TUNNELING?

Of course, the most unambiguous experimental evidence of quantum tunneling comes from spectroscopic measurements of coherent tunneling splittings in cases where a particle is transferred between symmetric (or nearly symmetric) potential wells. However, the interpretation of experimental measurements of $k(T)$ or kinetic isotope effects in irreversible reactions are not so clear-cut. In these cases, alternative explanations can often be presented to explain the results even in the absence of tunneling. In this section, we briefly discuss some possible ways to explain large deviations from the Arrhenius law and the existence of a low-temperature limit within the framework of CLTST. At $T \cong T_c$ the apparent prefactor decreases along with activation energy. In CLTST, small prefactors arise from large negative activation entropy $\Delta S^{\#}$ (see, e.g., Eyring et al. [1983]). In this language (2.6) formally gives

$$\Delta S^{\#} = -\frac{2k_B}{\hbar} S(E_a) \quad (2.94)$$

This entropy of activation is determined by the ratio of partition functions, which generally has a slight temperature dependence. The typical value of $k_c = 10^{-1}-10^{-5} \, \mathrm{s}^{-1}$ corresponds to a drop in $\Delta S^{\#}$ of 65–85 cal/mol·K. Since only vibrational degrees of freedom are involved in a solid-state reaction, the sole reason for this change may be the increase in their frequencies in the transition state:

$$\Delta S^{\#} = k_B \sum_{n=1}^{N} \ln\left(\frac{\omega_n}{\omega_n^{\#}}\right) \quad (2.95)$$

where N is the total number of transverse vibrational modes with frequencies ω_n and $\omega_n^{\#}$ in the initial and transition states. To obtain the indicated values of $\Delta S^{\#}$, one must suppose that $N > 10^2$ and $\omega_n^{\#} > \omega_n$. Even this somewhat artificial assumption cannot explain the temperature dependence of $\Delta S^{\#}$ shown in Figure 2.4, because CLTST at $\omega_n^{\#} > \omega_n$ predicts a rise of $\Delta S^{\#}$ instead of a decrease when the vibrations become quantum, i.e., when $\beta \hbar \omega_n / 2 \cong 1$. Therefore, an additional assumption is needed to explain a sharp increase in the number of vibrations involved in the reaction when the temperature approaches T_c. This model of classical cooperative transitions has been speculated about as an alternative to tunneling, but no confirmations to such a scheme exist.

Another possibility that has been considered is that $k(T)$ can be represented as

$$k(T) = k_0 \exp(-\beta E_a) + k_1 \qquad (2.96)$$

where the rate $k_1 \ll k_0$ is associated with passing through an activationless channel with a very small transition probability. That is, a small but real hole exists in the barrier, and the system must pass through a small volume of phase space to get to the product valley through the hole. The origin of such a channel can hardly be substantiated within the framework of CLTST, in view of the above entropy arguments. Although these hypotheses do not clash with any basic principles and they cannot be discarded a priori, they hardly pretend to be a universal explanation to the numerous experimental results that have been obtained near the low-temperature limit.

ONE-DIMENSIONAL MODELS

CONTENTS

In this chapter we expand on the problem of one-dimensional motion in a potential $V(x)$. Although it is a textbook example, we use here the less traditional Feynman path integral formalism, the advantage of which is a possibility of straightforward extension to many dimensions. In the following sections on tunneling theories we shall use dimensionless units, in which $\hbar = 1$, $k_B = 1$ and the particle has unit mass.

3.1. THE MAIN PATH INTEGRAL RELATIONS

The path integral formulation of quantum mechanics relies on the basic idea that the evolution operator of a particle is expressed in terms of the time-independent Hamiltonian, $\hat{H}(x, p) = p^2/2 + V(x)$ [Feynman and Hibbs, 1965]:

$$\langle x_f | \exp(-i\hat{H}t_0) | x_i \rangle \equiv K(x_f, x_i | t_0) = \int D[x(t)] \exp\{iS[x(t)]\} \quad (3.1)$$

with the normalization

$$K(x, x' | 0) = \delta(x - x') \quad (3.2)$$

where the path integral sums up all the paths connecting the points $x(0) = x_i$ and $x(t_0) = x_f$, each path having the weight $\exp(iS)$. If we make

55

time discrete by introducing the points $0 = t_1 < t_2 < \cdots < t_{\ell-1} < t_\ell = t_0$, $\ell \to \infty$, the symbol $D[x(t)]$, which indicates summation over all possible paths $x(t)$, may be thought of as $D[x(t)] = N_0 \, dx(t_2) \cdots dx(t_{\ell-1})$, where N_0 is the normalization factor providing the validity of (3.2). The action S is defined via the classical Lagrangian:

$$S = \int_0^{t_0} dt \, L(x, \dot{x}) \qquad L(x, \dot{x}) = \tfrac{1}{2}\dot{x}^2 - V(x) \tag{3.3}$$

The Fourier transform of the propagator (3.1) yields the energetic Green's function:

$$G(x, x' \mid E) = i \int_0^\infty dt \, K(x, x' \mid t) \exp(iEt) \tag{3.4}$$

which is the solution to the time-independent Schrödinger equation:

$$(\hat{H} - E)G = \delta(x - x') \tag{3.5}$$

The poles of the spectral function

$$g(E) = \mathrm{Tr}(\hat{H} - E)^{-1} = \int dx \, G(x, x \mid E) \tag{3.6}$$

correspond to the energy eigenvalues E_n^0. This function can also be represented as

$$g(E) = i \int_0^\infty dt \, \exp(iEt) \int dx_i \int_{x_i = x_f} D[x(t)] \exp\{iS[x(t)]\} \tag{3.7}$$

As seen from (3.7), the closed paths with $x(t_0) = x(0)$ fully determine the energy spectrum of the system. The propagator K, when expressed in terms of the energy eigenfunctions $|n\rangle$, has the form

$$K(x_f, x_i \mid t_0) = \sum \exp(-iE_n^0 t_0)\langle x_f \mid n\rangle\langle n \mid x_i\rangle \tag{3.8}$$

The statistical operator (density matrix)

$$\exp(-\beta\hat{H}) = \sum \exp(-\beta E_n^0)\langle x_f \mid n\rangle\langle n \mid x_i\rangle \tag{3.9}$$

is formally obtained from the quantum mechanical evolution operator by replacing it_0 by β. This change of variables $t = -i\tau$, called the Vick rotation [Callan and Coleman, 1977], turns the potential upside down and therefore replaces the Lagrangian in (3.3) by the classical Hamiltonian:

$$t \to -i\tau \qquad x \to x \qquad \dot{x} \to i\dot{x} \qquad V(x) \to -V(x)$$

$$L \to H \equiv \tfrac{1}{2}\dot{x}^2 + V(x) \qquad E \to -E \tag{3.10}$$

The resulting Euclidean action is

$$-is = S_E[x(\tau)] = \int_0^\beta d\tau \, H(x, \dot{x}) = \int_{x_i}^{x_f} p \, dx + E\beta \tag{3.11}$$

The density matrix is equal to [Feynman, 1972]

$$\rho(x_f, x_i \mid \beta) \equiv \langle x_f | \exp(-\beta H) | x_i \rangle$$

$$= \int_{x(0)=x_i}^{x(\beta)=x_f} D[x(\tau)] \exp\{-S_E[x(\tau)]\} \tag{3.12}$$

and, consequently, the partition function is

$$Z \equiv \exp(-\beta F) \equiv \mathrm{Tr}[\exp(-\beta \hat{H})]$$

$$= \int dx(0) \int_{x(0)=x(\beta)} D[x(\tau)] \exp\{-S_E[x(\tau)]\} \tag{3.13}$$

where F is the free energy. As in (3.7), only the closed paths enter into the expression (3.13). The density matrix $\rho(x_f, x_i \mid \beta)$ satisfies the differential equation

$$\frac{\partial \rho}{\partial \beta} = -\hat{H}\rho \tag{3.14}$$

with the initial condition $\rho(x, x' \mid 0) = \delta(x - x')$ (cf. (3.2)).

We now write out the expressions for the density matrix of a free particle and harmonic oscillator. In the first case $\rho(x, x' \mid \beta)$ is a Gaussian with a half-width equal to the thermal de Broglie wavelength:

$$\rho(x, x' \mid \beta) = (2\pi\beta)^{-1/2} \exp\left[-\frac{(x - x')^2}{2\beta}\right] \tag{3.15}$$

For a harmonic oscillator with frequency ω

$$\rho(x, x' \mid \beta) = \left(\frac{\omega}{2\pi \sinh \beta\omega}\right)^{1/2}$$

$$\times \exp\left\{\frac{\omega}{2 \sinh \beta\omega}[(x^2 + x'^2)\cosh \beta\omega - 2x'x]\right\} \quad (3.16)$$

and, finally,

$$Z = \exp(-\beta F) = \left[2\sinh\left(\frac{\beta\omega}{2}\right)\right]^{-1} \quad (3.17)$$

3.2. NUMERICAL PATH INTEGRAL METHODS

To complete the picture of quantum mechanics based on path integrals, in this section we briefly review the methods of its numerical realization. We do not present a full survey of this vast field, nor do we provide practical guidelines that should be strictly followed when doing path integrals. Our purpose is only to outline some general ideas of how these objects can be calculated on a computer and to show that, in principle, doing this is a task no more complicated than numerically solving a Schrödinger equation.

It should be noted that in one dimension, straightforward basis set methods implementing directly formulae (3.8) and (3.9) (where eigenenergies and eigenfunctions are found from diagonalization of the Hamiltonian in a sufficiently large basis) are very efficient in most cases and should be preferred. However, when the dimensionality of the problem is increased, the number of basis functions required grows exponentially, so the basis set approach applies to only a very few degrees of freedom. For path integral methods, the computational effort increases much slower—linearly in most cases. Further, the use of an influence functional [Feynman and Vernon, 1963] for the Hamiltonians of the system–bath form greatly facilitates the task because part of the work can be performed analytically.

The most straightforward way to compute a path integral numerically is to make time discrete for the propagator (3.1):

$$\langle x_f | \exp(-iHt)|x_i\rangle = \int dx_1 \cdots \int dx_{n-1} \prod_{k=1}^{n} \langle x_k|\exp(-iH\,\Delta t|x_{k-1}\rangle \,,$$

$$\Delta t = t/n \quad (3.18)$$

For short time propagators entering into this equation one usually exploits the well-known Trotter formula:

$$\langle x_k| \exp(-\mathrm{i}H\,\Delta t|x_{k-1}\rangle = \left(\frac{1}{2\pi\mathrm{i}\,\Delta t}\right)^{1/2} \exp\Big\{\mathrm{i}\,\frac{1}{2\,\Delta t}\,(x_k - x_{k-1})^2$$

$$-\frac{\Delta t}{2}\,[V(x_k) + V(x_{k-1})]\Big\} \tag{3.19}$$

Although in the limit $\Delta t \to 0$, $n \to \infty$ Eqs. (3.18) and (3.19) give the exact result, in practice it is desirable to reduce the dimensionality of the integral n, that is, to use time steps as large as possible. To this end, improved short time propagators are used, which enhance convergence of the method. In particular, Makri and Miller [1989a] proposed a systematic way of improving the convergence of path integral calculations by expressing the short time propagator as the exponential of a power series in time. For small enough numbers of time steps, that is, for short enough total times, one may even evaluate the integral (3.18) by numerical quadrature. Otherwise, one has nothing better to do than to invoke the Monte Carlo method to compute multidimensional integrals such as (3.18).

The most simple procedure can be carried out in the case of path integral for the equilibrium density matrix, i.e., with $\mathrm{i}t = \beta$. Specifically, the Metropolis algorithm [Metropolis et al., 1953] applies to n-dimensional integrals of the general form

$$I = \int \mathrm{d}x\ \rho(x)f(x) \tag{3.20}$$

where $\rho(x)$ is a normalized, positive definite weighting function, and $f(x)$ is a smooth function. According to the Monte Carlo algorithm, the function $f(x)$ is sampled at a large number of points with distribution ρ. The points subject to the distribution function are generated by performing a random walk in space x such that the moves leading to more favorable regions of space (with greater ρ) are always accepted, while other steps are accepted with a certain probability. The statistical error of this method is inversely proportional to the square root of the number of steps. For applications of Monte Carlo techniques to the tunneling problem, described by imaginary-time path integrals, and comparison with semiclassical methods see Alexandrou and Negele [1988]. Figure 3.1 shows typical trajectories of tunneling in a one-dimensional double-quadratic potential that are generated in the way described above. As seen from Figure 3.1, averaging over many such trajectories leads to a

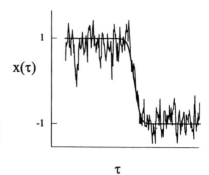

Figure 3.1. A typical Monte Carlo-generated tunneling path in the quartic double well. The smooth curve shows the quasiclassical solution. (From Alexandrou and Negele [1988].)

mean trajectory that is smooth because the fluctuations are canceled out. It is close to the classical trajectory in the upside-down barrier.

The Monte Carlo method is practical only when the integrand is dominantly positive. If the integrals of the positive and negative regions are separately much larger in absolute value than their sum, the statistical noise will be comparable to the integral itself. This difficulty, called, the *sign problem*, is encountered if the real-time dynamics is in question, when the exponential factors in (3.18) oscillate quickly and there is no normalizable positive weighting function. We are not aware of any universal prescription that would allow one to generally circumvent the sign problem. Several ways are mentioned below that proved to be successful for some particular problems.

Makri and Miller [1987b], Doll and Freeman [1988], Doll et al. [1988] (see also the review by Makri [1991b]) exploited the stationary phase approximation (i.e., the semiclassical limit) as an initial approximation to the path integral. For example, for the multidimensional integral of the form $\int dx \exp[iS(x)]$, one may obtain the following approximation [Makri, 1991b]:

$$\int dx \exp[iS(x)] \cong \int dx_0 \exp[iS(x_0)]\{\det[\hat{\mathbf{1}} + i\hat{\mathbf{c}}\mathbf{S}_2(x_0)]\}^{1/2}\rho(x_0) \quad (3.21)$$

where the weighting function is

$$\rho(x_0) = \exp[-\tfrac{1}{2}\mathbf{S}_1(x_0)\hat{\mathbf{c}}\mathbf{S}_1(x)] \quad (3.22)$$

Here, $\hat{\mathbf{c}}$ is an arbitrary positive matrix (for example, $\hat{\mathbf{c}} = c\hat{\mathbf{1}}$, $c > 0$), and the matrices $\mathbf{S}_1 = \partial S/\partial x$ and $\mathbf{S}_2 = \partial^2 S/\partial x\,\partial x$ are taken at the point x_0. At $c \rightarrow 0$ Eq. (3.21) is exact, but the distribution (3.22) tends to be uniform. Conversely, at $c \rightarrow \infty$ the distribution (3.22) has very sharp peaks at the stationary phase points, where $\mathbf{S}_1 = 0$, and (3.21) is simply the semiclassical approximation to the actual integral. The choice of intermediate values of c allows a significantly improved result compared with that given by the stationary phase approximation. This method is efficient when there is one or only a few classical paths. When the dynamics is dominated by many interfering classical trajectories, this method does not solve the problem.

Another direction that is pursued to make Monte Carlo evaluation of real-time path integrals feasible is constructing less oscillatory effective short time propagators (see Makri [1993] and references therein). The idea may be illustrated by considering formula (3.8) for the propagator. This expression is a sum of many oscillating terms. If we compare the exact propagator (i.e., the infinite sum) with a truncated expansion involving only a finite number of energy levels, we may not notice any resemblance, because adding higher energy levels makes the expression increasingly oscillatory. Note that this problem does not occur for the imaginary-time propagator (3.9), since high-energy terms there are associated with exponentially decreasing contributions. The way out is to note that we are usually not interested in the propagator itself, but in propagating a wave function using this propagator. If the initial wave function expanded in the eigenenergy basis set does not contain contributions from high energy levels, use of the truncated propagator and the exact one to obtain the wave function at time t will give the same answer, even though the propagators look different. This observation underlies the group of methods where effective propagators are constructed, which give the same result as the exact ones acting on a certain class of relevant wave functions and, at the same time, are much smoother, so that they allow Monte Carlo evaluation of the path integral. For example, one may use a truncated basis set of plane waves to eliminate large momentum components, which are responsible for the strongly oscillatory behavior of the propagator and are associated with insignificant contributions of the paths with high kinetic energies.

As an alternative to making time discrete, we mention the methods based on Fourier expansion of the path with subsequent integration over Fourier coefficients. These methods are mostly applied to calculate statistical properties at finite temperatures (see, e.g., Doll and Freeman [1984], Doll et al. [1985], Topper and Truhlar [1992], and Topper et al. [1992]).

All of the above methods are easily extended to more degrees of freedom. There are also approaches that rely on a certain form of multidimensional Hamiltonian which should be represented as some "system" coupled to a bath of harmonic oscillators, as described in Section 2.2. In Chapter 5 we see that integration over these bath oscillator modes can be performed exactly, leading to an influence functional. Integration over the system coordinate can be performed using a numerically constructed one-dimensional propagator (3.8) combined with the influence functional [Makri, 1992]. This method combines the feasibility of one-dimensional basis set methods with the attractive property of a harmonic bath which can be treated analytically. In some cases it also permits the use of large time steps so that instead of Monte Carlo evaluation, one can use direct quadrature.

3.3. TUNNELING TRAJECTORIES FOR A ONE-DIMENSIONAL BARRIER

Consider a tunneling event through a parabolic barrier [Doll et al., 1972]. The classical motion at energy $E < V_0$ to the left of the barrier obeys the equation

$$x(t) = -x_1 \cosh(\omega_{\#} t) \qquad (3.23)$$

where at $t = 0$ the particle is located at the turning point $x = x_1 = (2E/\omega_{\#}^2)^{1/2}$. At $t \to \infty$ the particle goes to minus infinity, having reflected from the barrier at $t = 0$. This picture holds for classical mechanics when overcoming the barrier is impossible at $E < V_0$. Formally, the tunneling trajectories appear when t is considered complex, i.e.,

$$t = t_1 + it_2 \qquad (3.24)$$

The solution (3.23) then oscillates as

$$x(t) = -x_1 \cosh(\omega_{\#} t_1) \cos(\omega_{\#} t_2) - ix_1 \sinh(\omega_{\#} t_1) \sin(\omega_{\#} t_2) \quad (3.25)$$

At

$$t_2 = \frac{\pi(2n + 1)}{\omega_{\#}} \qquad n = 0, \pm 1, \pm 2, \ldots \qquad (3.26)$$

Eq. (3.25) differs from (3.23) by only a sign. That is, a solution appears that describes the classical motion to the right of the barrier. Therefore, tunneling penetration through the barrier may be described by choosing the time contour in the complex t plane consisting of half the real axis $(-\infty, 0)$, a segment of the imaginary axis $(0, -i\pi/\omega_{\#})$, and a straight line parallel to the real axis $(-i\pi/\omega_{\#}, \infty - i\pi/\omega_{\#})$. The particle moves in real time until it strikes the barrier and then begins its motion in imaginary time. After leaving the barrier at $t_2 = -\pi/\omega_{\#}$, the motion proceeds only in real time (i.e., the imaginary part of t is constant). The value $n = -1$

corresponds to the time contour that is located closest to the real time. This contour is associated with a single pass through the barrier (kink), whose contribution to the action equals

$$W(E) = \frac{\pi}{\omega_{\#}} (V_0 - E) \tag{3.27}$$

It is readily seen that this quantity is independent of the kink's position on the real axis, t_1; i.e., whenever the kink is started, after performing half of the vibration with imaginary period $2\pi/\omega^{\#}$ the coordinate remains real and the action satisfies (3.27). The contours with $n = 1, \pm 2, \pm 3, \ldots$ correspond to multiple barrier crossings. It is clear from the semiclassical picture that the barrier transparency is proportional to $\exp[-2W(E)]$. The trajectories for which the imaginary increment is $2\pi n/\omega_{\#}$ with integer n are associated with reflection from the barrier.

As mentioned before, the parabolic shape of the barrier can be assumed only for sufficiently small x. In reality, the potential should be finite at $x \to \pm\infty$. As an illustration consider the Eckart barrier:

$$V(x) = V_0 \, \text{sech}^2(\alpha x) \tag{3.28}$$

If the particle has energy E and is located at the left turning point at $t = 0$, then its trajectory satisfies

$$x(t) = -\frac{1}{\alpha} \sinh^{-1}\{\lambda \cosh[\alpha(2E)^{1/2}t]\} \tag{3.29}$$

with $\lambda = (V_0/E - 1)^{1/2}$. A real-valued trajectory of classical motion with $x(\infty) = \infty$ occurs if time is incremented by an imaginary value such that

$$\lambda \cosh[\alpha(2E)^{1/2}t] = \pm\pi i \tag{3.30}$$

whence

$$t = \pm t_0 + i\frac{\pi}{2\alpha}\left(\frac{1}{2E}\right)^{1/2}(2n + 1) \qquad t_0 = \frac{1}{2}\left(\frac{1}{2E}\right)^{1/2}\tanh^{-1}\left(\frac{E}{V_0}\right)^{1/2} \tag{3.31}$$

These points are the branch points of the multivalued function $\sinh^{-1}z(t)$ in the complex t plane. The one-kink action equals

$$W(E) = \frac{\pi}{\alpha}[(2V_0)^{1/2} - (2E)^{1/2}] \tag{3.32}$$

However, this value is independent of the kink starting point only within the segment $(-t_0, t_0)$ i.e., near the barrier. Outside this segment (i.e., in asymptotic region $|t| > t_0$) the trajectory is reflected from the barrier. This example demonstrates a general property: The value of the action $W(E)$ is same for two time contours connecting two arbitrary points t_1 and t_2, if there is no branch point inside the region bordered by these contours.

To find the total transition probability, one must sum up the contributions from all multiple barrier crossings (all n). At low enough energies and high barriers the contribution from a single kink is dominant, and the other contributions can be neglected. The general case is discussed in Section 3.4.

Now consider tunneling of the particle with a separable two-dimensional Hamiltonian [Altkorn and Shatz, 1980]:

$$H = \tfrac{1}{2}(P_Q^2 + P_q^2) + \tfrac{1}{2}\omega_1^2 q^2 + V_0 - \tfrac{1}{2}\omega_\#^2 Q^2 \qquad (3.33)$$

The longitudinal mode Q has energy E, while the energy of the transverse mode q is quantized. Classical equations of motion have the solution

$$Q = -Q_0 \cosh(\omega_\# t) \qquad q = \left[\frac{2(n + \tfrac{1}{2})}{\omega_1}\right]^{1/2} \cos(\omega_1 t + \delta) \qquad (3.34)$$

where n is the quantum number of the q vibration. As described above, we choose the value Q_0 such that at $t = 0$ the particle strikes the left turning point:

$$Q_0 = -\omega_\#^{-1}\{2[V_0 - \omega_1(n + \tfrac{1}{2}) - E]\}^{1/2} \qquad (3.35)$$

In view of the separability of the potential, the time contour may be taken exactly the same as for the one-dimensional parabolic barrier considered above. For the q coordinate to be the same at both turning points, the phase δ should be equal to

$$\delta = \frac{i\pi\omega_1}{2\omega_\#} + \pi m \qquad m = 0, \pm 1, \ldots \qquad (3.36)$$

This trajectory is shown in Figure 3.2. The coordinate Q is real under the barrier, while both the momenta are complex.

3.4. DECAY OF A METASTABLE STATE

Consider a potential $V(x)$ having a single minimum separated from a continuum by a sufficiently large barrier satisfying (1.1), for example, a

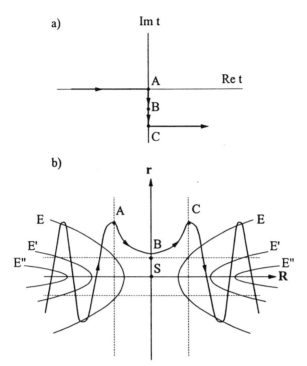

Figure 3.2. Time contour (a) and real part (b) of the tunneling trajectory for a separable system of a parabolic barrier–harmonic oscillator (schematic). Curves E, E', and E'' are equipotentials. Vertical and horizontal dashed lines show the loci of vibrational and translational turning points. Points A, B, and C indicate the corresponding times and positions along trajectory. (From Altcorn and Schatz [1980]).

cubic parabola (Figure 3.3):

$$V(x) = \frac{1}{2} \omega_0^2 x^2 (1 - x/x_0)$$
(3.37)

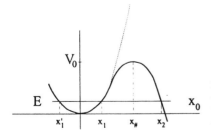

Figure 3.3. Cubic parabola potential. Turning points are shown. The dashed line indicates the harmonic potential having the same intrawell frequency.

Since the Hamiltonian is unbounded, the energy of each metastable state in the well is complex [Landau and Lifshitz, 1981]:

$$E_n = E_n^0 - \frac{i\Gamma_n}{2} \tag{3.38}$$

whence the wave function decays exponentially in time:

$$|\psi|^2 \propto \exp(-\Gamma_n t) \tag{3.39}$$

Metastability implies that $\Gamma_n \ll E_n^0$, so that the wave function inside the well is close to that of a stationary state with energy E_n^0. Strictly speaking, the energy spectrum in this case is quasidiscrete with the density of states:

$$\rho(E) = (2\pi)^{-1} \sum_n \frac{\Gamma_n}{(E - E_n^0)^2 + \Gamma_n^2/4} \tag{3.40}$$

We seek the poles of the spectral function $g(E)$ given by (3.7). In the WKB approximation the path integral in (3.7) is dominated by those classical trajectories that give an extremum to the action functional

$$0 = \frac{\delta S[x(t)]}{\delta x} = \frac{d^2 x}{dt^2} + \frac{dV(x)}{dx} \tag{3.41}$$

In fact, (3.41) is the usual stationary-phase approximation performed for an infinite-dimension path integral, which picks up the trajectories with classical action S_{cl}. Further, at fixed time t we take the integral over x_i again in the stationary-phase approximation, which gives

$$0 = \frac{\partial S_{cl}}{\partial x_i} = \frac{\partial S_{cl}}{\partial x(0)} + \frac{\partial S_{cl}}{\partial x(t)} = \dot{x}(t) - \dot{x}(0) \tag{3.42}$$

This equation shows that only the closed classical trajectories ($x(t) = x(0)$ and $\dot{x}(t) = \dot{x}(0)$) should be taken into account, and the energy spectrum is determined by these periodic orbits [Balian and Bloch, 1974; Gutzwiller, 1967; Miller, 1975b; Rajaraman, 1975]. Finally, the stationary-phase integration over time yields the identity

$$E = -\frac{\partial S_{cl}}{\partial t} = E_{cl}(t) \tag{3.43}$$

which means that the classical periodic orbits should have energy E. Therefore, with exponential accuracy the spectral function is represented by the sum over all periodic orbits:

$$g(E) \propto \sum_{\text{periodic orbits}} \exp\left\{i\left[\frac{W(E) - l\pi}{2}\right]\right\} \qquad (3.44)$$

where the short action equals

$$W(E) = S_{cl} + Et = \int p \, dx \qquad (3.45)$$

The additional phases $\ell\pi/2$, where ℓ is the number of turning points encountered along the trajectory, emerge because of the breakdown of the WKB approximation near the turning points [Gutzwiller, 1967; Levit et al., 1980; McLaughlin, 1972]. Vibrations with energy E in the well have a period

$$\tau_1 = 2\int_{x_1'}^{x_1} dx\{2[E - V(x)]\}^{-1/2} \qquad (3.46)$$

and short action (3.45)

$$W_1(E) = 2\int_{x_1'}^{x_1} dx\{2[E - V(x)]\}^{1/2} \qquad (3.47)$$

where x_1' and x_1 are the turning points (see Figure 3.3). If the state in the well were stable (the potential in Figure 3.3 shown by dashed line), these classical vibrations would be the only possible periodic orbits entering into (3.44). Summing up the series (3.44) over the number of vibrational periods n, we get

$$g(E) \propto \sum_{n=1}^{\infty} \exp\{in[W_1(E) - \pi]\} = (\exp\{-i[W_1(E) - \pi]\} - 1)^{-1} \quad (3.48)$$

Thus, $g(E)$ has an infinite number of poles at

$$W_1(E) = 2\pi(n + \tfrac{1}{2}) \qquad (3.49)$$

This expression is nothing but the Bohr-Sommerfeld quantization rule (see, e.g., Landau and Lifshitz [1981]). In the metastable potential of Figure 3.3 there are also imaginary-time periodic orbits satisfying (3.41) that develop between the turning points inside the classically forbidden region. It is these trajectories that are responsible for tunneling [Levit et al., 1980; McLaughlin, 1972]. They have imaginary period and action:

$$i\tau_2 = 2 \int_{x_1}^{x_2} dx \{2[E - V(x)]\}^{-1/2}$$

$$iW_2(E) = 2 \int_{x_1}^{x_2} dx \{2[E - V(x)]\}^{1/2}$$

(3.50)

The classically accessible region to the right of the turning point x_2 does not contain any closed trajectories. An arbitrary periodic orbit that involves n "swings" inside the well and m barrier passages has complex period $\tau_{nm}(E) = n\tau_1(E) + im\tau_2(E)$ and action $nW_1(E) + imW_2(E)$. Each orbit with given n and m $(m + n \geq 1)$ enters into the sum (3.44) with the combinatorial coefficient C_{n+m}^n, and the direct summation gives

$$g(E) \propto \frac{\exp[i(W_1 - \pi)] + \exp(-W_2)}{1 - \exp[i(W_1 - \pi)] - \exp(-W_2)}$$

(3.51)

whence the quantization condition is

$$\exp\{i[W_1(E) - \pi]\} + \exp[-W_2(E)] = 1$$

(3.52)

Since $\exp(-W_2) \ll 1$, this equation can be solved iteratively by using (3.38), whence

$$\Gamma_n = \left(\frac{\partial W_1}{\partial E}\right)_{E=E_n^0}^{-1} \exp[-W_2(E_n^0)] = \frac{1}{\tau_1(E_n^0)} \exp[-W_2(E_n^0)] \quad (3.53)$$

Formula (3.53) demonstrates that the decay rate for a metastable state is equal to the inverse period of classical vibrations in the well (*attempt frequency*) times the barrier transparency. The more traditional treatment of metastable-state decay using the one-dimensional WKB approximation is given in Appendix A.

3.5. THE Im F METHOD

We now consider the calculation of the rate constant for decay of a metastable state. In principle it can be done by statistical averaging of Γ_n from (3.53), but there is a more elegant and general way that relates the rate constant to the imaginary part of the free energy. Recalling (3.38) we write the rate constant as

$$k = Z_0^{-1} \sum \Gamma_n \exp(-\beta E_n^0) = 2\beta^{-1} \frac{\text{Im}\,Z}{Z_0} = 2\,\text{Im}\,F \qquad (3.54)$$

where Z_0 is the real part of the partition function in the well; it is calculated by neglecting the decay. This expression enables one to use the path integral expression (3.13). As in the previous section, we look for the stationary points of the integral, i.e., the trajectories that produce an extremum of the Euclidean action (3.11), and thus satisfy the classical equation of motion in the upside-down barrier

$$0 = \frac{\delta S[x(t)]}{\delta x} = -\frac{d^2 x}{d\tau^2} + \frac{dV(x)}{dx} \qquad (3.55)$$

with the periodic boundary condition $x(\tau + \beta) = x(\tau)$. These trajectories have the integral of motion

$$E = \tfrac{1}{2}\left(\frac{dx}{d\tau}\right)^2 - V(x) \qquad (3.56)$$

in accordance with prescription (3.10). There are two trivial solutions to (3.55), $x \equiv 0$ and $x \equiv x_\#$. The first of these corresponds to Z_0, while the second corresponds to the partition function of the transition state. Aside from the trivial solutions, there is a β-periodic upside-down barrier trajectory $x_{\text{ins}}(\tau)$ called an instanton, or "bounce"[1] [Callan and Coleman, 1977; Langer, 1969; Polyakov, 1977]. As $\beta \to \infty$, the instanton spends most of its time in the vicinity of the point $x = 0$, visiting the barrier region (near $x_\#$) only during some finite time $\tau_{\text{ins}} \sim \omega_\#^{-1}$ (Figure 3.4).

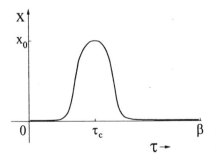

Figure 3.4. The bounce trajectory in the cubic potential at $\beta\omega_0 \to \infty$.

[1] Langer, who was the first to introduce the Im F method in his original paper [Langer, 1969] on nucleation theory called it a bubble.

When T is raised, the instanton amplitude decreases until the trajectory collapses to the point $x(\tau) \equiv x_{\#}$. As follows from the arguments of Section 2.1, this happens at $\beta_c = 2\pi/\omega_{\#}$. In other words, the crossover phenomenon is associated with disappearance of the instanton.

Unlike the trivial solution $x \equiv 0$, the instanton, as well as solution $x(\tau) \equiv x_{\#}$, is not the minimum of the action $S[x(\tau)]$, but rather a saddle point, because there is at least one direction in the space of functions $x(\tau)$ (i.e., toward the absolute minimum $x(\tau) \equiv 0$) in which the action decreases. Hence, if we were to try to use the steepest descent approximation in the path integral (3.13), we would obtain divergences from these two saddle points. This is not surprising because the partition function corresponding to the unbounded Hamiltonian also diverges. Langer [1969] proposed extending the integration to complex space and then using the steepest descent approximation. This procedure makes the partition function converge, and it acquires the sought-after imaginary part Im Z coming from the two saddle-point solutions. Accordingly, working with exponential accuracy, one has two contributions to Im Z, proportional $\exp[-S_{\text{ins}}(\beta)]$, where S_{ins} is the instanton action, and to $\exp(-\beta V_0)$. These contributions correspond to tunneling and thermally activated over-barrier transitions. Whereas the latter is dominant at high temperatures $(\beta < \beta_c)$, it becomes negligible compared with the contribution from the instanton in the region of the low-temperature plateau.

The uncertainty principle requires that any extremum path should be "spread," and the next step in our calculation is to find the prefactor in (3.54) by incorporating small fluctuations around the instanton solution, in the spirit of the usual steepest descent method. Following Callan and Coleman [1977], let us represent an arbitrary path in the form $x(\tau) = x_{\text{ins}}(\tau) + \delta x(\tau)$ and expand the action functional up to quadratic terms in $\delta x(\tau)$, assuming that these deviations are small:

$$S[x(\tau)] = S_{\text{ins}} + \frac{1}{2} \int_0^\beta d\tau \, \delta x(\tau) \left\{ -\partial_\tau^2 + \frac{d^2 V[x(\tau)]}{dx^2} \right\} \delta x(\tau) \qquad (3.57)$$

where ∂_τ^2 is the second derivative operator. The linear terms are absent from (3.57) due to the extremality condition (3.55). Suppose that we have found the basis of eigenfunctions of the differential operator $-\partial_\tau^2 + d^2 V[x(\tau)]/dx^2$

$$\left\{ -\partial_\tau^2 + \frac{d^2 V[x(\tau)]}{dx^2} \right\} x_n(\tau) = \epsilon_n x_n(\tau) \qquad (3.58)$$

Expanding an arbitrary path $x(\tau)$ in this basis

$$\delta x(\tau) = \sum c_n x_n(\tau) \tag{3.59}$$

we rewrite (3.57) as

$$S = S_{\text{ins}} + \sum \frac{1}{2} \epsilon_n c_n^2 \tag{3.60}$$

The path integration reduces to an integration over all coefficients c_n with the measure

$$D[x(\tau)] = N_0 \prod_n \frac{dc_n}{(2\pi)^{1/2}} \tag{3.61}$$

We do not have to worry about the normalizing factor N_0 because it will eventually cancel out in the ratio $\text{Im}\, Z/Z_0$. The integral (3.13) with the action (3.60) is Gaussian and equal to

$$\int N_0 \prod_N \frac{dc_n}{(2\pi)^{1/2}} \exp(-S_{\text{ins}}) \exp\left(-\sum \frac{1}{2} \epsilon_n c_n^2\right)$$

$$= \frac{i}{2} N_0 \left(\prod |\epsilon_n|\right)^{-1/2} \exp(-S_{\text{ins}})$$

$$\equiv \frac{i}{2} N_0 \left|\det\left(-\partial_\tau^2 + \frac{d^2 V}{dx^2}\right)\right|^{-1/2} \exp(-S_{\text{ins}}) \tag{3.62}$$

where we have defined the determinant of the differential operator as the product of its eigenvalues. Since the instanton is a saddle point, one of the eigenvalues, say ϵ_1, must be negative, and, according to Langer, the integration contour for c_1 should be distorted in the c_1 plane, as shown in Figure 3.5. This makes the integral (3.62) imaginary and provides an additional factor $\frac{1}{2}$. A more vexing point is that one ϵ_n in (3.58) equals zero. To see this, note that the function $\dot{x}_{\text{ins}}(\tau)$ can be readily shown to satisfy (3.58) with $\epsilon_0 = 0$. Since the instanton trajectory is closed, it can be

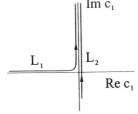

Figure 3.5. Integration contour in the complex plane for an unstable mode. Contours L_1 and L_2 are used to calculate $\text{Im}\, Z$ and the barrier partition function, respectively.

started from an arbitrary point. It is this *zero mode*, which is responsible for the time-shift invariance of the instanton solution. Therefore, the non-Gaussian integration over c_0 is expected to be the integration over positions of the instanton center τ_c (see Figure 3.4). The eigenfunction x_0 is the properly normalized $\dot{x}_{ins}(\tau)$

$$x_0(\tau) = N\dot{x}_{ins}(\tau) \tag{3.63}$$

The normalization condition $\int_0^\beta x_0^2(\tau)\, d\tau = 1$ yields

$$N = S_0^{-1/2} \qquad S_0 = \int_0^\beta d\tau\, \dot{x}_{ins}^2 \tag{3.64}$$

and the integration over c_0 is replaced by

$$(2\pi)^{-1/2}\, dc_0 = \left(\frac{S_0}{2\pi}\right)^{1/2} d\tau_c \tag{3.65}$$

where integration over τ_c simply gives the factor β. Excluding the zero mode from (3.62) by use of (3.65) and recalling that Z_0 may be represented in the same way as (3.62)

$$Z_0 = N_0[\det(-\partial_\tau^2 + \omega_0^2)]^{-1/2} = \left[2\sinh\left(\frac{\beta\omega_0}{2}\right)\right]^{-1} \tag{3.66}$$

one obtains finally

$$k = 2\,\mathrm{Im}\,F = \left(\frac{S_0}{2\pi}\right)^{1/2}\left|\frac{\det'(-\partial_\tau^2 + d^2V/dx^2)_{ins}}{\det(-\partial_\tau^2 + \omega_0^2)}\right|^{-1/2}\exp(-S_{ins}) \tag{3.67}$$

where the prime indicates that the zero mode is omitted from the determinant. Callan and Coleman [1977] have obtained this formula in the limit $\beta \to \infty$ by summing up all multi-instanton contributions to $\mathrm{Im}\,F$, i.e., by taking into account the trajectories that pass through the barrier more than once. These trajectories enter into $\mathrm{Im}\,F$ with factors $\exp(-nS_{ins})/n!$ (where n is the number of passes) and, therefore, they can be neglected when $S_{ins} \gg 1$.

Formula (3.67) has been applied to the cubic parabola (3.37) at $T = 0$

by Caldeira and Legett [1983]. The result is

$$k_c = 60^{1/2} \omega_0 \left[\frac{S_{ins}}{2\pi} \right]^{1/2} \exp(-S_{ins})$$

$$S_{ins} = \int_0^{x_0} [2V(x)]^{1/2} \, dx = \frac{36V_0}{5\omega_0}$$

(3.68)

and the crossover temperature is $T_c = \omega_0/2\pi$. The same result has been obtained with more traditional quantum mechanics [Schmid, 1986] by directly solving the Schrödinger equation in the WKB approximation (see Appendix A). The calculation of $\Gamma_0 = k_c$ with the aid of (3.52) and (3.53) for the same potential [Wartak and Krzeminski, 1989] gives a result that is ~3.34 times greater than (3.68). This discrepancy is due to use of the semiclassical quantization condition (3.52) for the ground state, whereas both the instanton method and Schmid's calculation essentially utilize the exact harmonic oscillator wave function in the well. However, formulae (3.52) and (3.53) may work better than the instanton approach (3.67) when anharmonicity is substantial even for the ground state [Hontscha et al., 1990]. As shown by Rajaraman [1975] and Waxman and Leggett [1985], the basic instanton relationship (3.67) is equivalent to Eqs. (2.6) and (2.7) obtained by extension of CLTST to the quantum regime. Comparison of (2.6) and (2.7) with (3.67) gives the semiclassical estimate for Im Z:

$$2 \, Im \, Z = \beta(2\pi)^{-1/2} \left| \frac{d\beta}{dE} \right|^{-1/2} \exp(-S_{ins})$$

$$= \beta(\pi)^{-1/2} \left| \frac{d^2 S_{ins}}{d\beta^2} \right|^{1/2} \exp(-S_{ins})$$

(3.69)

In particular, the high-temperature regime $T > T_c$, where the trivial trajectory $x(\tau) \equiv x_\#$ is the only solution contributing to Im F, is correctly described by the instanton formalism. Furthermore, the equivalence of (2.6) (at $T = 0$) and (3.68) for the cubic parabola is demonstrated in Appendix B. Although at first the infinite determinants in (3.67) might look less attractive than simple formulas (2.6) and (2.7) or the direct WKB solution by Schmid, it is the instanton approach that permits direct generalization to dissipative tunneling and to the multidimensional problem.

Gillan [1987], while studying tunneling of a particle interacting with a classical oscillator bath, proposed a physically transparent CLTST-based approach to the quantum rate constant using the centroid approximation.

This idea has been developed and tested by Voth et al. [1989b]. Following this work, we wish to modify the CLTST formula (2.12), which is rewritten as

$$k_{CLTST} = \tfrac{1}{2} u_{cl} Z_0^{-1} Z_\#(x_\#)$$ (3.70)

where the constrained transition-state partition function $Z_\#$ is

$$Z_\#(x_\#) = (2\pi\beta)^{-1/2} \int dx \, \exp[-\beta V(x)] \delta(x - x_\#)$$ (3.71)

so as to incorporate tunneling effects. The velocity factor $u_{cl} = \langle |\dot{x}| \rangle = (2/\pi\beta)^{1/2}$ in (3.70) represents the reactive flux for those trajectories that have reached the transition state. The original idea of approximating the quantum mechanical partition function by a classical one belongs to Feynman [Feynman and Hibbs, 1965; Feynman and Kleinert, 1986]. Expanding an arbitrary β-periodic orbit, entering into the partition function path integral, in Fourier series by Matsubara frequencies ν_n

$$x(\tau) = \beta^{-1} \sum_{n=-\infty}^{\infty} \sum_{-\infty}^{\infty} x_n \exp(i\nu_n\tau) \qquad \nu_n = \frac{2\pi n}{\beta}$$ (3.72)

it is easy to see that the kinetic energy term in the action $S[x(\tau)]$ takes the form

$$\int_0^\beta \frac{1}{2} \dot{x}^2 \, d\tau = \frac{1}{2} \beta \sum_{n=-\infty}^{\infty} \sum_{-\infty}^{\infty} \nu_n^2 x_n x_{-n}$$ (3.73)

For small β's the contribution of paths with large x_n's ($n \neq 0$) to the partition function Z is suppressed because they are associated with large kinetic energy terms proportional to ν_n^2. That is why the partition function actually becomes the integral over the zero Fourier component x_0. It is therefore plausible to conjecture that the quantum corrections to the classical TST formula (3.70) may be incorporated by replacing $Z_\#$ by

$$Z_{c\#}(x_\#) = \int D[x(\tau)] \exp\{-S[x(\tau)]\} \delta(x_c - x_\#)$$ (3.74)

where the *centroid* coordinate is

$$x_c = \frac{x_0}{\beta} = \beta^{-1} \int_0^\beta x(\tau) \, d\tau$$ (3.75)

To show that this guess is actually consistent with the Im F approach and to see what happens to the velocity factor u_{cl} at low temperature, let us examine the statistics of centroids. We introduce the centroid density

$$Z_c(R) = \langle \delta(R - x_c) \rangle = \int D[x(\tau)] \exp\{-S[x(\tau)]\}\delta(x_c - R)$$

$$= \int_{-\infty}^{\infty} \frac{d\lambda}{2\pi} \int D[x(\tau)] \exp\{-S[x(\tau)] - i\lambda(x_c - R)\} \tag{3.76}$$

where we have used the integral representation of the δ function. Apart from integration over λ and a constant factor, (3.76) is the partition function Z_λ in the complex potential $V_\lambda(x) = V(x) + i\lambda x$:

$$Z_\lambda = \int D[x(\tau)] \exp\{-S_\lambda[x(\tau)]\}$$

$$\tag{3.77}$$

$$S_\lambda[x(\tau)] = \int_0^\beta d\tau \left[\frac{\dot{x}^2}{2} + V_\lambda(x) \right]$$

so that $Z_c(R)$ is the Fourier transform of Z_λ

$$Z_c(R) = \int_{-\infty}^{\infty} \frac{d\lambda}{2\pi} \exp(i\lambda R) Z_\lambda \tag{3.78}$$

In fact, Z_λ is twice Im Z defined above,[2] taken for V_λ [Stuchebrukhov, 1991]; therefore,

$$Z_\lambda = \beta(2\pi)^{-1/2} \left| \frac{d^2 S_\lambda}{d\beta^2} \right|^{1/2} \exp(-S_{\lambda,\text{ins}}) \tag{3.79}$$

Expanding the action $S_{\lambda,\text{ins}}$ around $\lambda = 0$, one obtains

$$S_{\lambda,\text{ins}} = S_{\text{ins}} + \left(\frac{\lambda}{\beta} \right) S_1 + \left(\frac{\lambda}{\beta} \right)^2 S_2 + \cdots \tag{3.80}$$

where

$$S_1 = i\beta \langle x \rangle_{\text{ins}}$$

$$S_2 = \left(\frac{2 d^2 S_{\text{ins}}}{d\beta^2} \right)^{-1} \left[\frac{d}{d\beta} \beta \langle x^2 \rangle_{\text{ins}} + \left(\frac{dS_1}{d\beta} \right)^2 \right] \tag{3.81}$$

---

[2] In a semiclassical evaluation of the barrier partition function Z_λ the integration goes along the whole imaginary axis in the c_1 plane (see Figure 3.5).

and $\langle x \rangle_{ins}$ and $\langle x^2 \rangle_{ins}$ are imaginary time averages over the instanton trajectory. Neglecting higher-order terms in (3.80), one finds from (3.78) that the centroid density is Gaussian:

$$Z_c(R) = 2 \, \text{Im} \, Z \pi^{-1/2} (\Delta x)^{-1} \exp\left[\frac{(R - \langle x \rangle_{ins})^2}{(\Delta x)^2}\right] \quad (3.82)$$

where the characteristic length Δx is

$$\Delta x = -\frac{4S_2}{\beta^2} \quad (3.83)$$

It is obvious from (3.82) that the modified "transition-state" position in the quantum region is given by

$$\tilde{x}_\# = \langle x \rangle_{ins} \quad (3.84)$$

and comparison of (3.54) with (3.82) gives

$$k = \frac{\pi^{1/2} \, \Delta x \, \beta^{-1} Z_c(\hat{x}_\#)}{Z_0} \quad (3.85)$$

This formula looks like the CLTST formula (3.70), if we introduce the quantum velocity factor

$$u = 2\pi^{1/2} \, \Delta x \, \beta^{-1} \quad (3.86)$$

At high temperatures ($\beta \to 0$) the centroid (3.75) collapses to a point so that the centroid partition function (3.74) becomes a classical one (3.71) and the velocity (3.86) should approach the classical value u_{cl}. In particular, it can be shown directly [Voth et al., 1989b] that the centroid approximation yields the correct Wigner formula (2.11) for a parabolic barrier at $T > T_c$, if one uses the classical velocity factor u_{cl}. A direct calculation of Δx for a parabolic barrier at $T > T_c$ gives

$$\Delta x_{pb}^2 = \frac{2}{\beta \omega_\#^2} \quad (3.87)$$

That is, the centroids are distributed according to the classical function $Z_c(R) \propto \exp[\beta \omega_\#^2 (R - x_\#)^2 / 2]$. Use of (3.87) for (3.86), however, does not give the correct value of u_{cl}; . To fit smoothly the high- and low-temperature regions, Voth et al. [1989b] have written $u = 2\pi^{1/2} \, \Delta x \, \beta^{-1} \phi$, where $\phi = 1$ at low temperatures, and $\phi = T_c/T = \beta \omega_\# / 2\pi$ at $T > T_c$ (see also Affleck [1981]). This correction factor recovers the correct behavior of u at high temperatures, where formulae (3.81)–(3.83) do not work. For the Eckart barrier

$$V(x) = V_0 \operatorname{sech}^2\!\left(\frac{x}{x_0}\right) \tag{3.88}$$

a simple empirical formula for the velocity factor

$$u \cong u_{cl}\!\left(1 + \ln \frac{\beta\omega_\#}{2\pi}\right) \tag{3.89}$$

has been shown by Voth et al. [1989b] to work with 1% accuracy at $T_c/2 \le T \le T_c$. While being very attractive in view of their similarity to CLTST, on closer inspection formulae (3.84)–(3.86) reveal their deficiency at low temperatures. As $\beta \to \infty$, the characteristic length Δx from (3.83) becomes large, and the expansion (3.80) as well as the Gaussian approximation for the centroid density breaks down. Although the results of Voth et al. [1989b], have demonstrated the success of the centroid approximation for the Eckart barrier at $T \ge T_c/2$, no realistic case of a potential with bound initial state was considered at sufficiently low temperature, so there are no grounds to trust formula (3.85) as an estimate for k_c.[3] The situation becomes even worse for an asymmetric potential like that in (3.37), because at low temperatures nearly all the period β is spent on dwelling in the potential well (see Appendix B), so that $\lim_{\beta \to \infty} \langle x \rangle_{ins} = 0$. In other words, unless the potential is strictly symmetric,[4] the transition-state position $\tilde{x}_\#$ tends to the minimum of the initial state! It is natural to expect that the centroid approximation will work well when $\tilde{x}_\#$ does not deviate too far from $x_\#$. To summarize, the centroid method is an instructive way to describe in a unique TST-like manner both the high $(T > T_c)$ and fairly low $(T < T_c)$ temperature regions, but it does not give a reliable estimate for k_c.

3.6. TUNNELING SPLITTING IN A DOUBLE WELL

Coleman's method can be applied to finding the ground-state tunneling

---

[3] Strictly speaking, the concept of k_c itself makes no sense for a potential like the Eckart one, unless one artificially introduces Z_0 as the partition function of a bound initial state, which is not described by this potential. That is to say, it is reasonable to consider the combination kZ_0, rather than k alone.

[4] And it is the symmetric situation that has been studied by Gillan [1987] and Voth et al. [1989b].

splitting in a symmetric double well [Vainshtein et al., 1982], for example,

$$V(x) = \lambda(x^2 - x_0^2)^2 \tag{3.90}$$

It is expedient, as in the previous subsection, to study the partition function Z in the limit $\beta \to \infty$, which is real now. The trajectories satisfying (3.55) are composed of *single passes* or *kinks* and *antikinks*. During a single kink the particle spends an infinite time sliding from the upside-down potential top $x = -x_0$, crosses the barrier region $x \approx 0$ during some finite time, and then approaches the point $x = x_0$ during an infinite time. The same event in the reverse order is an antikink. For the potential (3.90) the kink (antikink) is described by

$$x_k(\tau) = \pm x_0 \tanh \frac{\omega_0}{2}(\tau - \tau_c) \tag{3.91}$$

where $(8\lambda x_0^2)^{1/2}$ is the frequency of classical vibrations in the well near $x = \pm x_0$, and τ_c is an arbitrary position of the kink center, $0 < \tau_c < \beta$. The single kink action

$$S_k = \int_{-x_0}^{x_0} dx[2V(x)]^{1/2} = \frac{\omega_0^3}{12\lambda} \tag{3.92}$$

is independent of τ_c. Each extremal trajectory includes n kink–antikink pairs, where n is an arbitrary integer, and the kink centers are placed at the moments $0 < \tau_1 < \cdots < \tau_{2n} < \beta$ forming the *instanton gas* (Figure 3.6). Its contribution to the overall path integral may be calculated in exactly the same manner as in the previous section, with the assumption that the instanton gas is dilute, i.e., the kinks are independent of each other:

$$Z_n = Z_0 \int_0^\beta d\tau_1 \cdots \int_{\tau_{2n-1}}^\beta d\tau_{2n} \left(\frac{\Delta}{2}\right)^{2n} = \frac{\beta^{2n}}{(2n)!} \left(\frac{\Delta}{2}\right)^{2n} Z_0 \tag{3.93}$$

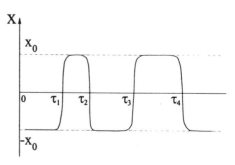

Figure 3.6. Dilute instanton gas.

where

$$\frac{\Delta}{2} = \left(\frac{S_k}{2\pi}\right)^{1/2} \left|\frac{\det'(-\partial_\tau^2 + d^2V/dx^2)_{\text{kink}}}{\det(-\partial_\tau^2 + \omega_0^2)}\right|^{-1/2} \exp(-S_k) \qquad (3.94)$$

and Z_0 is given by (3.66). Summation of all the terms in (3.93) gives

$$Z = Z_0 + \sum_{n=1}^{\infty} Z_n = Z_0 \cosh\left(\frac{\beta\Delta}{2}\right) \qquad (3.95)$$

What we have obtained in (3.95) is the partition function of a two-level system, each level having the energy $\omega_0/2 \pm \Delta/2$. Hence, (3.94) gives us the desired ground-state tunneling splitting Δ. The analytical continuation of relation (3.95) to real time (imaginary β) leads to a two-level system propagator oscillating as $\cos(\Delta t/2)$. This picture corresponds to coherent oscillations of the probability of finding a particle in one of the wells with frequency Δ.

Consider in some greater detail the calculation of the determinant in (3.94) for the potential (3.90). The eigenvalue equation

$$\left(-\partial_\tau^2 + \frac{d^2V}{dx^2}\right)x_n(\tau) = \frac{\partial^2 x_n}{\partial\tau^2} + \omega_0^2\left[1 - \frac{3}{2}\text{sech}^2\left(\frac{\omega_0\tau}{2}\right)\right]x_n$$

$$= \epsilon_n x_n(\tau) \qquad (3.96)$$

is formally equivalent to the stationary Schrödinger equation for the inverted Eckart potential, the exact solution of which is well known [Landau and Lifshitz, 1981]. The spectrum of bound states satisfies the relation $(\omega_0^2 - \epsilon_n)^{1/2} = \omega_0(1 - n/2)$, $n = 0, 1, \ldots$. Therefore, there are only two bound states ($n = 0, 1$) with eigenvalues $\epsilon_0 = 0$ and $\epsilon_1 = 3\omega_0^2/4$. The normalized zero mode with $\epsilon_0 = 0$ is given by

$$x_0(\tau) = \left(\frac{3\omega_0}{8}\right)^{1/2} \text{sech}^2\left[\frac{\omega_0(\tau - \tau_c)}{2}\right] \qquad (3.97)$$

Formula (3.97) may also be obtained from (3.63). The contribution of x_1 to the ratio of determinants in (3.94) is equal to $\frac{3}{4}$. Apart from the bound states, there is a continuous spectrum of eigenvalues ϵ_n whose contribution to (3.94) may be shown [Vainshtein et al., 1982] to be equal to $\frac{1}{12}$. Finally, (3.94) gives

$$\Delta = \frac{\omega_0}{\pi}\left(2\pi\frac{\omega_0^3}{\lambda}\right)^{1/2}\exp\left(-\frac{\omega_0^3}{12\lambda}\right) \qquad (3.98)$$

A disadvantage of this method is that it applies only to the lowest energy doublet. It is natural to calculate the tunneling splittings of highest energy levels with the same method as in Section 3.4. Following Miller [1979], we consider the more general situation of a double well that is not necessarily symmetric. The spectral function $g(E)$ in (3.44) comes from summation over closed classical trajectories with energy E. Each such trajectory is composed of vibrations in the classically accessible regions (1) and (3) and the imaginary-time orbits in the barrier region (2) (see Figure 2.1 replacing Q by x). Introducing the phases

$$W_1(E) = 2 \int_{x_1'}^{x_1} dx \{2[E - V(x)]\}^{1/2} - \pi$$

$$iW_2(E) = 2 \int_{x_1}^{x_2} dx \{2[E - V(x)]\}^{1/2} \tag{3.99}$$

$$W_3(E) = 2 \int_{x_2}^{x_2'} dx \{2[E - V(x)]\}^{1/2} - \pi$$

and, in view of the exponentially small value of $\exp(-W_2)$, neglecting the multiple crossings of the barrier, one obtains for $g(E)$

$$g(E) \propto \frac{\exp(iW_1)}{1 - \exp(iW_1)} - \exp(-W_2) \frac{\exp[i(W_1 + W_3)]}{[1 - \exp(iW_1)]^2 [1 - \exp(iW_3)]} \tag{3.100}$$

For a symmetric potential $W_1 = W_3 = W(E)$. Approximating $W(E)$ near the nth energy level E_n of an isolated well by

$$W(E) = 2\pi n + \left(\frac{dW}{dE}\right)_{E=E_n} (E - E_n) \tag{3.101}$$

so that the Bohr-Sommerfeld condition holds at $E = E_0$ (with the second term in (3.100) neglected), one obtains

$$g(E) \propto \frac{1}{E - E_n} + \left(\frac{dW}{dE}\right)_{E=E_n}^{-2} \frac{\exp(-W_2)}{(E - E_n)^3}$$

$$\cong \frac{1}{2} \left(\frac{1}{E - E_n - \Delta_n/2} + \frac{1}{E - E_n + \Delta_n/2} \right) \tag{3.102}$$

and the tunneling splitting of the nth energy level is

$$\Delta_n = 2\left(\frac{dW}{dE}\right)_{E=E_n}^{-1} \exp\left(-\frac{W_2}{2}\right) = \frac{2}{\tau_n} \exp\left(-\frac{W_2}{2}\right) \qquad (3.103)$$

where τ_n is the period of classical vibrations in the well with energy E_n. This formula resembles (3.53) and, as we shall show in due course, this similarity is not accidental. Note that at $n=0$ the short action $W_2(E_0)/2$ taken at the ground state energy E_0 is not equal to the kink action S_k (3.92). Since in the harmonic approximation for the well $\tau_0 = 2\pi/\omega_0$, this difference should be compensated by the prefactor in (3.98), but generally speaking, expressions (3.98) and (3.103) are not identical because Eq. (3.103) uses the semiclassical approximation for the ground state, whereas (3.98) does not.

Let us now consider an asymmetric potential. If E_1 and E_2 are the individual energy levels in each well so that $W_1(E_1) = 2\pi n_1$ and $W_2(E_2) = 2\pi n_2$, then (3.100) becomes

$$g(E) \propto \frac{1}{E - E_1} + \frac{\exp(-W_2)}{(E - E_2)^2(E_1 - E_2)}[W_1'(E_1)W_2'(E_2)]^{-1} \qquad (3.104)$$

Comparing this with $g(E) \propto (E - E_1 - \Delta E)^{-1}$ at $\Delta E \ll |E_1 - E_2|$, one obtains for the shift of the energy level

$$\begin{aligned}\Delta E &= [(E_1 - E_2)W_1'(E_1)W_2'(E_2)]^{-1} \exp(-W_2) \\ &= [(E_1 - E_2)\tau_1\tau_2]^{-1} \exp(-W_2) \\ &= \frac{\omega_1}{2\pi}\frac{\omega_2}{2\pi}(E_1 - E_2)^{-1} \exp(-W_2) \qquad (3.105)\end{aligned}$$

where ω_1 and ω_2 are the vibrational frequencies in the respective wells. The same procedure performed near $E = E_2$ gives $\Delta E_1 = -\Delta E_2$. To gain physical insight into the asymmetric situation let us compare Miller's result (3.105) with that obtained by considering a formal two-state problem with the Hamiltonian matrix

$$\begin{bmatrix} E_1 & V_{12} \\ V_{21} & E_2 \end{bmatrix} \qquad (3.106)$$

If $|V_{12}| \ll |E_1 - E_2|$, then the shift of an energy level appears in the second order in V_{12}, viz.,

$$\Delta E \cong \frac{|V_{12}|^2}{E_1 - E_2} \qquad (3.107)$$

Comparing this expression with (3.105), one obtains the formal definition of the quasiclassical tunneling matrix element

$$|V_{12}|^2 = \frac{\omega_1}{2\pi} \frac{\omega_2}{2\pi} \exp(-W_2) \qquad (3.108)$$

This definition is also valid in the symmetric case ($E_1 = E_2$) when $V_{12} = \Delta/2 = (2\pi)^{-1}\omega_1 \exp(-W_2/2)$.

In time-dependent perturbation theory [Landau and Lifshitz, 1981] the transition probability from the state 1 to state 2 is related to the perturbation by the golden rule:

$$\Gamma = 2\pi|V_{12}|^2\rho_2 \qquad (3.109)$$

Substituting V_{12} from (3.108) into this expression and taking into account the fact that the density of energy levels in the final state ρ_2 is ω_2^{-1}, one obtains the formula (3.53) for decay of the metastable state. Strictly speaking, the solution for the time-dependent Schrödinger equation for a double well would exhibit coherent oscillations of the probability of finding the particle in any given well, rather than an exponential decay. Expression (3.109) comes about either for a metastable state or when there is an additional mechanism destroying the phase coherence (e.g., resulting from interaction with other degrees of freedom (the bath)). We discuss this problem in Chapter 5.

The calculation of propagator (3.1) for an asymmetric double well was carried out by Cesi et al. [1991].

3.7. NONADIABATIC TUNNELING

The results of previous sections in this chapter relied on the assumption that the transition occurred on a single potential energy surface $V(x)$ characterized by a barrier separating two wells. In reality, this potential is usually created from two different (initial and final) electronic states. The separation of electron and nuclear coordinates in each of these states gives rise to a diabatic basis and a nondiagonal Hamiltonian matrix

$$\hat{H} = \frac{P^2}{2}\hat{1} + \begin{bmatrix} V_i(x) & V_d(x) \\ \bar{V}_d(x) & V_f(x) \end{bmatrix} \qquad (3.110)$$

where $\hat{1}$ is the 2×2 unit matrix. The off-diagonal matrix elements are significant in the vicinity of the crossing point x_c such that $V_i(x_c) = V_f(x_c)$. Diagonalization of the potential energy matrix in (3.110) gives the adiabatic terms

$$V_\pm = \tfrac{1}{2}(V_i(x) + V_f(x)) \pm \tfrac{1}{2}\{[V_i(x) - V_f(x)]^2 + 4V_d^2(x)\}^{1/2} \quad (3.111)$$

which are separated at the crossing point by the adiabatic splitting $2|V_d|$ (Figure 3.7). The usual tunneling problem, which we considered previously, is associated with the situation when this splitting is so large that the influence of the upper term V_+ can be neglected, and tunneling occurs in the potential V_-. It is obvious, however, that when $V_d = 0$, no transitions are possible between the noninteracting terms. Therefore, the relationship between adiabatic splitting and the tunneling probability requires special consideration. For a classical transition, this is the well-known Landau-Zener-Stueckelberg problem [Landau, 1932; Stueckelberg, 1932; Zener, 1932], which has been investigated in detail in molecular collision theory (see, e.g., Child [1974] and Nakamura [1991]). With the assumption that near x_c the diabatic terms are linear and V_d is independent of x, the probability of transition between the diabatic terms at $E > V_+(x_c)$ depends on the parameter

$$\delta = \frac{|V_d|^2}{v|F_1 - F_2|} \quad (3.112)$$

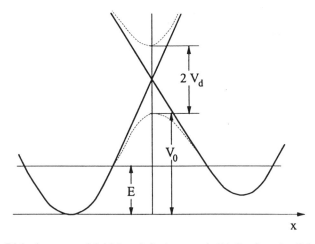

Figure 3.7. Diabatic terms of initial and final states (solid lines) and adiabatic terms (dashed lines). Adiabatic splitting is $2V_d$.

where F_1 and F_2 denote the slopes of the diabatic terms, and v is the classical velocity of the particle at $x = x_c$. The transition probability is

$$B_0(\delta) = 1 - \exp(-2\pi\delta) \qquad (3.113)$$

At $\delta > 1$, $B_0 \approx 1$ and transition occurs along the lower adiabatic term V_-. At $\delta \ll 1$, $B_0 = 2\pi\delta$, and there is a large probability that the particle passes the crossing point remaining on the initial diabatic term. This is the case in which perturbation theory in V_d is valid, yielding the golden rule formula. With increasing energy, the parameter δ decreases, thus enhancing nonadiabatic effects. In essence, the system traverses the crossing region so rapidly that it hardly feels the effect of the second diabatic term, so the transition probability is small.

The problem of nonadiabatic tunneling in the Landau-Zener approximation has been solved by Ovchinnikova [1965]. For further refinements of the theory and the ways to go beyond this approximation, see Laing et al. [1977], Holstein [1978], Coveney et al. [1985], and Zhu and Nakamura [1992]. The nonadiabatic transition probability for a more general case of dissipative tunneling is found in Appendix C at the end of Chapter 5. We cite here only the result of the dissipationless case, which is commensurate with the results of the papers cited above. When $E < V_-(x_c)$, the total transition probability is the product of the adiabatic tunneling rate, calculated in previous sections, and a factor resembling the Landau-Zener-Stueckelberg expression:

$$B = \frac{2\pi\delta^{-1} e^{-2\delta}\delta^{2\delta}}{\Gamma^2(\delta)} \qquad (3.114)$$

where δ is defined by (3.112) but the classical velocity is replaced by the absolute value of imaginary-time instanton velocity,

$$v = [2(V_-(x_c) - E)]^{1/2} \qquad (3.115)$$

As can be seen from (3.115), nonadiabaticity increases with decreasing energy, as opposed to the classical case. This is a straightforward consequence of the Vick rotation (3.10).

According to the general ideas formulated in Sections 2.3 and 3.6 (see especially formulas (3.108) and (3.109) and Appendix C) the incoherent tunneling rate is proportional to the square of the tunneling splitting. Therefore, when the diabatic terms are symmetric with respect to the crossing point, the tunneling splitting is the product of the factor $B^{1/2}$ and the splitting in the lower adiabatic potential V_-, Δ_{ad}

$$\Delta = \frac{(2\pi)^{1/2}\delta^{-1/2+\delta}\,e^{-\delta}}{\Gamma(\delta)}\Delta_{ad} \tag{3.116}$$

In the nonadiabatic regime, Δ is proportional to the adiabatic splitting $2|V_d|$. The instanton trajectory crosses the barrier twice, each passage bringing the factor Δ/Δ_{ad} associated with the probability to cross the nonadiabaticity region remaining on the same adiabatic term (and thus jumping from one diabatic term to the other). These two crossings result in one prefactor (3.114) when tunneling is incoherent.

The problem of two crossing points was considered by Shimshoni and Gefen [1991]. In this case the resulting value of the transition probability is dominated by an interference effect, which becomes important when the separation between crossing points is comparable with the characteristic length of interaction region for a single Landau-Zener transition.

3.8. QUANTUM TRANSITION STATE THEORY

So far we have considered two limiting cases of transition, the decay of a metastable state and quantum probability oscillations in a double well, of which only the first case permits the use of the concept of a rate constant. The latter, in general, describes the relaxation of a chemical system initially deviated from its equilibrium between reactants and products. According to the Onsager regression hypothesis (see, e.g., Chandler [1987]), this relaxation may be expressed in terms of fluctuations of the equilibrium system. The regression hypothesis, together with the fluctuation–dissipation theorem (FDT) leads to the Kubo linear response theory [Kubo, 1957; Kubo and Nakajima, 1957], which was first applied to chemistry by Yamamoto [1960]. The decay of reactant concentration $a(t)$ is characterized by the time-correlation function

$$\frac{[a(t)-a(0)]}{a(0)} = \frac{\langle \delta a(t)\,\delta a(0)\rangle}{\langle [\delta a(0)]^2\rangle} = \exp(-kt) \tag{3.117}$$

Since the decay is associated with passing through the barrier, the quantity $a(t)$ is nothing but the step function $a = \theta(x^{\#} - x)$. Differentiating (3.117) and finally setting $t = 0$, one obtains [Chandler, 1987] the expression for the rate constant:

$$k = \langle a\rangle^{-1}\langle \dot{x}(0)\delta(x - x^{\#})\theta(x^{\#} - x)\rangle \tag{3.118}$$

The CLTST rate constant is obtained from (3.118) by the total neglect of correlations between the flux $\dot{x}(0)$ and the concentration, viz.,

$$k_{\text{CLTST}} = \tfrac{1}{2}\langle a\rangle^{-1}\langle \dot{x}(0)\rangle\langle\delta(x - x^{\#})\rangle \qquad (3.119)$$

where the identity $\langle \dot{x}(0)\theta(x^{\#} - x))\rangle \equiv -\tfrac{1}{2}\langle \dot{x}(0)\rangle$ has been used. This illustrates that the θ function selects the direction of the flux. Relation (2.2), which we cited before, is nothing but (3.119) written for a canonical ensemble.

A consistent quantal TST (QTST) has been worked out by Miller and coworkers [Miller, 1974; Miller et al., 1983; Tromp and Miller, 1986; Voth et al., 1989a]. In quantum mechanics the classical flux $\dot{x}$ is replaced by the symmetrized flux operator:

$$\hat{F} = \tfrac{1}{2}[\hat{p}\delta(x - x^{\#}) + \delta(x - x^{\#})\hat{p}] \qquad \hat{p} = -i\partial/\partial x \qquad (3.120)$$

It can be readily verified that the matrix elements of the flux operator between two states are

$$|\langle n|F|n'\rangle|^2 = \tfrac{1}{4}|\psi_n(x^{\#})\psi'_{n'}(x^{\#}) - \psi'_n(x^{\#})\psi_{n'}(x^{\#})|^2 \qquad (3.121)$$

Further, the step function $\theta[x(t) - x_{\#}]$ is replaced by the projection operator $\hat{\mathbb{P}}$ selecting the states that evolve finally to the product valley at $t \to \infty$:

$$\hat{\mathbb{P}} = \lim_{t\to\infty}[\exp(i\hat{H}t)\theta(x - x^{\#})\exp(-i\hat{H}t)] \qquad (3.122a)$$

Its equivalent representation is [Miller et al., 1983]

$$\hat{\mathbb{P}} = \lim_{t\to\infty}[\exp(i\hat{H}t)\theta(p)\exp(-i\hat{H}t)] \qquad (3.122b)$$

The identity of (3.122a) and (3.122b) means that the particle hits the product valley only having crossed the dividing surface $x = x^{\#}$ from left to right. If we were to use simply the step function $\theta(x - x^{\#})$, we would be neglecting the recrossings of the dividing surface. Finally, the formally exact quantal expression for the rate constant is

$$k = Z_0^{-1}\lim_{t\to\infty}\{\text{Tr}[\exp(-\beta\hat{H})\hat{F}\hat{\mathbb{P}}]\} \qquad (3.123)$$

Using the fact that the symmetrized flux operator commutes with the density matrix, and representing the latter as $\exp(-\beta\hat{H}) \equiv \exp[-(\beta -$

$\lambda)\hat{H}]\exp(-\lambda\hat{H})$ one may rewrite (3.123) as

$$k = Z_0^{-1} \lim_{t\to\infty} [\text{Tr}(\hat{F}\exp[i\hat{H}(t+i\lambda)]\theta(x-x^{\#})$$

$$\times \exp\{-i\hat{H}[t-i(\beta-\lambda)]\})] \qquad (3.124)$$

It is customary to take $\lambda = \beta/2$, and, after some manipulations, the following expressions for k can be obtained:

$$k = Z_0^{-1}\int_0^\infty dt\, C_f(t) = z_0^{-1}\lim_{t\to\infty}C_{fx}(t) = z_0^{-1}\lim_{t\to\infty}\frac{d}{dt}C_x(t) \qquad (3.125)$$

where the flux–flux correlation function C_f is defined by

$$C_f(t) = \text{Tr}[\hat{F}\exp(i\hat{H}t_c^*)\hat{F}\exp(-i\hat{H}t_c)] \qquad (3.126)$$

where $t_c = t - i\beta/2$. The cross correlation function of position and flux C_{fx}

$$C_{fx}(t) = \text{Tr}[F\exp(i\hat{H}t_c^*)\theta(x-x^{\#})\exp(-i\hat{H}t_c)] \qquad (3.127)$$

and the left–right spatial correlation function C_x

$$C_x(t) = \text{Tr}[\theta(x^{\#}-x)\exp(i\hat{H}t_c^*)\theta(x-x^{\#})\exp(-i\hat{H}t_c)] \qquad (3.128)$$

are related with C_f by

$$C_f(t) = \frac{dC_{fx}}{dt} = \frac{d^2C_x}{dt^2} \qquad (3.129)$$

According to (3.119), CLTST approximates $C_f(t)$ by the δ function, being, in a sense, a zero-time limit of the first of the equations (3.125). In the representation of eigenfunctions $|n\rangle$, $|n'\rangle$ the flux–flux correlation function is

$$C_f(t) = \sum_{n,n'}\exp\left[-\frac{\beta}{2}(E_n+E_{n'})\right]\cos[(E_n-E_{n'})t]|\langle n|\hat{F}|n'\rangle|^2 \qquad (3.130)$$

so that the rate constant is

$$k = \pi Z_0^{-1}\sum_{n,n'}\exp(-\beta E_n)|\langle n|\hat{F}|n'\rangle|^2\delta(E_n-E_{n'}) \qquad (3.131)$$

The correlation functions (3.126)–(3.128) may be written out explicitly in terms of the propagator $K(x_f, x_i | t_c)$ [Miller et al., 1983]. For the particular case of a parabolic barrier one has

$$C_f = \frac{1}{2\pi\beta} \frac{\beta\omega_\#/2}{\sin(\beta\omega_\#/2)} \exp(-\beta V_0) \frac{\omega_\# \sin^2(\beta\omega_\#/2)\cosh(\omega_\# t)}{[\sin^2(\beta\omega_\#/2) + \sinh^2(\omega_\# t)]^{3/2}}$$

(3.132)

which after integration over time gives the Wigner formula (2.11). The function C_f exponentially decays with a characteristic time close to $1/\omega_\#$.

The correlation functions introduced above may be expressed with the use of path integrals. Consider, for one, the function C_x. Its explicit form is

$$C_x(t) = -\text{Tr}\{\exp(-\beta\hat{H})\theta(x - x^\#)\exp[i\hat{H}(t + i\lambda)]$$

$$\times \theta(x - x^\#)\exp[-i\hat{H}(t + i\lambda)]\}$$

$$= -\langle h(0)h(t + i\lambda)\rangle$$

(3.133)

where $h(t)$ is the $\theta(x - x^\#)$ operator in the Heisenberg representation $h(t) = \exp(i\hat{H}t)\theta(x - x^\#)\exp(-i\hat{H}t)$. Since the result does not depend on λ, it can be rewritten by use of the Kubo transform:

$$C_x(t) = -\beta^{-1}\int_0^\beta d\lambda\langle h(0)h(t + i\lambda)\rangle$$

(3.134)

For purely imaginary time $t = i\tau$ the correlator in (3.134) is expressible via the path integral

$$\langle h(0)h(\tau)\rangle = Z_0^{-1}\int\int\int D[x(\tau)]\exp\{-S[x(\tau)]\}\theta[x(0) - x^\#]$$

$$\times \theta[x(\tau) - x^\#]$$

(3.135)

where the integration is performed over all closed β-periodic paths. The complex-time correlation function then may be found by analytically continuing (3.135). For example, this correlator is readily calculated for a symmetric double well by use of the instanton method of Section 3.6 to give [Gillan, 1987]

$$\langle h(0)h(\tau)\rangle = \frac{\cosh[\beta\Delta/2 - i(t + i\lambda)\Delta]}{\cosh(\beta\Delta/2)}$$

(3.136)

This correlation function oscillates in real time, thus providing no rate constant, as expected for coherent tunneling. This reveals a deficiency of the one-dimensional tunneling model, in which the rate constant, strictly speaking, can be obtained only for an unbound initial or final state, i.e., for a gas-phase reaction. In general, the analytical continuation of the imaginary-time correlator is hardly justifiable, especially for long times ($\beta \to \infty$). The real-time correlation function, represented by a triple path integral, was given in Voth et al. [1989a,b].

Although the correlation function formalism provides formally exact expressions for the rate constant, only the parabolic barrier has proven to be analytically tractable in this way. It is difficult to consistently follow the relationship between the flux–flux correlation function expression and the semiclassical Im F formulae at $\beta \to \infty$. So far, the correlation function approach has been used for fairly high temperatures to accurately study the quantum corrections to CLTST. The behavior of functions C_f, C_{fx}, and C_x far below T_c has not yet been studied.

APPENDIX A

Decay of a Metastable State and Tunneling Splitting in Terms of the One-Dimensional WKB Approximation

Following Schmid [1986], consider the decay of a metastable state described by a potential of rather general form:

$$
V(x) = \begin{cases} \dfrac{\omega_0^2 x^2}{2} & |x| < x_1 \\ -F(x - x_1) & |x - x_1| < x_1 \end{cases} \tag{A.1}
$$

In the vicinity of x_1, a shift in energy means only a redefinition of x_1, so one may suppose without the loss of generality that $E = 0$. Then the solution of the Schrödinger equation, near the right turning point is expressed via the Airy functions:

$$
\psi(x) = N\left[Bi\left(\frac{x_1 - x}{a}\right) + iAi\left(\frac{x_1 - x}{a}\right) \right] \qquad a = (2F)^{-1/3} \tag{A.2}
$$

with the asymptotic form at $x - x_1 \gg a$:

$$
\psi \cong N\pi^{-1/2}\left(\frac{x - x_1}{a}\right)^{-1/4} \exp\left[\frac{2}{3}i\left(\frac{x - x_1}{a}\right)^{3/2} + \frac{i\pi}{4}\right] \tag{A.3}
$$

This function describes an outgoing wave. To the left of x_1, the wave function is a sum of two contributions, the principal and reflected waves

$$\psi = \psi_0 + \psi_1$$

$$\psi_0 = N\pi^{-1/2}\left(\frac{x_1 - x}{a}\right)^{-1/4} \exp\left[\frac{2}{3}\left(\frac{x_1 - x}{a}\right)^{3/2}\right] \tag{A.4}$$

$$\psi_1 = \frac{1}{2} N\pi^{-1/2}\left(\frac{x_1 - x}{a}\right)^{-1/4} \exp\left[-\frac{2}{3}\left(\frac{x_1 - x}{a}\right)^{3/2}\right]$$

Near the minimum of the potential $x = 0$ the wave function is the parabolic cylinder function

$$\psi(x) = D_\nu\left(-\frac{x}{\delta_0}\right) \qquad \delta_0 = (2\omega_0)^{-1/2} \tag{A.5}$$

The solution is chosen such that ψ vanishes far to the left. At $x \gg \delta_0$ the solution takes the asymptotic form

$$\psi = \psi_0 + \psi_1 \tag{A.6}$$

$$\psi_0 = \left|\frac{x}{\delta_0}\right|^\nu \exp\left(-\frac{x^2}{4\delta_0^2}\right)$$

$$\psi_1 = -\nu(2\pi)^{1/2}\left(\frac{\delta_0}{x}\right)^{1+\nu} \exp\left(\frac{x^2}{4\delta_0^2}\right) \tag{A.7}$$

To the left of the origin $x = 0$ the asymptotic behavior of the wave function is that of ψ_0. The wave function (A.5) corresponds to the energy

$$E_0 = \omega_0(\nu + \tfrac{1}{2}) \tag{A.8}$$

whence it is seen that the decay rate equals

$$\Gamma = 2 \operatorname{Im} E_0 = 2\omega_0|\nu| \tag{A.9}$$

To match the asymptotic solutions (A.4) and (A.6) we employ the standard semiclassical ansatz[1]:

$$\psi_0(x) = \exp\left\{-\frac{1}{\hbar}[W(x) + \hbar W_1(x)]\right\} \tag{A.10}$$

---

[1] Although $\hbar = 1$ in our units, we include the Planck constant explicitly in the following expansion to emphasize its asymptotic nature.

which gives the eikonal equation in the leading order in $\hbar$,

$$\frac{\partial W}{\partial x} = [2V(x)]^{1/2} \tag{A.11}$$

and the continuity equation in the next order,

$$W_1'W' - \tfrac{1}{2}W'' + \omega_0(\tfrac{1}{2} + \nu) = 0 \tag{A.12}$$

From (A.11) we have

$$W(x) = \begin{cases} \dfrac{x^2}{4\delta_0^2} & |x| < x_1 \\[3mm] \dfrac{S}{2} - \dfrac{2}{3}\left(\dfrac{x_1 - x}{a}\right)^{3/2} & |x_1 - x| < x_1 \end{cases} \tag{A.13}$$

where S is the total Euclidean action in the barrier, corresponding to twofold crossing of the barrier (for example, $S = \frac{8}{15}\omega_0 x_1^2$ for the cubic parabola potential). Solution (A.13) agrees with the asymptotic forms (A.4) and (A.6). The prefactor is obtained by a straightforward integration of (A.12), which yields

$$\exp[-W_1(x)] \cong \begin{cases} \left|\dfrac{x}{\delta_0}\right|^{\nu} & |x| < x_1 \\[3mm] \left(\dfrac{4x_1}{\delta_0}\right)^{\nu}\left(\dfrac{x_1 - x}{16x_1}\right)^{-1/4} & |x_1 - x| < x_1 \end{cases} \tag{A.14}$$

Comparison of (A.13) and (A.14) with the asymptotic expressions (A.4), (A.6), and (A.7) gives the value of the normalization constant

$$N = 2\pi^{1/2}\left(\frac{x_1}{a}\right)^{1/4}\left(\frac{4x_1}{\delta_0}\right)^{\nu}\exp\left(-\frac{S}{2}\right) \tag{A.15}$$

For the reflected wave ψ_1 one writes the same expansion, changing the sign in front of $W(x)$. The prefactor equals

$$\exp[-W_1(x)] \cong \begin{cases} -(2\pi)^{1/2}\nu\left(\dfrac{\delta_0}{x}\right)^{1+\nu} & |x| < x_1 \\[3mm] -(2\pi)^{1/2}\nu\left(\dfrac{\delta_0}{4x_1}\right)^{1+\nu}\left(\dfrac{x_1 - x}{16x_1}\right)^{-1/4} & |x_1 - x| < x_1 \end{cases} \tag{A.16}$$

Comparing this again with the asymptotic expressions, one obtains the imaginary part of the energy:

$$\nu = -i(8\pi)^{-1/2}\left(\frac{4x_1}{\delta_0}\right)^{1+2\nu} \exp(-S) \qquad (A.17)$$

which confirms the initial assumption that the energy is complex. The decay rate Γ is related to ν by (A.9). One may infer also the value of Γ by directly calculating the probability flux at $x \to \infty$, which, by definition, is equal to

$$J = Im\left(\psi * \frac{\partial \psi}{\partial x}\right) \qquad (A.18)$$

It follows from (A.3) and (A.15) that this influx is independent of x, which confirms that it corresponds with a genuine decay rate constant. For the cubic parabola potential $(a = (\omega_0^2 x_1)^{-1/3})$ it equals

$$J = 4\omega_0 x_1 \exp(-S) \qquad (A.19)$$

The rate constant is the flux normalized by the probability density in the initial well, that is,

$$\Gamma = \frac{J}{\int dx|\psi_0(x)|^2} = \frac{(2\pi)^{1/2}J}{\delta_0} \qquad (A.20)$$

Remembering that $\Gamma = 2\omega_0|\nu|$ and using (A.2) it is easy to check that (A.17) and (A.20) are identical when $\Gamma \ll \omega_0$. In particular, for the cubic parabola potential from (A.19) and (A.20) we get

$$\Gamma = \omega_0 60^{1/2}\left(\frac{S}{2\pi}\right)^{1/2} \exp(-S) \qquad (A.21)$$

Expression (A.20) clarifies the origin of a large prefactor compared to ω_0, which was pointed out in Section 2.1. The radius of the region in which the wave function is localized is of order of δ_0, while the size of "source" for the outgoing wave equals $x_1 \gg \delta_0$, so $\Gamma \sim \omega_0 x_1/\delta_0$.

The tunneling splitting in a symmetric double well may be considered in a similar way. For infinitely high barrier the ground state with energy E_0 is twice degenerate, with wave functions $\psi_0(x)$ and $\psi_0(-x)$. Tunneling removes this degeneracy, leading to the tunneling splitting $\Delta \ll \omega_0$ between the energy levels of the symmetric and antisymmetric states with

$\psi_{s,a} = 2^{-1/2}[\psi_0(x) \pm \psi_0(-x)]$:

$$E_{a,s} = E_0 \pm \frac{\Delta}{2} \qquad (A.22)$$

The value of Δ may be determined in terms of the wave functions of localized states $\psi_0(x)$ (see Landau and Lifshitz [1981] and Dekker [1986]) and is described by a formula similar to (A.18):

$$\Delta = \frac{\psi_0(0)\psi_0'(0)}{\int dx |\psi_0(x)|^2} \qquad (A.23)$$

This formula demonstrates that the tunneling splitting is determined—like the imaginary part of metastable state energy (A.20)—as a normalized probability flux through the dividing line. In the present case this flux corresponds to coherent probability oscillations between the wells rather than exponential decrease of the survival probability in the well, so Δ is a real value.

The semiclassical wave function (A.10) in the barrier region may be written in the form

$$\psi_0(x) = N[2V(x)]^{-1/4} \exp\left[-\frac{W(x_0)}{2} + \frac{W(x)}{2} - \frac{\omega_0\tau(x)}{2} \right]$$

$$W(x) = \int_{-x}^{x} dx' [2V(x')]^{1/2} \qquad \tau(x) = \int_0^x dx' [2V(x')]^{-1/2} \qquad (A.24)$$

The term including ω_0 in the exponent accounts for the zero-point energy $\omega_0/2$. In the classically accessible region near the minima $\pm x_0$ one may use the harmonic oscillator approximation. The factor N is determined in a manner similar to the route described previously for a metastable potential. The tunneling splitting calculated from (A.23) is

$$\Delta = \frac{\omega_0}{\pi} (2\pi)^{1/2} \frac{x_0}{\delta_0} \exp\left(-\frac{S}{2}\right) \qquad (A.25)$$

Comparing (A.25) with (A.20) one sees that the exponent in (A.20) is twice as large as that in (A.25). This difference between coherent and incoherent processes has already been discussed. As in (A.20), the prefactor is increased by the factor of $\sim x_0/\delta_0$.

APPENDIX B

Equivalence of the Instanton Approach to Semiclassical TST

Consider a metastable potential well like that in Figure 3.3. In the vicinity of the parabolic well $(\tau \to 0)$ at $\beta \gg \omega_0^{-1}$ the instanton solution (3.55) corresponding to a harmonic oscillator in imaginary time is given by

$$x = r \cosh(\omega_0 \tau) . \tag{B.1}$$

The instanton energy is equal to

$$E = \tfrac{1}{2} \omega_0^2 r^2 \tag{B.2}$$

(We use mass-scaled coordinates and set $\hbar = 1$, $k_B = 1$.) Let $r_0 \gg r$ be the coordinate of an arbitrary point still lying in the region where the harmonic approximation for $V(x)$ is correct with the (arbitrary) given accuracy. Then, if the instanton starts out at $\tau = 0$, the time it takes to reach the point r_0 equals

$$\tau_1 = \omega_0^{-1} \ln\left(\frac{2r_0}{r}\right) \tag{B.3}$$

where we have replaced the hyperbolic cosine in (B.1) with an exponential. The total instanton period may be represented as

$$\beta = 2\tau_1 + \tau_2 \tag{B.4}$$

where τ_2 is the time during which the particle crosses the barrier and returns to the point r_0. The gist of the derivation is that, when $\beta \to \infty$ and, consequently, $r \to 0$, the time of lingering near the point $x = 0$, τ_1 goes to infinity, while the barrier passage time τ goes to some finite value, so that $\tau_1 \gg \tau_2$ for large enough β. Neglecting τ_2 in (B.4) and differentiating $\beta \cong 2\tau_1$ with respect to E, we immediately obtain from (B.2) and (B.3)

$$\frac{\partial \beta}{\partial E} = (\omega_0 E)^{-1} \tag{B.5}$$

The solution to (B.5)

$$E = E_0 \exp(-\beta \omega_0) \tag{B.6}$$

depends on the constant E_0, which is determined by the actual shape of $V(x)$. To find E_0 we study first the auxiliary problem of a cusp-shaped harmonic potential with a wall placed at $x = x_p$ (see Figure 2.5):

$$V(x) = \begin{cases} \frac{1}{2}\omega_0^2 x^2 & x < x_p \\ -\infty & x > x_p \end{cases} \tag{B.7}$$

For this particular potential Eq. (B.1) is exact throughout the whole range of β and thus we can take $r_0 = x_p$ so that $\tau_2 \equiv 0$. From (B.2)–(B.4) it is easy to see now that

$$E_0 = 2\omega_0^2 x_p^2 = 4V_0 \tag{B.8}$$

Then, given an arbitrary potential, we compare it to the reference potential (B.7), adjusting the width of the latter x_p such that the instanton periods β are equal when the energies E are the same. In other words we require

$$\lim_{E \to 0} \left[\int_r^{r'} dx(2V(x) - E)^{-1/2} - \int_r^{x_p} dx(\omega_0^2 x^2 - 2E)^{-1/2} \right] = 0 \tag{B.9}$$

where r is related to E by (B.2) and r' is the second turning point for the potential $V(x)$ at given E. Equation (B.9) inexplicitly defines x_p for which the value of E_0 is the same for the both potentials. Thus, for an arbitrary potential we have $E_0 = 2\omega_0^2 x_p^2$ with x_p defined by (B.9). In particular, for the cubic parabola $V(x) = \frac{1}{2}\omega_0^2 x^2(1 - x/x_0)$ direct use of (B.9) leads to $x_p = 4x_0$, so that $E_0 = 32\omega_0^2 x_0^2$. Insertion of this into (2.6) and (2.7) gives (3.68).

TWO-DIMENSIONAL TUNNELING

CONTENTS

Our discussion so far has concerned one-dimensional models, which, as a rule, do not directly apply to real chemical systems for the reasons discussed in Chapter 1. In this chapter we discuss how one-dimensional models can be extended to many dimensions. For clarity and simplicity, we confine the treatment to two dimensions, although the generalization to more dimensions is straightforward.

From the very simple WKB considerations it is clear that the barrier transparency is proportional to the Gamov factor $\exp[-2 \int ds\, (2\{V[s(\mathbf{Q})] - E)\}^{1/2}]$, where $s(\mathbf{Q})$ is a path in two dimensions ($\mathbf{Q} = \{Q_1, Q_2\}$) starting in the initial well and crossing the barrier. The *most probable tunneling path*, or instanton, which maximizes the Gamov factor, represents a compromise of two competing factors: the barrier height and its width. That is, one has to optimize the instanton path not only in time, as in the previous section, but also in space. This complicates the problem so that an analytical solution is usually impossible.

4.1. DECAY OF A METASTABLE STATE

Again we use the Im F method in which the tunneling rate is determined by the nontrivial instanton paths that extremize the Euclidean action in the barrier. Suppose, for example, we let the potential $V(\mathbf{Q})$ have a single minimum at $\mathbf{Q} = 0$, $V(0) = 0$, separated from the continuous spectrum by a high barrier. The extrema of the action

$$S[\mathbf{Q}(\tau)] = \int_0^\beta d\tau [\tfrac{1}{2}\dot{Q}_1^2 + \tfrac{1}{2}\dot{Q}_2^2 + V(Q_1, Q_2)] \tag{4.1}$$

are the instanton trajectories subject to

$$\frac{d^2 Q_i}{d\tau^2} = \frac{\partial V}{\partial Q_i} \tag{4.2}$$

These equations form a fourth-order system of differential equations that cannot be solved analytically in most interesting nonseparable cases. Further, according to these equations, the particle slides from the "hump" of the upside-down potential $-V(\mathbf{Q})$ (see Figure 4.1), and, unless the initial conditions are specially chosen, it exercises an infinite aperiodic motion. In other words, the instanton trajectory with the required periodic boundary conditions

$$\mathbf{Q}(\tau) = \mathbf{Q}(\tau + \beta) \tag{4.3}$$

is unstable with respect to all deviations except the time shift.

Once the instanton trajectory has been found numerically, one proceeds to the calculation of the prefactor, which amounts to finding determinants of differential operators. The direct two-dimensional

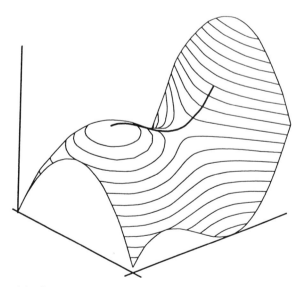

Figure 4.1. Instanton trajectory on a two-dimensional upside-down PES.

generalization of (3.67) is

$$k = \left(\frac{S_0}{2\pi}\right)^{1/2} \left[\frac{\det'(-\partial_\tau^2 \mathbf{1} + \partial^2 V/\partial Q_i \partial Q_j)_{\text{ins}}}{\det(-\partial_\tau^2 \mathbf{1} + \mathbf{K}_0)}\right]^{-1/2} \exp(-S_{\text{ins}}) \quad (4.4)$$

where $\partial_\tau^2 \mathbf{1}$ is the second-derivative operator multiplied by the unit 2×2 matrix, and $\mathbf{K}_0$ is a matrix of the form

$$\mathbf{K}_0 = \begin{bmatrix} \omega_+^2 & 0 \\ 0 & \omega_-^2 \end{bmatrix} \quad (4.5)$$

with the squares of the well eigenfrequencies, $\omega_+^2 > \omega_-^2$, on the diagonal. This matrix is simply the diagonal form of the matrix $\partial V/\partial Q_i \partial Q_j$, taken at the point $\mathbf{Q} = 0$. The same matrix enters into the numerator in (4.4) where it is taken on the instanton trajectory. The action S_0 is equal to

$$S_0 = \int_0^\beta d\tau (\tfrac{1}{2}\dot{Q}_1^2 + \tfrac{1}{2}\dot{Q}_2^2) \quad (4.6)$$

We pause now to highlight some features of a two-dimensional instanton. Let us introduce the local normal coordinates Q_+ and Q_- about the potential minimum, which correspond to the higher and lower vibrational frequencies ω_+ and ω_-. The asymptotic behavior of the instanton solution at $\beta \to \infty$, $\tau \to 0$, is described by $Q_+ \cong Q_\pm^0 \cosh(\omega_\pm \tau)$ (see Appendix B), where Q_+^0 goes to zero faster than Q_-^0 so as to keep Q_+ finite. Therefore, the asymptotic instanton direction coincides with Q_-.

The conclusion we have arrived at is what may be called a *generalized Fukui theorem*, which states that the reaction path near the minimum goes along the lowest eigenfrequency direction (see Tachibana and Fukui [1979]). Further consideration of the Fukui theorem is addressed in Section 8.3. Here we note only that this theorem is not valid when the instanton path is constrained by the symmetry of potential to lie totally along the direction of one of initial state modes.

When the temperature is raised, the period decreases, and the instanton amplitude decreases until the trajectory collapses to a point (the saddle point). At the saddle point $\mathbf{Q}_\#$ the potential has two normal frequencies, the imaginary longitudinal $i\omega_\#$, and transverse $\omega_{t\#}$. Thus we expect the crossover temperature to be

$$T_c = \frac{\omega_\#}{2\pi} \quad (4.7)$$

This formula, however, tacitly supposes that the instanton period depends monotonically on its amplitude so that the zero-amplitude vibrations in the upside-down barrier possess the smallest possible period $2\pi/\omega_{\#}$. This is obvious for sufficiently nonpathological one-dimensional potentials, but in two dimensions this is not necessarily the case. Benderskii et al. [1993] found that there are certain cases of strongly "bent" two-dimensional PES's in which the instanton period has a minimum at a finite amplitude. Therefore, the crossover temperature, formally defined as the lowest temperature at which the instanton still exists, turns out to be higher than that predicted by (4.7). At $T > T_c$ the trivial solution $\mathbf{Q} \equiv \mathbf{Q}_{\#}$ ($\mathbf{Q}_{\#}$ is the saddle point coordinate) replaces the instanton; the action is $S = \beta V_{\#}$ (where $V_{\#}$ is the barrier height at the saddle point) and the Arrhenius dependence $k \propto \exp(-\beta V_{\#})$ holds.

Although formula (4.4) gives the formal instanton expression for the rate constant, it is difficult to apply directly, because the numerical evaluation of an infinite product of eigenvalues of the differential operator is a rather challenging problem. It is very tempting to reduce the ratio of determinants in (4.4) to something like (3.69). However, that formula is written solely for one dimension. The way out is to divide (4.4) into transverse and longitudinal parts and to deal with each of them separately. We introduce the coordinates s, which runs along the instanton path, and x, which measures the deviation away from the instanton path. The fluctuations around the instanton trajectory are of two types. The *temporal* or *longitudinal* fluctuations affect only the s motion (i.e., they maintain $x = 0$). It is this sort of fluctuation that is present in one dimension, leading to (3.67). The transverse fluctuations introduce a finite spatial width of the instanton trajectory, spreading it to a *channel*. If this channel is narrow enough, one may use the local harmonic expansion around the points of the instanton trajectory, and it is this expansion that results in the ratio of determinants in (4.4). In the new coordinates the action, when expanded up to quadratic terms around the instanton solution, is

$$S = S_{\text{ins}} + \frac{1}{2} \int_0^\beta d\tau \, \delta s(\tau) \left(-\partial_\tau^2 + \frac{\partial^2 V}{\partial s^2} \right) \delta s(\tau)$$

$$+ \frac{1}{2} \int_0^\beta d\tau \, \delta x(\tau) \{ -\partial_\tau^2 + \omega_t^2[s(\tau)] \} \delta x(\tau) \qquad (4.8)$$

where ω_t is an s-dependent *transverse vibrational frequency*, which results from the canonical transformation from the coordinates $\{Q_1, Q_2\}$ to $\{s, x\}$. Its explicit expression in terms of the original potential $V(\mathbf{Q})$ can be obtained in principle by the method of Someda and Nakamura [1991], but, as will be seen later, one does not actually need it. There is also an important case in which the coordinates s and x are known in advance, without solving the instanton equations of motion, so that the potential is approximately separable, and ω_t is the actual transverse frequency with respect to the reaction path s, $\omega_t(s) = (\partial^2 V/\partial x^2)^{1/2}$.

The most remarkable thing about expression (4.8) is that it does not contain any cross-terms $\delta x\, \delta s$. This is a consequence of the fact that the equation of motion for s, $d^2s/d\tau^2 = dV(s)/ds$, does not include any x-dependent terms. In the new coordinates the determinants break up into longitudinal and transverse parts and (4.4) become

$$k = B_t k_{1D} \qquad (4.9)$$

where the one-dimensional rate constant equals

$$k_{1D} = \left(\frac{S_0}{2\pi}\right)^{1/2} \left| \frac{\det'(-\partial_\tau^2 + \partial^2 V/\partial s^2)_{ins}}{\det(-\partial_\tau^2 + \omega_-^2)} \right|^{-1/2} \exp(-S_{ins}) \qquad (4.10)$$

and the transverse prefactor is

$$B_t = \left[\frac{\det(-\partial_\tau^2 + \omega_t^2)}{\det(-\partial_\tau^2 + \omega_+^2)} \right]^{-1/2} \qquad (4.11)$$

The zero mode, which is associated with the longitudinal fluctuations, is now included in (4.10), and when $\omega_t > 0$, the determinants in (4.11) do not suffer from the zero-mode problem. The value k_{1D} is simply the rate of tunneling (3.67) in the dynamical one-dimensional barrier $V(s)$ along the instanton trajectory. As for B_t, it incorporates the effect of transverse vibrations around the instanton trajectory. To calculate (4.10), one may employ the apparatus of Chapter 3 designed for one-dimensional tunneling. In particular, now it is possible to make use of (3.69) together with (3.66), which gives

$$k_{1D} = \left(\frac{2}{\pi}\right)^{1/2} \sinh\left(\frac{\omega_-\beta}{2}\right) \left| \frac{\partial E}{\partial \beta} \right|^{1/2} \exp(-S_{ins}) \qquad (4.12)$$

We proceed now to the calculation of B_t, following Benderskii et al. [1992a]. The denominator in (4.11) (apart from normalization) is equal to the harmonic oscillator partition function $[2\sinh(\omega_+\beta/2)]^{-1}$. The numerator is the product of ϵ_n's satisfying the equation of Schrödinger type

$$\{-\partial_\tau^2 + \omega_t^2[s(\tau)]\}x_n(\tau) = \epsilon_n x_n(\tau)$$

$$x_n(\tau + \beta) = x_n(\tau) \qquad (4.13)$$

Consider a more general eigenvalue equation without the periodic boundary condition

$$\{-\partial_\tau^2 + \omega_t^2[s(\tau)]\}x(\tau) = \epsilon(z)x(\tau) \qquad (4.14)$$

According to the Floquet theorem [Arnold, 1978], this equation has a pair of linearly independent solutions of the form $x(z, \tau) = u(z, \pm\tau)$ $\exp(\pm 2\pi i z\tau/\beta)$, where the function u is β-periodic. The solution becomes periodic at integer $z = \pm n$, so that the eigenvalues ϵ_n we need are $\epsilon_n = \epsilon(\pm n)$. To find the infinite product of ϵ_n's we exploit the analytical properties of the function $\epsilon(z)$. It has two simple zeros in the complex plane such that

$$\{-\partial_\tau^2 + \omega_t^2[s(\tau)]\}x(\tau) = 0 \qquad (4.15)$$

$$x(\tau + \beta) = \exp(\pm\lambda)x(\tau) \qquad (4.16)$$

Note that Eq. (4.15) is simply the linearized equation of motion for the classical upside-down barrier ($\delta S/\delta x = 0$) for the new coordinate x. Therefore, while $x \equiv 0$ corresponds to the instanton, the nonzero solution to (4.15) describes how the trajectory "escapes" from the instanton solution, when deviated from it. The parameter λ, referred to as the stability angle [Gutzwiller, 1967; Rajaraman, 1975], generalizes the harmonic oscillator phase $\omega_t\beta$, which would stand in (4.16), if ω_t were constant. The fact that λ is real is a reminder of the aforementioned instability of the instanton in two dimensions. Guessing that the determinant $\det(-\partial_\tau^2 + \omega_t^2)$ is a function of λ only, and using the Poisson summation formula, we are able to write

$$\frac{d}{d\lambda}\ln[\det(-\partial_\tau^2 + \omega_t^2)] = \sum_{-\infty}^{\infty} \epsilon_n^{-1}\frac{d\epsilon_n}{d\lambda}$$

$$= \sum_{-\infty}^{\infty} \int_{-\infty}^{\infty} dz\, \epsilon(z)^{-1}\frac{d\epsilon(z)}{d\lambda}\exp(2\pi i n z) \qquad (4.17)$$

which reduces to

$$\det(-\partial_\tau^2 + \omega_t^2) = \left[2 \sinh\left(\frac{\lambda}{2}\right)\right]^2 \tag{4.18}$$

Finally, we arrive at

$$B_t = \frac{\sinh(\omega_+ \beta/2)}{\sinh(\lambda/2)} \tag{4.19}$$

Formula (4.9) together with (4.12) and (4.19) is the semiclassical TST result first obtained by Miller [1975a] and developed later in Chapman et al. [1975] and Hanggi and Hontscha [1988, 1991].

As a simple illustration of this technique, consider the case of high-frequency ω_t, viz., $\omega_t^{-2}\partial\omega_r/\partial\tau \ll 1$ for the instanton trajectory (but $\omega_t/2$ is still small compared to the total barrier height $V_\#$). Then the quasi-classical approximation can be invoked to solve Eq. (4.15), which yields for λ

$$\lambda = \int_0^\beta \omega_t[s(\tau)]\, d\tau \tag{4.20}$$

It is readily seen that when β is sufficiently large so that the hyperbolic sines in (4.19) can be replaced by exponentials, the effect of the prefactor B_t is to replace the potential $V(s)$ by the vibrationally adiabatic potential:

$$V_{vad}(s) = V(s) + \frac{\omega_t(s)}{2} \tag{4.21}$$

Since the transverse vibration frequency at the barrier top is usually lower than ω_+, the vibrationally adiabatic barrier is lower than the bare one.

The deviation from the vibrationally adiabatic approximation can be illustrated by consideration of the instantaneous switch of transverse frequency from ω_1 to ω_2 ($\omega_1 > \omega_2$). If τ_1 and $\beta - \tau_1$ are the times of the frequencies ω_1 and ω_2, respectively, then the stability parameter is given by

$$\sinh\left(\frac{\lambda}{2}\right) = \frac{1}{2}\left[4 \sinh^2\left(\frac{\phi_1}{2} + \frac{\phi_2}{2}\right) + \left(\frac{\omega_1}{\omega_2} + \frac{\omega_2}{\omega_1} - 1\right) \sinh \phi_1 \sinh \phi_2\right]^{1/2}$$

where $\phi_i = \omega_i \tau_i$. When $\omega_1 \gg \omega_2$ this result differs from the adiabatic approximation, $\sinh(\frac{1}{2}\lambda_{ad}) = \sinh(\phi_1/2 + \phi_2/2)$ by a factor $\sim \frac{1}{2}(\omega_1/\omega_2)^{1/2} > 1$. In other words, nonadiabacity caused by stepwise frequency dependence reduces B_t by this factor as compared to the vibrationally adiabatic approximation (Benderskii et al., 1993).

As another illustration, note that above the crossover point T_c the temperature dependence $k(T)$ exhibits an activation energy equal to

$$E_a \cong V^{\#} + \frac{\omega_t^{\#}}{2} - \frac{\omega_+}{2} \qquad (4.22)$$

where $V_{\#}$ and $\omega_t^{\#}$ are the barrier height and transverse vibration frequency at the saddle point, and we have neglected the effect of the zero-point energy of the low-frequency vibration ω_-. Unlike Eq. (4.21) vibrational adiabaticity is not required for (4.22) to be valid, and the only requirement is that ω_t be quantal (i.e., $\beta\omega_t \gg 1$). In particular, this situation is usually realized in hydrogen transfer reactions because the intramolecular vibration for hydrogen remains quantal at room temperatures and higher.

The functional (4.8) permits us to study the collection of paths that actually contribute to the partition function path integral thus leading to the determinant (4.18). Namely, the symmetric Green's function for the deviation $x(\tau)$ from the instanton path is given by [Benderskii et al., 1992a, 1993b]

$$G(\tau, \tau') = \langle x(\tau)x(\tau')\rangle$$

$$= \left[2W \sinh\left(\frac{\lambda}{2}\right)\right]^{-1}\left[x(\tau)\tilde{x}(\tau')\exp\left(\frac{\lambda}{2}\right) + x(\tau')\tilde{x}(\tau)\exp\left(-\frac{\lambda}{2}\right)\right]$$

$$0 < \tau < \tau' < \beta \qquad (4.23)$$

where the functions $x(\tau)$ and $\tilde{x}(\tau)$ satisfy (4.15) and (4.16), and W is their Wronskian. The equal-time correlator then gives the *fluctuational width of the tunneling channel*, inside which the dominant paths lie:

$$\langle x^2(\tau)\rangle = W^{-1}x(\tau)\tilde{x}(\tau)\coth\left(\frac{\lambda}{2}\right) \qquad (4.24)$$

In the vibrationally adiabatic limit this formula reduces to the familiar form:

$$\langle x^2(\tau)\rangle = [2\omega_t(\tau)]^{-1}\coth\left\{\frac{1}{2}\int_0^\beta d\tau\, \omega_t[s(\tau)]\right\} \qquad (4.25)$$

Equation (4.25) indicates that the quantum number $\langle n\rangle$ of the transverse x vibration is an adiabatic invariant of the trajectory. At $T = 0$, $\langle x^2(\tau)\rangle$

becomes the instantaneous zero-point spread of transverse vibration $(2\omega_t)^{-1}$, in accord with the uncertainty principle.

The practical method of calculating of λ is different from that used in derivation of (4.19). Since λ is invariant with respect to canonical transformations, it is preferable to seek it in the initial coordinate system. Writing the linearized equation for deviations from the instanton solution $\delta \mathbf{Q}$

$$\left[-\partial_\tau^2 \mathbf{1} + \left(\frac{\partial^2 V}{\partial Q_i \partial Q_j} \right)_{\text{ins}} \right] \delta \mathbf{Q} = 0 \qquad (4.26)$$

we define the monodromy matrix $\mathbf{M}$ through which the solution to (4.26) is transformed over the period

$$\begin{bmatrix} \delta Q_1(\tau + \beta) \\ \delta Q_2(\tau + \beta) \\ \delta \dot{Q}_1(\tau + \beta) \\ \delta \dot{Q}_2(\tau + \beta) \end{bmatrix} = \mathbf{M} \begin{bmatrix} \delta Q_1(\tau) \\ \delta Q_2(\tau) \\ \delta \dot{Q}_1(\tau) \\ \delta \dot{Q}_2(\tau) \end{bmatrix} \qquad (4.27)$$

This matrix has two unit eigenvalues corresponding to the zero mode, and the other two eigenvalues are $\exp(\pm\lambda)$ entering into (4.19). The general formulae remain the same when the number of degrees of freedom N is greater than 2. The $2N \times 2N$ monodromy matrix has $2N - 2$ eigenvalues $\exp(\pm\lambda_j)$ and a doubly degenerate unit eigenvalue resulting from the time-shift invariance of the instanton. The transverse prefactor then becomes

$$B_t = \prod_j \frac{\sinh(\omega_j \beta / 2)}{\sinh(\lambda_j / 2)} \qquad (4.28)$$

where ω_j are the normal frequencies in the well except the smallest one.

In Benderskii et al. [1993a] a numerical instanton analysis has been carried out for tunneling escape from the metastable well defined by the Hamiltonian

$$H(Q, q) = V_0 \left[\frac{1}{2} \dot{Q}^2 + \frac{1}{2} \dot{q}^2 + \frac{1}{2} Q^2 \left(1 + \frac{C^2}{\Omega^2} - Q^n \right) + CQq + \frac{1}{2} \Omega^2 q^2 \right] \qquad (4.29)$$

The coordinates, *coupling parameter* C and frequency Ω in (4.29), are dimensionless, and time is measured in dimensionless units $\omega_0 \tau$, where ω_0 is the frequency of small vibrations in the well for the *adiabatic* potential

$V_a(Q)$, taken along the MEP $q_a(Q)$, defined by

$$\frac{\partial V}{\partial q} = 0 \qquad q_a = -\frac{CQ}{\Omega^2} \qquad V_a = \frac{1}{2}V_0 Q^2(1 - Q^n) \qquad (4.30)$$

For convenience of notation we accept from here on that each frequency of the problem ω_i has a dimensionless counterpart denoted by a capital Greek letter, so that $\omega_i = \omega_0 \Omega_i$. The model (4.29) may be thought of as a particle in a one-dimensional cubic parabola potential coupled to the q vibration. The saddle point coordinates, defined by $\partial V/\partial Q = \partial V/\partial q = 0$, are

$$Q_\# = \left(\frac{2}{n+2}\right)^{1/n} \qquad q_\# = -\frac{C}{\Omega^2}\left(\frac{2}{n+2}\right)^{1/n} \qquad (4.31)$$

and the barrier height is

$$V_\# = V(Q_\#, q_\#) = V_0 \frac{n}{2(n+2)}\left(\frac{2}{n+2}\right)^{2/n} \qquad (4.32)$$

In the limit of large n the potential (4.29) becomes a harmonic well with an absorbing wall located at $Q = 1$. This problem was discussed in Section 2.5.

Figure 4.2 demonstrates the instanton trajectories at different temperatures for $C = 0.5$, $\Omega = 0.5$, $n = 2$. For temperatures close to the T_c the trajectory runs near the saddle point, and it deviates from the saddle point with increasing β. The Hamiltonian (4.29) with $n = 1$ has recently been studied numerically within the complex scaling method [Hontscha et al. 1990]. Using these data we can estimate the accuracy of the instanton

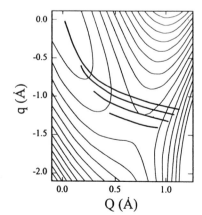

Figure 4.2. Contour plot and instanton trajectories at $\beta\omega_0 = 6.4$, 7.8, 10.2, and 25.0 for PES (4.29) with $C = \Omega = 0.5$, $n = 2$. Greater length of trajectory corresponds to greater β. (From Benderskii et al. [1993].)

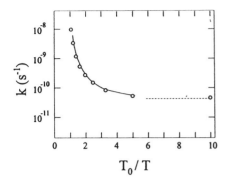

Figure 4.3. Plot of k against T_0/T for the PES (4.29) with $\Omega = 0.1$, $C = 0.0357$, $n = 1$, and $V^\#/V_0 = 3$. The solid curve shows the instanton result, the dashed line corresponds to the low-temperature limit using (4.9) and (3.53), and the circles represent exact numerical calculations from Hontscha et al. [1990]. (From Benderskii et al. [1993].)

method. Note, however, that the method of Hontscha et al. [1990] was applicable only to moderately high barriers, namely $V_\#/\omega_0 \leq 3$, whereas the instanton method is expected to work best for higher barriers. The ratio $\omega_0/V_\#$ may be regarded as the parameter of "quantumness"; the smaller it is, the better the instanton approximation works. In Figure 4.3 the Arrhenius plot $\ln[k(T)/\omega_0]$ versus $T_0/T = \beta\omega_0/2\pi$ is shown for $V_\#/\omega_0 = 3$, $\omega = 0.1$, and $C = 0.0357$. The discrete points are the data of Hontscha et al. [1990]. The solid line is obtained with the two-dimensional instanton method. One sees that the agreement between the instanton result and the exact quantal calculations is essentially perfect. The low-temperature limit obtained with the use of the periodic orbit theory expression for k_{1D} (dashed line) is also in excellent agreement with the exact result. Figure 4.4 shows the dependence of $\ln(k_c/\omega_0)$ on the

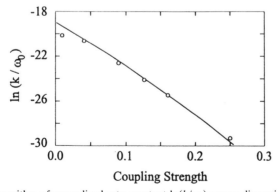

Figure 4.4. Logarithm of normalized rate constant $\ln(k/\omega_0)$ versus dimensionless coupling strength C^2/Ω^2 for PES (4.29) with $\Omega = 0.1$, $n = 1$, and $V^\#/V_0 = 3$. The discrete points and solid curve correspond to the instanton result [Benderskii et al., 1993] and exact results [Hontscha et al., 1990], respectively.

coupling strength defined as C^2/Ω^2. The solid line corresponds to the exact result from Hontscha et al. [1990], and the discrete points are obtained with the instanton method. For all practical purposes the instanton results may be considered exact.

Figure 4.5 shows an instanton trajectory for the parameters chosen in Figure 4.3. Unlike in Figure 4.2, the trajectory consists of two fairly straight segments that meet at an angle of nearly 90°. This sharp "bend" of the tunneling path becomes more prominent with decreasing Ω. The phenomenon can be understood on the basis of the sudden theory of tunneling, developed by Benderskii et al. [1980], Pollak [1986a], and Levine et al. [1989]. This theory exploits the fact that when $\Omega \ll 1$ the tunneling event may be considered instantaneous on the time scale of the q vibration. More accurately, one rewrites the problem in terms of the coordinates Q_+ and Q_-. The probability of finding the particle at a certain point Q_- is given by the diagonal element of the density matrix $P(Q_-) = \rho(Q_-, Q_-, \beta)$, which in the harmonic approximation is described by (3.16), $\rho_h(Q_-, Q_-, \beta) \propto \exp[-\omega_- Q_-^2 \tanh(\beta\omega_-/2)]$. Having reached the point Q_-, the particle is assumed to suddenly tunnel along the fast coordinate Q_+ with the probability $k_{1D}(Q_-)$, which is described in terms of the usual one-dimensional instanton. The rate constant comes from averaging of the one-dimensional tunneling rate over positions of the slow vibration mode:

$$k = \int dQ_- \rho(Q_-, Q_-, \beta) k_{1D}(Q_-) \qquad (4.33)$$

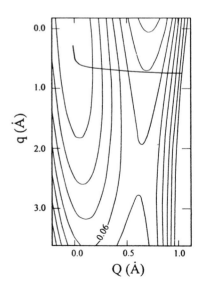

Figure 4.5. Contour plot and instanton trajectory for PES (4.29) with $\omega_0\beta = 35$, $\Omega = 0.1$, $C = 0.0357$, and $n = 1$. (From Benderskii et al. [1993].)

In the light of the path integral representation for $\rho(Q_-, Q_-, \beta)$, the latter may be semiclassically represented as proportional to $\exp[-S_1(Q_-)]$. In this expression, $S_1(Q_-)$ is the Euclidean action on the β-periodic trajectory that starts and ends at the point Q_- and visits the potential minimum $Q_- = 0$ for $T = 0$. The one-dimensional tunneling rate, in turn, is proportional to $\exp[-S_2(Q_-)]$, where S_2 is the action in the barrier for the closed straightforward trajectory which goes along the line with constant Q_-.

The integral in (4.33) may be evaluated by the steepest descent method, which leads to an optimum value of $Q_- = Q_-^*$. This amounts to minimization of the total action $S_1 + S_2$ over the positions of the "bend point" Q_-. In fact, in the sudden approximation one looks for the minimum of the barrier action taken on a certain class of paths, each consisting of two straightforward segments. If the actual extremal path is close to one of the paths from this class—and this is indeed the case for low enough Ω—then the sudden approximation provides accurate results. In particular, the sudden approximation permits calculation of the rate constant to an accuracy of 10% at $V_\# / \omega_0 = 3$, $\Omega = 0.1$, $C \leq 0.05$ [Hontscha et al., 1990]. For the cubic parabola ($n = 1$ in (4.29)) at small coupling parameter the rate constant in the sudden approximation may be evaluated analytically by using the one-dimensional instanton result (3.68) for k_{1D}:

$$k_{c,sudden} = 60^{1/2} \omega_0 \left(\frac{S_{1D}}{2\pi}\right)^{1/2} \left(1 + \frac{7C^2}{3\omega^2}\right) \exp\left\{S_{1D}\left[1 + \frac{5C^2}{2\omega^2}(1 - 3\Omega)\right]\right\}$$

(4.34)

where S_{1D} is the one-dimensional instanton action defined in (3.68).

4.2. TUNNELING SPLITTING

The formula for the tunneling splitting in two dimensions is a simple generalization of (3.94):

$$\Delta = 2\left(\frac{S_k}{2\pi}\right)^{1/2} \left[\frac{\det'(-\partial_\tau^2 \mathbf{1} + \partial^2 V/\partial Q_j \partial Q_j)_{kink}}{\det(-\partial_\tau^2 \mathbf{1} + \mathbf{K}_0)}\right]^{-1/2} \exp(-S_k) \quad (4.35)$$

where S_k is the one-kink action:

$$S_k = \int_{-\infty}^{\infty} d\tau\left[\tfrac{1}{2}\dot{Q}_1^2 + \tfrac{1}{2}\dot{Q}_2 + V(Q_1, Q_2)\right] \quad (4.36)$$

and $\mathbf{K}_0$ is the eigenfrequency matrix in the well. To apply the Floquet apparatus of previous subsection, it is expedient to "double" the kink, i.e., to use the periodic instanton trajectory with the action $S_{ins} = 2S_k$ at $\beta \rightarrow \infty$. When the kink and antikink on this trajectory are separated by an infinite time, each eigenvalue of the operator $-\partial_\tau^2 + \partial^2 V/\partial Q_i \partial Q_j$ will be doubled in the spectrum of the same operator taken for the whole instanton. Thus, we may rewrite (4.36) as

$$\Delta = \left(\frac{S_{ins}}{\pi}\right)^{1/2} \left[\frac{\det''(-\partial_\tau^2 1 + \partial^2 V/\partial Q_i \partial Q_j)_{ins}}{\det(-\partial_\tau^2 1 + \mathbf{K}_0)}\right]^{-1/4} \exp\left(-\frac{S_{ins}}{2}\right) \quad (4.37)$$

where the operators are taken now for the full periodic trajectory, and the double prime indicates that two zero eigenvalues are omitted. Now we are in position to use the results of the previous section to get

$$\Delta = \Delta_{1D} \tilde{B}_t \quad (4.38)$$

where Δ_{1D} is the one-dimensional tunneling splitting given by (3.103) or (3.94):

$$\Delta_{1D} = \left(\frac{S_{ins}}{\pi}\right)^{1/2} \left[\frac{\det''(-\partial_\tau^2 + \partial^2 V/\partial s^2)}{\det(-\partial_\tau^2 + \omega_-^2)}\right]^{-1/4} \exp\left(-\frac{S_{ins}}{2}\right) \quad (4.39)$$

and $\tilde{B}_t$ is the transverse prefactor

$$\tilde{B}_t = \lim_{\beta \rightarrow \infty} \{\exp[\tfrac{1}{4}(\beta \omega_+ - \lambda(\beta)]\} = \lim_{\beta \rightarrow \infty} B_t^{1/2} \quad (4.40)$$

An example of a numerically calculated trajectory in a symmetric double well is presented in Figure 4.6 for the Hamiltonian

$$H(Q, q) = V_0\left(\frac{1}{2}\dot{Q}^2 + \frac{1}{2}\dot{q}^2 + Q^4 - 2Q^2 - CQ^2q + \frac{1}{2}\Omega^2\right.$$
$$\left. \times \left(q + \frac{C}{\Omega^2}\right)^2 + 1 - \frac{C^2}{2\Omega^2}\right) \quad (4.41)$$

with $\Omega = 0.163$, $C = 0.185$. All the parameters except V_0 are dimensionless here. The time is gauged in dimensionless units $\tau = \omega_0 t$, where $8^{1/2}\omega_0$ is the vibrational frequency in the well of the one-dimensional potential taken at $q \equiv 0$

$$V_{1D}(Q) = V_0(Q^2 - 1)^2 \quad (4.42)$$

Since we are going to use the Hamiltonian (4.41) rather extensively in

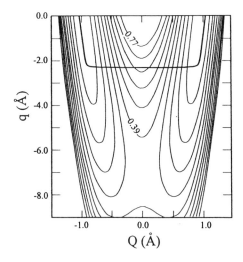

Figure 4.6. Contour plot and instanton trajectory for PES (4.41) with $\omega_0\beta \to \infty$, $C = 0.185$, and $\Omega = 0.163$. (From Benderskii et al. [1993].)

subsequent sections, as a simple two-dimensional model for an exchange chemical reaction, it is beneficial to note some of its salient features in advance. The minimum energy path (MEP) is determined by $\partial V/\partial q = 0$, whence

$$q_{\mathrm{a}} = \frac{C}{\Omega^2}(Q^2 - 1) \tag{4.43}$$

and the adiabatic barrier along this path is

$$V_{\mathrm{a}}(Q) = V_0(1 - b)(Q^2 - 1)^2 \qquad b = \frac{C^2}{2\Omega^2} \tag{4.44}$$

The dimensionless upside-down barrier frequency is $\Omega_\# = 2(1 - b)^{1/2}$, and the transverse frequency $\Omega_{\mathrm{t}\#} = \Omega$. The instanton action at $\beta = \infty$ in the one-dimensional potential (4.42) is given by (cf. Eq. (3.92))

$$S_{\mathrm{ins,1D}} = \frac{2^{7/2}V_0}{3\omega_0} \tag{4.45}$$

The situation presented in Figure 4.6 corresponds to the sudden limit, as we have already seen. Having reached a bend point at the expense of the low-frequency vibration, the particle then cuts straight across the angle between the reactant and product valley, tunneling along the Q direction. The sudden approximation holds when the vibrational frequency Ω is less than the characteristic instanton frequency, which is of the order of $\Omega_\#$. In particular, proton transfer reactions, which are characterized by high intramolecular vibrational frequencies are usually studied in this approximation [Ovchinnikova, 1979; Babamov and Marcus, 1981].

When the transverse frequency is high ($\Omega \gg \Omega_{\#}$) there is another way to avoid having to actually solve the instanton equations. In this case the factor of the barrier height becomes prevalent over that of its width, because any deviations of the trajectory from the MEP (4.43) entail a great rise of the barrier. Therefore, the q vibration adiabatically follows the motion along the Q coordinate, according to Eq. (4.43) and the trajectory of tunneling is the MEP. This is the adiabatic (small-curvature) approximation [Miller, 1983]. Projected onto the Q axis, this motion looks like that of a particle with kinetic energy $\frac{1}{2}\dot{Q}^2(1 + \alpha Q^2)$, $\alpha = 4C^2/\Omega^4$, and thus with variable effective dimensionless mass $M^* = 1 + \alpha Q^2$. The evaluation of the instanton action is straightforward and gives [Benderskii et al., 1990, 1991a]

$$S_{ad} = \frac{3}{4}S_{ins,1D}(1 - b)^{1/2}\left\{(1 + \alpha)^{1/2}\left(\frac{1}{2} - \frac{1}{\alpha}\right) + \alpha^{-1/2}\right.$$

$$\left. \times [1 + (4\alpha)^{-1}]\sinh^{-1}(\alpha^{1/2})\right\} \tag{4.46}$$

It is not difficult to show that the inequality $\Omega \ll \Omega_{\#}$, which should be met for the sudden approximation to hold, is equivalent to (2.86) if we introduce the angle 2φ between the reactant and product valleys where $\tan \varphi = \Omega^2/C$. The borders of the regions of validity of the sudden and adiabatic approximations in the (C, Ω) plane are symbolically drawn in Figure 4.7. The only physically sensible parameters are those for which

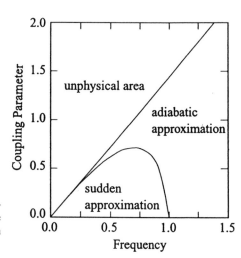

Figure 4.7. Domains of validity of sudden and adiabatic approximations in the (C, Ω) plane for PES (4.41). (From Benderskii et al. [1993].)

$b < 1$ in (4.44). Note that even for low vibration frequencies Ω, the adiabatic limit may hold for large enough coupling parameter C (see the "bill" of the adiabatic approximation domain in Figure 4.7). This situation was referred to as the strong fluctuation limit by Benderskii et al. [1991a,c], and it actually takes place for heavy particle transfer, as described in Chapter 9. In Chapter 5 we shall describe how both the sudden and adiabatic limit may be viewed from a unique perspective.

As we have seen, the role of phonons in two-dimensional tunneling can be elucidated by considering linearly and symmetrically coupled double-well potentials. In both cases the bending of the reaction path is caused by coupling to a vibration. The pure effect of the vibration-induced "squeezing" of the reactive channel (without bending) may be conventionally studied using the potential

$$V(Q, q) = V_0[(Q^2 - 1)^2 + \tfrac{1}{2}[\Omega^2 + C(Q^2 - 1)]q^2\} \qquad (4.47)$$

This is quite different from the first two. Due to the reflection symmetry of this potential, the instanton path always remains directed along the Q axis. The transverse q vibration changes only the width of the reactive channel according to Eqs. (4.23) and (4.24). When $C > 0$, the vibrationally adiabatic squeezed barrier is greater than the bare one. This case of dynamically induced formation of the barrier was studied by Auerbach and Kivelson [1985] in context of nuclear physics. The opposite case $C < 0$, corresponding to the vibration-assisted tunneling, will be considered in Section 8.3.

4.3. PERIODIC ORBITS IN A SYMMETRIC DOUBLE WELL

We now describe how the periodic orbit theory of Section 3.6, which relates the energy levels with the poles of the spectral function $g(E)$, can be extended to two dimensions. For simplicity we shall illustrate this extension by the simplest model, in which the total PES is constructed of two paraboloids that cross at some dividing plane. Each paraboloid is characterized by two eigenfrequencies, ω_+, ω_-. As explained in Section 2.5 (see Figure 2.15), the paraboloids are placed symmetrically either with respect to the dividing plane (symmetric case) or with respect to a point (antisymmetric case). Our discussion draws on Benderskii et al. [1992b].

In accordance with the one-dimensional periodic orbit theory, any orbit contributing to $g(E)$ is supposedly constructed from closed classical orbits in the well and subbarrier imaginary-time trajectories. These two classes of trajectories are bordered by the turning points. For the present model the classical motion in the well is separable, and the harmonic approximation for classical motion is quite reasonable for more realistic potentials if only relatively low energy levels are involved.

Consider, for example, the antisymmetric case. We choose the origin of the coordinate system in one of the wells, and the center of symmetry has the coordinates $Q_{+\#}$, $Q_{-\#}$ (Figure 4.8). Inside the well, the classical trajectories are Lissajous figures bordered by the rectangle formed by the lines $Q_{\pm} = Q^t_{\pm}$, and $Q_{\pm} = -Q^t_{\pm}$, where $Q^t_{\pm}$ are the turning point coordinates:

$$Q_{\pm} = Q^t_{\pm} \cos(\omega^t_{\pm} + \psi_{\pm}) \tag{4.48}$$

The effect of sufficiently weak anharmonicities of the potential on this picture will be to distort the rectangle comprising the classical trajectories so that the motion occurs on a two-dimensional torus belonging to the three-dimensional constant energy subspace of the total four-dimensional phase space of the system [Arnold, 1978].

The energy is

$$E = E_1 + E_2 = \frac{\omega^2_+(Q^t_+)^2}{2} + \frac{\omega^2_-(Q^t_-)^2}{2} \cong E^0_+ + E^0_-$$

$$= \left(n_+ + \frac{1}{2}\right)\omega_+ + \left(n + \frac{1}{2}\right)\omega_- \tag{4.49}$$

where tunneling is neglected in the zeroth approximation. Equation (4.48) does not describe periodic orbits regardless of the phases $\Psi_{\pm}$ unless the frequencies $\omega_{\pm}$ are commensurate. Thus, the first question to be addressed is how to semiclassically quantize a separate well. Furthermore, because of symmetry a tunneling orbit should pass through the

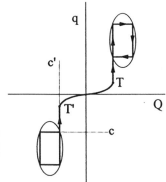

Figure 4.8. Two-dimensional periodic orbits for a vibration coupled antisymmetrically to the reaction coordinate. Caustics c and c' and take-off points T and T' are indicated. (From Benderskii et al. [1992b].)

point $\mathbf{Q}_{\#} = (Q_{+\#}, Q_{-\#})$. However, if it sets out from the turning point $\mathbf{Q}_t$ at $\tau = 0$, it will not necessarily hit the point $\mathbf{Q}_{\#}$, because it is impossible to satisfy simultaneously two imaginary time equations of motion

$$Q_{\pm} = Q_{\pm}^t \cosh(\omega_{\pm}\tau) \qquad (4.50)$$

when setting $\mathbf{Q} = \mathbf{Q}_{\#}$.

The solution is to sacrifice the classical trajectories. The crux of the matter is that we are trying to solve the variational problem for the action S with a given energy and fixed ends, while in the instanton theory the ends of a path are let loose. The solution of the problem at hand is not a classical trajectory, but a path that consists of segments of classical trajectories and caustics. The caustics are the envelopes of families of classical trajectories [Arnold, 1978]. Both the classical trajectories and caustics, and only they, possess a property that distinguishes them from all other paths. Namely, if we find for each point $\mathbf{Q}$ the momentum $\mathbf{P}(\mathbf{Q})$ of a trajectory that passes through this point, then, for a caustic or a classical trajectory, $\mathbf{P}$ is directed along its tangent. Moreover, the manifold $\mathbf{P}(\mathbf{Q})$ is single-valued in the regions bordered by caustics. The absolute value of momentum is obviously equal to $[2(V - E)]^{1/2}$ and, therefore, the Euclidean action along the above-mentioned path is $S = E\tau + \int ds [2(V - E)]^{1/2}$, where s is the coordinate along the path.

In the parabolic model, the equations for caustics are simply $Q_{+} = Q_{+}^t$, and $Q_{-} = Q_{-}^t$. The periodic orbits inside the well are not described by (4.48), but they run along the borders of the rectangle formed by caustics. It is these trajectories that correspond to topologically irreducible contours on a two-dimensional torus [Arnold, 1978] and lead to the quantization condition (4.49).

The semiclassical picture of tunneling then looks as follows (Figure 4.8). The particle starts out from one of the turning points[1] and runs along a subbarrier caustic, say that with $Q_{+} = Q_{+}^t$, until it reaches a "takeoff" point $Q_{-} = Q_{-}^t$, from which it begins to exercise the classical (upside-down barrier) motion. The coordinate and momenta are continuous along the whole path. From each turning point a pair of caustics comes out, and the correct caustic and the position of the takeoff point are picked up such that the path hits the point $\mathbf{Q}_{\#}$. When moving along the caustic, the "faster" degree of freedom is frozen, so as to eventually

---

[1] There are actually four turning points in each well. It does not matter from which one to start, because the straight-line segments between the turning points are incorporated into that part of $g(E)$ that is responsible for the single-well quantization.

"synchronize" the arrival at the point $\mathbf{Q}_\#$ for both degrees of freedom. The explicit equation for the takeoff point coordinate is

$$Q'_- = Q^t_- \cosh[\omega_-(\tau_- - \tau_+)] \quad \tau_\pm = \omega_\pm^{-1} \cosh^{-1}\left(\frac{Q^\#_\pm}{Q^t_\pm}\right) \quad (4.51)$$

where $\tau_\pm$ is the time of motion from the turning point to $Q_{\pm\#}$ for each coordinate. In the symmetric case the requirement to cross the point $\mathbf{Q}_\#$ is replaced by the requirement to cross the dividing line at a right angle.

For an N-dimensional paraboloid in the space $\mathbf{Q} = \{Q_1, \ldots, Q_N\}$ let us order the times to reach the point $\mathbf{Q}_\#$, $\tau_i = \omega_i^{-1} \cosh^{-1}(Q_{i\#}/Q_i^t)$, as follows: $\tau_1 > \tau_2 > \cdots > \tau_N$. Then the tunneling path consists of segments separated by takeoff points. At the kth segment, classical motion in the k-dimensional space of "slow" coordinates occurs, while the other $N - k$ degrees of freedom are "frozen." The genuine classical motion takes place only at the last segment, when all the times τ_i are equalized.

In conclusion we note that for a sufficiently dense energy spectrum the caustic segments have been shown to disappear after statistical averaging, which brings one back to the instanton and, for the present model, gives formulae (2.85).

CHEMICAL DYNAMICS IN THE PRESENCE OF A HEAT BATH

CONTENTS

When a system has a very large number of degrees of freedom, the multidimensional method, which treats them all equivalently, is impractical. Usually just one or a few reactive coordinates (i.e., the reactive coordinate and closely coupled modes) are of primary interest. The others may be considered as a heat bath. For example, in the $OH \cdots O$ fragment drawn in Figure 1.2 the primary degree of freedom is the proton coordinate, i.e., the position of the H atom relative to the O–O center. The O atoms, in turn, may participate in various intermolecular vibrations in a condensed medium. These intermolecular vibrations are nonreactive and they may be considered to be a heat bath. The choice of the bath is somewhat arbitrary because, for example, one could consider the two-dimensional reaction complex with one more relevant coordinate, say the O–O distance, and relegate the rest of the degrees of freedom to the bath. However, once the total Hamiltonian has a system–bath form, the main objective of the theory is to eliminate the bath degrees of freedom in order to write down the problem in terms of the system coordinates only. This is done by introducing the so-called influence functionals.

As seen from our discussion in Chapter 3, which dealt with one-dimensional problems, in many relevant cases one actually does not need the knowledge of the behavior of the system in real time to find the rate constant. As a matter of fact, the rate constant is expressible solely in terms of the equilibrium partition function imaginary-time path integrals. This approximation is closely related to the key assumptions of TST, and it is not always valid, as mentioned in Section 2.3. The general real-time description of a particle coupled to a heat bath is the Feynman-Vernon influence functional theory [Feynman and Vernon, 1963] which expresses the particle's reduced density matrix as a double path integral over the paths that develop in real time.[1] The only dissipative tunneling problem that has been thoroughly studied in real time is that of a two-level system coupled to a harmonic oscillator bath [Leggett et al., 1987]. Those results have provided an opportunity to estimate the accuracy of much more practical imaginary-time methods. So long as the system can be described by the rate constant—which rules out the case of localization and coherent tunneling—it may with reasonable accuracy be considered in the imaginary-time framework. For this reason we rely on the Im F approach in the main part of this chapter. In a separate subsection, the real-time dynamics of a TLS is analyzed, but on a simpler and less rigorous basis of Heisenberg equations of motion. A systematic and exhaustive discussion of this problem may be found in the review by Leggett et al. [1987].

5.1. THE QUASIENERGY METHOD

The total space of system coordinates consists of a tagged coordinate Q (conjugate momentum P) and a set of mass-scaled bath coordinates $\mathbf{q}$ (conjugate momenta $\mathbf{p}$). The Hamiltonian reads

$$H(P, Q, \mathbf{p}, \mathbf{q}) = H_0(P, Q) + H_b(\mathbf{p}, \mathbf{q}) + V_{int}(Q, \mathbf{q})$$

$$H_0(P, Q) = P^2/2 + V(Q)$$

$$H_b(\mathbf{p}, \mathbf{q}) = \sum_j \frac{p_j^2}{2} + V(\mathbf{q}) \qquad (5.1)$$

Following the Im F method one looks for the partition function of the system

$$Z = \int dq_i \, dQ_i \int D[Q(\tau)] D[q(\tau)] \exp\left[-\int_0^\beta (H_0 + H_b + V_{int}) \, d\tau \right] \quad (5.2)$$

where the path integral is taken over the closed paths having the periodic

---

[1] The influence functional theory, as it was formulated by Feynman and Vernon, relies on the additional assumption that the total (system and bath) density matrix can be factorized in the past. Without this assumption the theory requires a triple path integral, with one "thermal" integration over the imaginary-time axis [Grabert et al., 1988].

boundary conditions:

$$Q(\tau) = Q(\tau + \beta)$$

$$\mathbf{q}(\tau) = \mathbf{q}(\tau + \beta) \tag{5.3}$$

$$Q(0) = Q_i \qquad \mathbf{q}(0) = \mathbf{q}_i$$

If we fix a realization of the path $Q(\tau)$, then, when performing the path integration over $\mathbf{q}$, the particle may be treated as if it were subject to a time-dependent potential $V_{int}[Q(\tau), \mathbf{q}]$. From traditional quantum mechanics it is clear that this integration is equivalent to the solution of the time-dependent Schrödinger equation in imaginary time:

$$-\frac{\partial \phi(\mathbf{q}, \tau)}{\partial \tau} = (H_b + V_{int}[Q(\tau), \mathbf{q}])\phi(\mathbf{q}, \tau) \tag{5.4}$$

which allows one to find the bath propagator

$$K_b(\mathbf{q}_f, \mathbf{q}_i | -i\beta) = \hat{T} \exp\left[-\int_0^\beta d\tau(H_b + V_{int}) \right]$$

$$= \int D[\mathbf{q}(\tau)] \exp\left[-\int_0^\beta d\tau(H_b + V_{int}) \right] \tag{5.5}$$

where the time-ordering operator $\hat{T}$ appears in the expression for the propagator because the Hamiltonian is now time-dependent, unlike the problems we have discussed so far. The trace of K_b may be expressed in an elegant form if we exploit the periodicity of the potential V_{int} in (5.4). Namely, there is the so-called Floquet basis ϕ_n of solutions to (5.4) (see, e.g., Casati and Molinari, 1989), which satisfy the condition

$$\phi_n(\tau + \beta) = \exp(-\tilde{\epsilon}_n \beta)\phi_n(\tau) \tag{5.6}$$

The parameters $\tilde{\epsilon}_n$, which are called quasienergies, play the same role for periodic motion as the usual energies for time-independent Hamiltonians. By using definition (5.6) it is easy to obtain [Benderskii and Makarov, 1992]

$$\text{Tr } K_b = \int dq_i K_b(q_i, q_i | -i\beta) \equiv \tilde{Z}[Q(\tau)] = \sum_n \exp\{-\beta\tilde{\epsilon}_n[Q(\tau)]\} \tag{5.7}$$

This equation defines the quasienergy partition function functional. Its use results in the total partition function written in terms of the coordinate Q alone:

$$Z = \int dQ_i \int D[Q(\tau)] \exp\{-S_{eff}[Q(\tau)]\}$$

$$S_{eff}[Q(\tau)] = \int_0^\beta H_0 \, d\tau - \ln \tilde{Z}[Q(\tau)] \qquad Q(\tau) = Q(\tau + \beta) \qquad (5.8)$$

If the bath Hamiltonian is expressed in the form of an $N \times N$ matrix, then formulae (5.4), (5.6), and (5.7) may be represented as follows. Let $\mathbf{F}$ be a fundamental $N \times N$ matrix of the equation

$$-\frac{\partial \mathbf{F}}{\partial \tau} = (H_b + V_{int})\mathbf{F} \qquad (5.9)$$

Then define the monodromy matrix $\mathbf{M}$ [Shirley, 1965]

$$\mathbf{F}(\tau + \beta) = \mathbf{M}\mathbf{F}(\tau) \qquad \mathbf{M} = \mathbf{F}(\tau + \beta)\mathbf{F}(\tau)^{-1} \qquad (5.10)$$

Matrices $\mathbf{F}$ and $\mathbf{M}$ can be found by the straightforward integration of (5.9) with the initial conditions being N linearly independent vectors. Then the quasienergy partition function equals

$$\tilde{Z}[Q(\tau)] = \operatorname{Tr} \mathbf{M} \qquad (5.11)$$

Note also that $\mathbf{M}$ meets the Wronsky theorem:

$$\operatorname{Det} \mathbf{M} = \exp\left[-\int_0^\beta \operatorname{Tr}(H_b + V_{int}) \, d\tau\right] \qquad (5.12)$$

Thus, for Hamiltonians of finite dimensionality the effective action functional can be found immediately by integrating a system of ordinary differential equations. The most simple yet very important case is a bath of two-level systems:

$$H_b + V_{int} = \tfrac{1}{2}\Delta(Q)\sigma_x + \tfrac{1}{2}\epsilon(Q)\sigma_z \qquad (5.13)$$

The quasienergy partition function equals

$$\tilde{Z} = 2\cosh(\beta\tilde{\epsilon}) \qquad (5.14)$$

where $\pm\tilde{\epsilon}$ correspond to the Floquet solutions of the Schrödinger equation:

$$-\frac{2\partial c_1}{\partial \tau} = \epsilon c_1 + \Delta c_2$$

$$-\frac{2\partial c_2}{\partial \tau} = -\epsilon c_2 + \Delta c_1 \tag{5.15}$$

such that $c_{1,2}(\tau + \beta) = c_{1,2}(\tau) \exp(\pm\beta\tilde{\epsilon})$. In fact, when the temperature is less than the separation between two lowest energy levels of the bath, the latter can be approximated by a set of two-level systems [Caldeira and Leggett, 1983; Mermin, 1991].[2]

Generally speaking, the calculation of the quasienergies is itself a complex problem. There are, however, several limiting cases for which the quasienergy partition function may be evaluated at little cost. Two of these cases are considered below.

5.1.1. Adiabatic Approximation

Suppose that the characteristic frequency of the environment is greater than that of the Q system. Then Eq. (5.4) can be solved readily in the adiabatic approximation. The quasienergies are given by

$$\tilde{\epsilon}_n = \beta^{-1} \int_0^\beta \epsilon_n[Q(\tau)] \, d\tau \tag{5.16}$$

where $Q(\tau)$ is considered a parameter, that is, ϵ_n are the energy levels corresponding to (5.4) at constant values of Q. At sufficiently low temperatures when only the ground state (with energy $\epsilon_0(Q)$) survives in the partition function, the effective potential is renormalized from $V(Q)$ to

$$V_{\text{eff}}(Q) = V(Q) + \epsilon_0(Q) \tag{5.17}$$

At arbitrary temperatures and in the perturbative limit the energy can be estimated as the matrix element taken over unperturbed ϕ functions $\tilde{\epsilon}_n[Q(\tau)] \cong \epsilon_n + \beta^{-1} \int_0^\beta \{V_{\text{int}}[Q(\tau)]\}_{nn} \, d\tau$. Inserting this equality into the

---

[2] Therefore, at extremely low temperatures one may equally choose between the TLS and oscillator bath. The latter is usually supposed to be easier to handle, but the TLS bath model, in addition to its apparent applicability to the theory of glasses, has some very attractive features [Mermin, 1991; Shimshoni and Gefen, 1991; Suarez and Silbey, 1991b].

partition function one obtains to first order in V_{int}

$$\ln\left\{\frac{\tilde{Z}[Q(\tau)]}{Z_0}\right\} = -\int_0^\beta \langle V_{int}[\mathbf{q}, Q(\tau)]\rangle_{bath}\, d\tau \qquad (5.18)$$

where the partition function Z_0 and averaging correspond to the unperturbed bath. Thus, we arrive at the adiabatically renormalized effective potential

$$V_{eff}(Q) = V(Q) + \langle V_{int}[\mathbf{q}, Q(\tau)]\rangle_{bath} \qquad (5.19)$$

The adiabatic approximation in the form (5.17) or (5.19) allows one to eliminate the high-frequency modes and to concentrate only on the low-frequency motions. The most frequent particular case of adiabatic approximation is the vibrationally adiabatic potential:

$$V_{vad}(Q) = V(Q) + \sum \frac{\omega_i(Q)}{2} \qquad (5.20)$$

where ω_i are the frequencies of transverse vibrations.

5.1.2. Classical Low-Frequency Heat Bath

In the opposite case of very low-frequency bath degrees of freedom there is a wide range of temperatures in which the bath is classical while the Q coordinate is quantum. If ω_c is the characteristic frequency of the bath, it remains classical as long as $\beta\omega_c \ll 1$. The right side of the Schrödinger equation (5.4) now contains a rapidly oscillating potential V_{int} and the most obvious way to treat it is simply to average it over the period. A more accurate approach is based on the Kapitsa's method of generating an effective potential for fast oscillations (see, e.g., Bas' et al. [1971]). By expanding V_{int} in a Fourier series in the Matsubara frequencies $\nu_n = 2\pi n/\beta$

$$V_{int}[Q(\tau), \mathbf{q}] = \beta^{-1} \sum_{n=-\infty}^{\infty} V_n\{[Q(\tau)], \mathbf{q}\} \exp(i\nu_n\tau) \qquad (5.21)$$

we obtain the effective interaction potential $V_{eff}(\mathbf{q})$ as follows:

$$V_{eff}(\mathbf{q}) = \beta^{-1} \int_0^\beta V_{int}[Q(\tau), \mathbf{q}]\, d\tau - \sum_{n=1}^{\infty} \frac{|(\partial V_n/\partial \mathbf{q})|^2}{(2\pi n)^2} \qquad (5.22)$$

The classical bath "sees" the quantum particle potential as averaged over the characteristic time, which (if we recall that in conventional units it is

$\hbar/k_B T)$ vanishes in the classical limit $\hbar \to 0$. The quasienergy partition function for the classical bath now simply turns into an ordinary integral in configuration space:

$$\tilde{Z}[Q(\tau)] \propto \int d\mathbf{q} \exp[-\beta V_{\mathrm{eff}}(\mathbf{q})] \qquad (5.23)$$

where V_{eff} is a functional of path $Q(\tau)$. We have omitted the integration over momenta as constant factor which is canceled out when the ratio of Im Z and Re Z is taken. Inserting the latter expression into (5.8) one obtains

$$Z = \int d\mathbf{q} \int dQ_i \oint D[Q(\tau)]$$

$$\times \exp\left[-\int_0^\beta (H_0 + V_{\mathrm{int}}(Q,\mathbf{q})\, d\tau + \beta \sum_{n=1}^\infty \frac{|(\partial V_n/\partial \mathbf{q})|^2}{(2\pi n)^2} \right] \qquad (5.24)$$

Except for the nonlocal last term in the exponent, this expression is recognized as the average of the one-dimensional quantum partition function over static configurations of the bath. This formula without the last term was used by Dakhnovskii and Nefedova [1991] to handle a bath of classical anharmonic oscillators. The integral over $\mathbf{q}$ was evaluated with the steepest descent method leading to the most favorable bath configuration.

5.2. BATH OF HARMONIC OSCILLATORS

There are only a few Hamiltonians for which the path integration can be carried out exactly, the most illustrious case being the driven harmonic oscillator [Feynman, 1972; Feynman and Hibbs, 1965]:

$$H_b(\mathbf{p}, \mathbf{q}) = \sum_j \left(\frac{p_j^2}{2} + \frac{\omega_j^2 q_j^2}{2} \right)$$

$$V_{\mathrm{int}}(Q, \mathbf{q}) = \sum_j q_j f_j(Q) \qquad (5.25)$$

A common, but not always valid, assumption about f_j is $f_j(Q) = C_j Q$. A lot of the literature is devoted to analysis of this Hamiltonian, both classical and quantum mechanical. Two appealing features of this model draw so much attention to it. First, although microscopically one has very little information about the parameters entering into (5.25), it is known

[Caldeira and Leggett, 1983] that when the bath responds linearly to the particle motion, the operators q and p satisfying (5.25) can always be constructed, and the only quantity entering into various observables obtained from the model (5.25) is the spectral density (2.22):

$$J(\omega) = \frac{\pi}{2} \sum \omega_j^{-1} C_j^2 \delta(\omega - \omega_j) \qquad (5.26)$$

Second, the classical dynamics of this model are governed by the generalized Langevin equation of motion in the adiabatic barrier [Hanggi et al., 1990; Schmid, 1983]:

$$\frac{d^2 Q}{dt^2} + \int_{-\infty}^{t} dt' \eta(t - t') \left(\frac{dQ}{dt'} \right) + \frac{dV_{ad}}{dQ} = f(t)$$

$$\eta(t) = 2\pi^{-1} \int_0^{\infty} d\omega J(\omega) \frac{\cos(\omega t)}{\omega} \qquad (5.27)$$

$$V_{ad}(Q) = V(Q) - \sum \frac{C_j^2 Q^2}{2\omega_j^2}$$

where the fluctuating force $f(t)$ satisfies the usual fluctuation–dissipation relation:

$$\frac{1}{2} \langle f(t)f(0) + f(0)f(t) \rangle = \pi^{-1} \int_0^{\infty} d\omega \, J(\omega) \coth\left(\frac{\beta\omega}{2} \right) \cos(\omega t) \quad (5.28)$$

From a quantum mechanical point of view the Langevin equation (5.27) describes the evolution of the Heisenberg operator $Q(t)$. However, the simplicity of this equation is deceptive. For example, it is usually impossible to write down the equation of motion for the mean position $\langle Q \rangle$ in closed form, because averaging of (5.27) leads to the term $\langle dV_{ad}/dQ \rangle$, which is not equal to $dV_{ad}(\langle Q \rangle)/d\langle Q \rangle$, unless $V_{ad}(Q)$ is a harmonic oscillator potential. This is the reason why the quantum Langevin equation, although superficially similar to the classical Langevin equation, may describe tunneling, i.e., penetration into classically forbidden regions.

The situation is simplified when $V(Q)$ is a parabola, since the mean position of the particle now behaves as a classical coordinate. For a parabolic barrier (1.5), the total system, consisting of particle and bath, is represented by a multidimensional harmonic potential, and all one needs do is diagonalize it. In doing so, one finds a single unstable mode with imaginary frequency $i\lambda_\#$ and a spectrum of normal modes orthogonal to this coordinate. The quantity $\lambda_{\#}$ is the renormalized parabolic barrier frequency, which replaces $\omega_{\#}$ in a multidimensional theory. To calculate it we rewrite the Langevin equation (5.27) in terms of its Fourier components. Namely, if we take $Q = \Sigma Q_\nu \exp(i\nu t)$, then

$$-(\nu^2 + \omega_{\#}^2)Q_\nu + i\nu\eta_\nu Q_\nu = f_\nu \qquad (5.29)$$

where the subscript ν indicates the corresponding Fourier component. If we further define the susceptibility χ of the system as

$$\chi(\nu) \equiv \frac{\langle Q_\nu \rangle}{\langle f_\nu \rangle} = [-(\nu^2 + \omega_{\#}^2) + i\nu\eta_\nu]^{-1} \qquad (5.30)$$

then the eigenfrequency $\lambda_{\#}$ will correspond to the resonance, i.e., to the pole of χ. Setting $\nu = i\lambda_{\#}$ we thus find

$$\lambda_{\#} = \frac{\omega_{\#}^2}{\lambda_{\#} + \hat{\eta}(\lambda_{\#})} \qquad (5.31)$$

where $\hat{\eta}(\lambda_{\#})$ is the Laplace transform of $\eta(t)$ [Pollak, 1986a,b; Ford et al. 1988]. Use of conventional TST now gives the following result for the classical rate constant [Grote and Hynes, 1980; Pollak, 1986b]:

$$k = \left(\frac{\omega_0}{2\pi}\right)\left(\frac{\lambda_{\#}}{\omega_{\#}}\right)\exp(-\beta V_0) \qquad (5.32)$$

For ohmic friction $\eta(t) = \eta\delta(t)$, $\lambda_{\#} = [\omega_{\#}^2 + (\eta/2)^2]^{1/2} - \eta/2$, and (5.32) becomes the celebrated Kramers' formula for classical escape out of a metastable well in the case of moderate and strong damping [Kramers, 1940]. In accord with the predictions of multidimensional theory, the crossover temperature should be

$$T_c = \frac{\lambda_{\#}}{2\pi} \qquad (5.33)$$

as we pointed out in Section 1.2 (see formula (1.12)).

If the potential is parabolic, it seems credible that the inverted barrier frequency $\lambda_{\#}$ should be substituted for the parabolic barrier transparency to give the dissipative tunneling rate as

$$k \propto \left[1 + \exp\left(\frac{2\pi E}{\lambda_{\#}}\right)\right]^{-1} \qquad (5.34)$$

Such a treatment, while accurate above T_c, suffers from the fact that the actual form of the potential near the well is totally neglected. However, it can be used as the basis for a variational procedure with a parabolic reference [Pollak, 1986a].

Proceeding now to the instanton treatment of the Hamiltonian (5.25), we observe that the spectrum of quasienergies differs from that of unperturbed harmonic oscillator ($f(Q) = 0$) only by a shift that is independent of n [Bas' et al., 1971]:

$$\tilde{\epsilon}_n^i = \left(n + \frac{1}{2}\right)\omega_i + (2\beta)^{-1} \int_0^\beta f_i[Q(\tau)]\xi(\tau)\, d\tau \qquad (5.35)$$

where ξ is the periodic solution to the classical equation

$$-\frac{d^2\xi}{d\tau^2} + \omega_i^2 \xi = -f_i[Q(\tau)] \qquad (5.36)$$

As shown by Benderskii and Makarov [1992], one could consider an even more general problem with oscillator frequencies ω_i that are dependent on Q. The result would be

$$\tilde{\epsilon}_n = \frac{(n + \frac{1}{2})\lambda_i}{\beta} + (2\beta)^{-1} \int_0^\beta f_i[Q(\tau)]\xi(\tau)\, d\tau \qquad (5.37)$$

where λ_i is the stability parameter of the ith oscillator defined precisely as in the previous section. In particular, when $f_j \equiv 0$, the bath partition function would be $\tilde{Z} = \prod [2 \sinh(\lambda_j/2)]^{-1}$. For sufficiently small values of λ_j/β, the dependence of the quasienergy partition function is relatively weak and the factor $\tilde{Z}$ could be taken out of the path integral. This approach provides another derivation of the prefactor (4.28) in the multidimensional instanton theory.

Using (5.35) and (5.36) in (5.8) and dropping the constant partition function of the unperturbed harmonic oscillator in $\tilde{Z}$, we obtain the nonlocal effective action derived by Feynman (see also Caldeira and Leggett [1983]):

$$S_{eff} = S_0 - \frac{1}{2}\int_0^\beta \sum_j d\tau\, d\tau'\, G_j(\tau - \tau')f_j[Q(\tau)]f_j[Q(\tau')] \qquad (5.38)$$

where S_0 is the action of the particle in the bare potential $V(Q)$, $G_j(\tau - \tau')$ are the phonon Green's functions, whose Fourier components are $(\omega_j^2 + \nu_n^2)^{-1}$ in the expansion in the Matsubara frequencies $\nu_n = 2\pi n/$

β. For linear functions $f_j(Q) = C_j Q$, Eq. (5.38) takes the form

$$S_{\text{eff}} = \int_0^\beta d\tau \left[\frac{\dot{Q}^2}{2} + V_{\text{ad}}(Q) + \frac{1}{2} \int_0^\beta d\tau' K(\tau - \tau') Q(\tau) Q(\tau') \right]$$

(5.39)

$$K(\tau) = \beta^{-1} \sum_j \sum_n \frac{C_j^2 \nu_n^2}{\omega_j^2 (\omega_j^2 + \nu_n^2)} \exp(i\nu_n \tau)$$

$$= \beta^{-1} \sum_n \hat{\eta}(|\nu_n|) |\nu_n| \exp(i\nu_n \tau)$$

In the derivation of (5.39) we have extracted the δ-function term from the phonon Green's function, which, in turn, renormalized the bare potential V to the adiabatic one V_{ad}. An expression similar to (5.38) can be obtained for an arbitrary bath whenever the coupling is sufficiently weak and the functional $Z[Q(\tau)]$ can be expanded into the series

$$\tilde{Z}[Q(\tau)] \cong Z_0 + \int \left(\frac{\delta \tilde{Z}}{\delta Q} \right) Q(\tau) \, d\tau + \frac{1}{2} \int \int \left[\frac{\delta^2 \tilde{Z}}{\delta Q(\tau) \delta Q(\tau')} \right] Q(\tau) Q(\tau') \, d\tau \, d\tau'$$

The first-order term in this expansion renormalizes the potential $V(Q)$ while the bilinear term is analogous to the last term in (5.39). This is the linear response theory for the bath. In fact, it shows that to the extent that the bath quasienergy partition function is approximated by a quadratic functional, any bath can be represented as a set of effective harmonic oscillators. However, the Green's function of the bath $G_b(\tau, \tau') = \delta^2 Z / \delta Q(\tau) \delta Q(\tau')$ may have a form quite different from a single-phonon Green's function and may exhibit a strong temperature dependence. It would be interesting to see explicitly how this kernel reduces to the phonon Green's function, but with temperature-dependent parameters so that the friction coefficient η depends on temperature (e.g., for a liquid). To the authors' knowledge this has not been done as yet.

Returning to tunneling, we assume a metastable state like that in Figure 3.3 and use the Im F method. The extremal trajectories for S_{eff} satisfy the instanton equation of motion ($\delta S_{\text{eff}} / \delta Q = 0$)

$$-\frac{d^2 Q}{d\tau^2} + \frac{dV_{\text{ad}}(Q)}{dQ} + \int_0^\beta d\tau' K(\tau - \tau') Q(\tau') = 0 \qquad (5.40)$$

In the same way as it is done in the absence of dissipation, one obtains the instanton formula for the rate constant:

$$k = 2 \operatorname{Im} F = \left(\frac{S_0}{2\pi}\right)^{1/2} \left| \frac{(\det' \hat{\mathscr{L}})_{\text{ins}}}{(\det \hat{\mathscr{L}})_0} \right|^{-1/2} \exp(-S_{\text{ins}}) \qquad (5.41)$$

where $\hat{\mathscr{L}}$ is the integro-differential operator defined by

$$\hat{\mathscr{L}}Q \equiv \left(-\partial_\tau^2 + \frac{d^2 V_{\text{ad}}}{dQ^2}\right)Q(\tau) + \int_0^\beta d\tau' K(\tau - \tau')Q(\tau') \qquad (5.42)$$

and the subscripts "ins" and "0" indicate that the operators are taken on the instanton trajectory and on the static path $Q \equiv 0$, respectively. The action S_{ins} is the action from (5.39) on the instanton trajectory, and

$$S_0 = \int_0^\beta \dot{Q}^2 \, d\tau$$

At temperatures above T_c there is no instanton, and escape out of the initial well is accounted for by the static solution $Q \equiv Q_\#$ with action $S_{\text{eff}} = \beta V_0$ (where V_0 is the adiabatic barrier height here) which does not depend on friction. This follows from the fact that the zero Fourier component of $K(\tau)$ equals zero; hence the dissipative term in (5.39) vanishes if Q is constant. The dissipative effects come about only through the prefactor that arises from small fluctuations around the static solution. Decomposing the trajectory into a Fourier series

$$Q(\tau) = Q_\# + \beta^{-1} \sum_{n=-\infty}^{\infty} Q_n \exp(i\nu_n \tau) \qquad (5.43)$$

and using harmonic approximation for the potential near $Q = Q_\#$ one obtains the action

$$S_{\text{eff}} = \beta V_0 + \frac{1}{2\beta} \sum_{n=-\infty}^{\infty} [\nu_n^2 - \omega_\#^2 + \hat{\eta}(|\nu_n|)|\nu_n|]Q_n^2 \qquad (5.44)$$

where we have recalled that $Q_n = Q_{-n}$ because $Q(\tau)$ is real-valued. A similar expansion can be written in the vicinity of $Q = 0$. Path integration amounts to a Gaussian integration over Q_n's, whereas the integration over the unstable mode Q_0 is understood as described in Section 3.5. Also in that section we have motivated the correction factor $\phi = T_c/T = \beta\lambda_\#/2\pi$ which should be multiplied by the $\operatorname{Im} F$ result in order to reproduce the correct high-temperature behavior.

Direct use of the Im F formula finally yields

$$k = 2\beta^{-1}\phi \, \frac{\text{Im } Z}{\text{Re } Z} = \frac{\omega_0}{2\pi} \frac{\lambda^{\#}}{\omega^{\#}} \exp(-\beta V_0) \prod_{n=1}^{\infty} \frac{\nu_n^2 + \nu_n \hat{\eta}(\nu_n) + \omega_0^2}{\nu_n^2 + \nu_n \hat{\eta}(\nu_n) - \omega^{\#2}} \tag{5.45}$$

The last term in (5.45) accounts for quantum corrections to the classical escape rate (5.32) [Dakhnovskii and Ovchinnikov 1985; Grabert and Weiss, 1984; Melnikov and Meshkov, 1983; Wolynes, 1981]. In the case of ohmic dissipation the product in (5.45) can be calculated explicitly and one obtains for the quantum correction factor

$$\frac{k}{k_{\text{Kramers}}} = \frac{\Gamma(1 - \beta\Lambda_+^{\#}/2\pi)\Gamma(1 - \beta\Lambda_-^{\#}/2\pi)}{\Gamma(1 - \beta\Lambda_+/2\pi)\Gamma(1 - \beta\Lambda_-/2\pi)} \tag{5.46}$$

where k_{Kramers} is the Kramers' rate constant (5.32) and

$$\Lambda_{\pm} = -\frac{\eta}{2} \pm \left[\left(\frac{\eta}{2}\right)^2 - \omega_0^2\right]^{1/2}$$

$$\Lambda_{\pm}^{\#} = -\frac{\eta}{2} \pm \left[\left(\frac{\eta}{2}\right)^2 + \omega_{\#}^2\right]^{1/2} \tag{5.47}$$

Interestingly, the correction factor is practically the same as that given by the Wigner formula (2.15), and it is practically independent of friction [Hanggi, 1986]. This is in contrast with the relatively strong dependence of T_c on η.

The situation changes when we consider the behavior at low temperature. Friction affects not only the prefactor but the instanton action itself, and the rate constant depends strongly on η. In what follows we restrict ourselves to action alone, and for calculation of the prefactor we refer the reader to the original papers cited.

For the cusp-shaped harmonic potential

$$V(Q) = \begin{cases} \dfrac{\omega_0^2 Q^2}{2} & Q \leq Q_{\#} \\ -\infty & Q > Q_{\#} \end{cases} \tag{5.48}$$

the instanton equation (5.40) can be solved exactly in the Fourier representation for $Q(\tau)$ to give [Grabert et al., 1984a]

$$k \propto \exp\left(-\frac{Q_0^2}{2\delta^2}\right) \tag{5.49}$$

where $\delta(\eta, \beta)$ is the zero-point spread of damped harmonic oscillator:

$$\delta^2(\eta, \beta) = \beta^{-1} \sum_{n=-\infty}^{\infty} [\nu_n^2 + \omega_0^2 + |\nu_n|\hat{\eta}(|\nu_n|)]^{-1} \tag{5.50}$$

For ohmic friction this sum reduces to

$$\delta^2(\eta, \beta) = 2(\omega_0^2\beta)^{-1} + (2\pi)^{-1}(\Lambda_- - \Lambda_+)^{-1}$$
$$\times \left[\psi\left(1 - \frac{\Lambda_+\beta}{2\pi}\right) - \psi\left(1 - \frac{\Lambda_-\beta}{2\pi}\right) \right] \tag{5.51}$$

where Ψ is the digamma function, and $\Lambda_\pm$ are defined in (5.47). In the case of an undamped oscillator $\delta^2 = (2\omega_0)^{-1} \coth(\beta\omega_0/2)$. The asymptotical formulae can be written as

$$\delta^2 = \begin{cases} \dfrac{T}{\omega_0^2} & T \gg \omega_0 \\[2ex] (2\omega_0)^{-1}\left(1 - \dfrac{\eta}{\pi\omega_0}\right) & T = 0 \quad \eta \gg \omega_0 \\[2ex] 2(\pi\eta)^{-1} \ln\dfrac{\eta}{\omega_0} & T = 0 \quad \eta \ll \omega_0 \end{cases} \tag{5.52}$$

These formulae are cited in Section 2.2.

As seen from (5.49) and (5.52), at high temperatures the leading exponential term in the expression for k is independent of η and displays Arrhenius dependence with activation energy $E_a = V_0 = \omega_0^2 Q_\#^2/2$. Formally, because of the cusp, the instanton in this model never disappears and the crossover temperature defined by (5.33) is infinite. Practically, however, it is natural to define T_c as the temperature at which the dependence $\ln k(1/T)$, or $\delta^2(1/T)$ levels off. That is, $T_c \cong \omega_0/2$ in the absence of dissipation, and T_c decreases with increasing η. At strong friction and zero temperature $\delta^2 \propto 1/\eta$, apart from the weak logarithmic dependence, and the instanton action increases linearly with increasing η. This behavior is universal for different barrier shapes, as it can be shown for more realistic potentials from the scaling properties of the instanton solution [Grabert et al., 1984b]. Namely, let us rewrite (5.40) at $\beta = \infty$, having integrated the dissipative term by parts:

$$-\frac{d^2Q}{d\tau^2} + \frac{dV_{ad}(Q)}{dQ} + \int_0^\infty d\tau' \, g(\tau - \tau')\dot{Q}(\tau') = 0 \tag{5.53}$$

where the Fourier components of the new kernel $g(\tau - \tau')$ are $g_n = -i\eta(|\nu_n|)\,\mathrm{sign}(n)$. In the ohmic case the explicit form for g is

$$g(\tau) = \eta(\pi\tau)^{-1} \tag{5.54}$$

If we scale time as $\tau = \tilde{\tau}\eta$, then the first term in (5.53) decreases as $1/\eta^2$, while the other two are independent of friction. Therefore, at large η the second derivative term in (5.53), as well as the kinetic energy term in action, can be neglected, and the only effect of friction is to change the time scale. That is, the solution to (5.53) is $Q(\tau) = \tilde{Q}(\tau/\eta)$ where $\tilde{Q}$ is an η-independent function. The instanton velocity scales as $\dot{Q} \propto \eta^{-1}$, and the action (5.39) grows linearly with η, $S_{\mathrm{ins}} \propto \eta$.

The exact solution of the instanton equation in the large ohmic friction limit was found by Larkin and Ovchinnikov [1984] for the cubic parabola (3.37). At $T = 0$

$$Q_{\mathrm{ins}}(\tau) = \frac{4}{3}Q_0\left[1 + \left(\frac{\omega_0^2\tau}{\eta}\right)^2\right]^{-1}$$

$$S_{\mathrm{ins}} = \frac{2}{9}\pi\eta Q_0^2 \tag{5.55}$$

The instanton action behaves in accordance with the scaling predictions and is independent of ω_0. Loosely speaking, the frequency ω_0 is replaced by the friction coefficient η. Grabert et al. [1984b] studied the energy loss ΔE_{diss} in the dissipative tunneling process and found that ΔE_{diss} is saturated at large friction and becomes independent of η. For a cubic parabola, the maximum energy loss was found to be $\Delta E_{\mathrm{diss}} = 4V_0$.

5.3. DYNAMICS OF A DISSIPATIVE TWO-LEVEL SYSTEM

When the potential $V(Q)$ is symmetrical (or its asymmetry is smaller than the level spacing ω_0), then at low temperature ($T \ll \omega_0$) only the lowest energy doublet is occupied, and the total energy spectrum can be truncated to that of a TLS. If $V(Q)$ is coupled to vibrations whose frequencies are less than ω_0 and $\omega_{\#}$, then the system can be described by the spin-boson Hamiltonian:

$$H_{\mathrm{sb}} = \tfrac{1}{2}\Delta_0\sigma_x + \tfrac{1}{2}\epsilon\sigma_z + \sum \tfrac{1}{2}\omega_j^2 q_j^2 + \tfrac{1}{2}p_j^2 + Q_0\sigma_z C_j q_j \tag{5.56}$$

where $\Delta_0/2$ is the tunneling matrix element and ϵ the energy bias. The tunneling matrix element may be found from the instanton analysis, as described in Section 3.6. The extended coordinate Q is replaced in the spin-boson model by the matrix $Q_0\sigma_z$, which has two eigenvalues, $\pm Q_0$, corresponding to the minima of the original potential. If there are high-frequency vibrations among the bath oscillators ($\omega_j \gg \omega_0, \omega_\#$), their effect is to renormalize Δ_0 [Leggett et al., 1987], so we can assume that (5.56) contains only the low-frequency vibrations. Since we expect the coherence effects to show up in a symmetrical potential, it would not be obvious to use the Im F method for the spin-boson model. Yet, the truncation we did simplified the model to such an extent that the explicit real-time dynamics can be investigated.

There is a vast field in chemistry where the spin-boson model can serve practical purposes, namely, proton exchange reactions in condensed media [Borgis and Hynes, 1991; Borgis et al., 1989; Morillo et al., 1989; Morillo and Cukier, 1990; Suarez and Silbey, 1991]. The early approaches to this model used a perturbative expansion for weak coupling [Silbey and Harris, 1983]. Generally speaking, perturbation theory allows one to consider a TLS coupled to an arbitrary bath via the term $\hat{f}\sigma_z$, where $\hat{f}$ is an operator that acts on the bath variables. The equations of motion in the Heisenberg representation for the $\hat{\sigma}$ operators, $\partial\hat{\sigma}/\partial t = i[\hat{H}, \hat{\sigma}]$, have the form

$$\dot{\sigma}_x = -2\sigma_y\hat{f}$$

$$\dot{\sigma}_y = -\Delta_0\sigma_z + 2\sigma_x\hat{f} \qquad (5.57)$$

$$\dot{\sigma}_z = \Delta_0\sigma_y$$

Note that since the von Neumann's equation for evolution of the density matrix, $\partial\hat{\rho}/\partial t = -i[\hat{H}, \hat{\rho}]$, differs from the equation for $\hat{\sigma}$ only by a sign, similar equations can be written out for $\hat{\rho}$ in the basis of the Pauli matrices, $\hat{\rho} = \sigma_x\rho_x + \sigma_y\rho_y + \sigma_z\rho_z + \frac{1}{2}\mathbf{1}$. In the incoherent regime this leads to the master equation [Blum, 1981; Zwanzig, 1964]. For this reason the following analysis can be easily reformulated in terms of the density matrix.

From (5.57) one can obtain an integro-differential equation for the operator σ_z. What we need is the mean particle position, $\langle\sigma_z\rangle$, and to find it two approximations are made. First, in taking the bath averages we assume free bath dynamics. Second, we decouple the bath and pseudo-spin averages, guided by the perturbation theory. The result is a Langevin-like equation for the expectation $\langle\sigma_z\rangle$ [Dekker, 1987a; Meyer and Ernst, 1987; Waxman, 1985]

$$\frac{d^2\langle\sigma_z\rangle}{dt^2} + \Delta_0^2\langle\sigma_z\rangle + 4\int_0^t dt' \langle\hat{f}(t)\hat{f}(t')\rangle_s\langle\dot{\sigma}_z(t')\rangle = 0 \qquad (5.58)$$

where $\langle \hat{f}(t')\hat{f}(t')\rangle_s$ is the symmetrized autocorrelation function for $\hat{f}$, taken for the free bath. A damped oscillator equation of this type has been obtained for the first time by Nikitin and and Korst [1965] for the gas-phase model of a stochastically interrupted TLS (see Section 2.3). For the spin-boson Hamiltonian the correlator becomes

$$\langle \hat{f}(t)\hat{f}(t')\rangle_s = Q_0^2 \pi^{-1} \int_0^\infty d\omega\, J(\omega) \coth\left(\frac{\beta\omega}{2}\right) \cos[\omega(t-t')] \quad (5.59)$$

In the limit of extremely weak coupling, Eq. (5.58) becomes Markovian and takes the form of (2.43) and (2.44). As discussed in Section 2.3, for strong enough coupling, $\langle \sigma_z \rangle$ exhibits an exponential falloff with the rate constant proportional to Δ_0^2, and prefactor dependent on the bath spectrum. However, perturbative treatment cannot be counted on in the strong friction case. Instead, it is more applicable to the spectroscopic problem of broadening of the tunneling splitting spectral line due to dissipation.

The weak coupling scheme breaks down when the reorganization energy of the jth oscillator $E_{rj} = 2Q_0^2 C_j^2 \omega_j^{-2}$ exceeds its levels spacing ω_j. If $E_{rj} > \omega_j$, when the tunneling particle changes its position, the oscillator equilibrium positions shift through a considerable distance so that the bath cannot be thought of as unperturbed by the particle. It is then impossible to write down the perturbation series by using coupling as a small parameter. This obstacle may be circumvented with the aid of the so-called polaron transformation, which partially diagonalizes the Hamiltonian in the shifted oscillator basis [Leggett et al., 1987; Silbey and Harris, 1984]. The new Hamiltonian is

$$\hat{H}' = \hat{U}\hat{H}\hat{U}^{-1} = \frac{1}{2}\Delta_0(\sigma_+ e^{-i\hat{\Omega}} + \sigma_- e^{i\hat{\Omega}}) + \frac{1}{2}\epsilon\sigma_z + \sum \frac{1}{2}\omega_j^2 q_j^2 + \frac{1}{2}p_j^2$$
$$(5.60)$$

$$\hat{U} = \exp\left(-\frac{1}{2}i\sigma_z\hat{\Omega}\right) \qquad \hat{\Omega} = 2\sum \frac{p_j Q_0 C_j}{\omega_j^2} \qquad \sigma_\pm = \frac{1}{2}(\sigma_x \pm i\sigma_y)$$

The operator $\hat{U}$ shifts the q_j oscillator coordinate to its equilibrium through the distance $\pm Q_0 C_j / \omega_j^2$, the sign depending on the state of the TLS. All the coupling now is put into the term proportional to the tunneling matrix element and the small parameter of the theory is Δ_0 rather than C_j. To better understand the origin of the first term in (5.60) we separate from the Hamiltonian the part proportional to σ_x and average it over equilibrium oscillators. This gives rise to an effective tunneling splitting Δ_{eff}:

$$\frac{1}{2}\Delta_{\text{eff}}\sigma_x = \frac{1}{4}\Delta_0\langle e^{-i\hat{\Omega}} + e^{i\hat{\Omega}}\rangle = \frac{1}{2}\Delta_0\sigma_x \exp\left(-\frac{\langle\hat{\Omega}^2\rangle}{2}\right)$$

$$= \frac{1}{2}\Delta_0\sigma_x \exp\left(-\sum_j \frac{\phi_j}{2}\right) = \frac{1}{2}\Delta_0\sigma_x \exp\left(-\frac{\phi}{2}\right)$$

$$\phi = \sum_j \phi_j = 2\sum_j Q_0^2 C_j^2 \omega_j^{-3} \coth\left(\frac{\omega_j\beta}{2}\right)$$

$$= 4Q_0^2\pi^{-1}\int_0^\infty d\omega J(\omega)\omega^{-2} \coth\left(\frac{\beta\omega}{2}\right)$$

(5.61)

The term proportional to σ_y after averaging goes to zero. It is easy to verify that $\exp(-\Phi_j/2)$ is the statistically averaged overlap integral for the jth oscillator (cf. (2.43))

$$\exp\left(-\frac{\phi_j}{2}\right) = Z_j^{-1}\sum_n \exp[-\beta\omega_j(n + \tfrac{1}{2})]$$

$$\times \int dq_j\psi_n\left(q_j + \frac{2Q_0C_j}{\omega_j^2}\right)\psi_n(q_j)$$

(5.62)

where Z_j is its partition function. In the strong coupling case (in which we are presently interested) the absolute value of the overlap integral increases with increasing n (see Section 2.5). Nevertheless, as seen from (5.61), the average $\exp(-\Phi_j/2)$ decreases with increasing temperature because of the alternating sign of the summand in (5.62). Yet the transition probability should increase with increasing temperature, because, according to the golden rule it is proportional to the thermal average of a positive quantity, the *square* of the overlap integral (Franck-Condon factor).

Solving now the Heisenberg equations of motion for the σ operators perturbatively the same way as in the weak coupling case, one arrives (at $\epsilon = 0$) at the celebrated *noninteracting blip approximation* [Aslangul et al., 1985; Dekker, 1987b]:

$$\frac{d\langle\sigma_z(t)\rangle}{dt} + \int_0^t f(t-t')\langle\sigma_z(t')\rangle\,dt' = 0$$

(5.63)

where

$$f(t) = \Delta_0^2 \cos\left(\frac{4Q_0^2 R_1(t)}{\pi}\right) \exp\left[-\frac{4Q_0^2 R_2(t)}{\pi}\right]$$

$$R_1(t) = \int_0^\infty \omega^{-2} J(\omega) \sin(\omega t)\, d\omega \qquad (5.64)$$

$$R_2(t) = \int_0^\infty \omega^{-2} J(\omega)[1 - \cos(\omega t)] \coth\left(\frac{\beta\omega}{2}\right) d\omega$$

This approximation was originally derived and extensively explored by the path integral techniques (see the review by Leggett et al. [1987]). Most of the results cited in Section 2.3 can be obtained from (5.63) and (5.64). Equation (5.63) makes it obvious that the rate constant can be defined only when the integrand $f(t)$ falls off sufficiently fast, viz.,

$$k = \int_0^\infty dt\, f(t) \qquad (5.65)$$

Moreover, Eq. (5.65) is nothing but the omnipresent golden rule. To see this, just notice that the density of final states is identically equal to

$$\rho_f = \sum_f \delta(E_i - E_f) = (2\pi)^{-1} \sum_f \int_{-\infty}^\infty dt\, \exp(i(E_i - E_f)t) \qquad (5.66)$$

Substitution of this equation into the golden rule expression (1.14) together with the renormalized tunneling matrix element from (5.61) gives (5.65) after thermally averaging over initial energies E_i. In the case of a biased potential the expression for the forward rate constant is

$$k_{\text{fwd}} = \int_0^\infty dt\, \cos(\epsilon t) f(t) \qquad (5.67)$$

So far we have taken the tunneling matrix element Δ_0 to be independent of vibrational coordinates. In terms of our original model with extended tunneling coordinate Q, this assumption means that the vibrations asymmetrize the instantaneous potential $V(Q, \{q_j\})$ but do not modulate its height or width. This model does not describe the effect of vibration on tunneling (fluctuational barrier preparation) dealt with in Section 2.5. For example, consider the OH$\cdots$O fragment shown in Figure 1.2. The relative O–O distance is clearly the same in the initial and final states, and hence the O–O vibration cannot be considered linearly coupled to the reaction coordinate. Such a mode (call it q_1) is not associated with any reorganization energy, and this necessitates $C_1 = 0$. However, the O–O vibration, changing the tunneling distance, strongly modulates the barrier transparency and facilitates tunneling. For an asymmetric potential one also has to give up the condition $C_1 = 0$. To describe the effect of such *promoting modes* within the spin-boson Hamiltonian, the latter can be modified by replacing Δ_0 with

$$\Delta = \Delta_0 \exp(\gamma q_1) \tag{5.68}$$

[Borgis et al., 1989; Suarez and Silbey, 1991a], where q_1 is a particular "coupling" coordinate from the set $\{q_j\}$ that modulates the barrier. (Here we assume for simplicity that there is only one such coordinate.) The exponential form of Δ is accounted for by the Gamov-factor nature of this term. A similar approach that does not use the spin-boson Hamiltonian explicitly but exploits the assumption that tunneling is sudden on the time scale of the bath vibrational period was developed in the quantum diffusion theory [Flynn and Stoneham, 1970; Kagan and Klinger, 1974] and in chemical reaction theory [Goldanskii et al., 1989a,b; Siebrand et al., 1984] within the framework of radiationless transition theory [Kubo and Toyazawa, 1955]. Carrying out the same polaron transformation, one obtains the effective tunneling matrix element for the case (5.68):

$$\Delta_{\text{eff}} = \Delta_0 \exp\left[\frac{\gamma^2}{4\omega_1} \coth\left(\frac{\beta\omega_1}{2}\right)\right] \exp\left(-\frac{\phi}{2}\right) \tag{5.69}$$

The promoting effect of the q_1 vibration is represented in this formula by the first exponent, which has the sense of the tunneling matrix element (5.68) averaged over the Gaussian distribution of q_1 with the spread equal to $\langle q_1^2 \rangle = (2\omega_1)^{-1} \coth(\beta\omega_1/2)$. The effect of the reorganization of the heat bath in the transition, which always hinders tunneling, is described by the second exponent. Integrals like (5.65) and (5.67) are usually calculated with the steepest descent method by deforming the integration contour toward the imaginary axis. The analytical results for the spin-boson Hamiltonian with fluctuating tunneling matrix element (5.68) are detailed by Suarez and Silbey [1991a]. Here we discuss only the situation in which the q_1 vibration is quantum, i.e., $\omega_1\beta \gg 1$.

When the bath is classical, $\omega_j\beta \ll 1$, $j \neq 1$, the rate constant for the transition from left to right is given by

$$k = \frac{1}{4} \Delta_0^2 \exp\left(\frac{\gamma^2/2 - E_r'}{\omega_1}\right)\left(\frac{\pi\beta}{E_r'}\right)^{1/2} \exp\left[-\frac{\beta(E_r' - \epsilon)^2}{4E_r'}\right] \tag{5.70}$$

where E'_r is the reorganization energy for all oscillators except q_1, $E'_r = \Sigma_{j \neq 1} E_{rj}$. The rate constant (5.70) exhibits Arrhenius behavior associated with activation of classical degrees of freedom, and the tunneling rate is enhanced by the factor $\exp(\gamma^2 / 2\omega_1)$. At $\gamma = 0$ the result (5.70) is readily recognized as the well-known Holstein formula [Holstein, 1959], and it is formally equivalent to the Marcus formula (2.66) for radiationless transition, except that the matrix element V in (2.66) corresponds with diabatic coupling between the terms rather than with tunneling in adiabatic potential. This analogy suggests that (5.70) is equally valid for electronically adiabatic and nonadiabatic chemical reactions, once the matrix element is properly defined. At low temperatures, when the bath is quantum ($\beta\omega_j \gg 1$), the rate expression, expanded in series over coupling strength, breaks up into contributions from different processes involving the bath phonons

$$k = k_{1P} + k_{2P} + k_R + k'_R + \cdots \qquad (5.71)$$

where k_{1P} corresponds to one-phonon emission or absorption, k_{2P} to two-phonon emission and absorption, k_R to two-phonon Raman processes, and k'_R to the Raman processes involving one phonon and one quantum of the q_1 vibration. For example, the one-phonon contribution for an exoergic reaction ($\epsilon > 0$) at zero temperature is

$$k_{1P} = \frac{2\epsilon\eta Q_0^2 \Delta_{\text{eff}}^2}{\omega_c^2} \qquad (5.72)$$

in the deformational potential approximation $J(\omega) = \eta\omega^3\omega_c^{-2}\exp(-\omega/\omega_c)$. With increasing temperature, k_{1P} increases linearly with T at $\beta\epsilon > 1$, while k_{2P} and k_R exhibit T^2 and T^3 dependences, respectively. The low-temperature limit k_c is proportional to Δ_{eff}^2, whereas the prefactor depends on the particular spectral properties of the bath.

A disadvantage of the two-state methods is that modeling of a real potential energy surface (PES) by a TLS cannot always be done.[3] Moreover, this truncated treatment does not capture the high-temperature regime, since the truncation scheme does not hold at $T > \omega_0$. With the assumption that the transition is incoherent, similar approximations

---

[3] Leggett et al. [1987] have set forth a rigorous scheme that reduces a symmetric (or nearly symmetric) double well, coupled linearly to phonons, to the spin-boson problem, if the temperature is low enough. However, in the case of nonlinear coupling (which must be introduced to describe the promoting vibrations), no such scheme is known, and the use of the spin-boson Hamiltonian together with (5.68) relies rather on intuition, and is not always justifiable.

can be worked out immediately from the nonlocal effective action, as shown by Sethna [1981] and Chakraborty et al. [1988] for $T = 0$ and by Gillan [1987] for a classical heat bath.

Consider the $T = 0$ case. Integrating the nonlocal term in (5.39) by parts, we recast it in the form

$$S_{\text{eff}} = \int_{-\infty}^{\infty} d\tau \left\{ \frac{\dot{Q}^2}{2} + V_{\text{ad}}(Q) + \frac{1}{2} \int_{-\infty}^{\infty} d\tau' \sum_j \right.$$

$$\left. \times \left[\frac{C_j^2}{2\omega_j^3} \exp(-\omega_j |\tau - \tau'|) \right] \dot{Q}(\tau) \dot{Q}(\tau') \right\} \qquad (5.73)$$

In case of a symmetric (or just slightly asymmetric) potential, the instanton trajectory consists of a kink and antikink, which are separated by infinite time and do not interact with each other.[4] In other words, we may change the boundary conditions, such that the time spans from $-\infty$ to $+\infty$ for a single kink, and then multiply the action in (5.73) by factor 2. Sethna [1981] considered two limiting cases. The calculation of the action in the fast flip (sudden) approximation ($\omega_j \ll \omega_{\#}$) proceeds by utilizing the expansion $\exp(-\omega_i |\tau|) \cong 1 - \omega_i |\tau|$. The first term unity, after substitution into (5.73), gives precisely the quantity $\Phi/2$, which yields the Franck-Condon factor in the rate constant. The next term cancels the adiabatic renormalization and changes $V_{\text{ad}}(Q)$ back to the bare potential $V(Q)$. Thus, with exponential accuracy one finds that the rate constant is proportional to the Franck-Condon factor times the tunneling rate in the potential $V(Q)$, in agreement with (5.61) and (5.72). In the opposite case (i.e., the slow flip (adiabatic) limit, $\omega_j \gg \omega_{\#}$), the exponential kernel can be approximated by the delta function, $\exp(-\omega_j |\tau|) \cong 2\delta(\tau)/\omega_j$, thus renormalizing the kinetic energy and, consequently, multiplying the particle's effective mass by the factor $M^* = 1 + \Sigma\, C_j^2/\omega_j^4$. The rate constant is equal to the tunneling probability of a particle with the renormalized mass M^* in the adiabatic barrier $V_{\text{ad}}(Q)$:

$$k \propto \exp\left[-S_{\text{ins,1D}} \left(1 + \sum \frac{C_j^2}{\omega_j^4} \right)^{1/2} \right] \qquad (5.74)$$

---

[4] This approximation is not true, say, for the ohmic case, when the bath spectrum contains too many low-frequency oscillators. The nonlocal kernel falls off according to the power law, and the kink interacts with the antikink even for large time separations. We assume here that the kernel falls off sufficiently fast. This requirement also ensures the convergence of the Franck-Condon factor, and it is satisfied in most cases relevant for chemical reactions.

As discussed before, the mass renormalization is a reflection of the fact that the particle traces a distance longer than $2Q_0$ in the total multi-dimensional coordinate space.

The promoting vibrational modes, like q_1 in the above spin-boson treatment, cannot be introduced within the Hamiltonian (5.25) with linear coupling functions $f_j = C_j Q$, because such couplings suppress tunneling via the Franck-Condon factor. To study vibration-assisted tunneling in symmetric potentials it is necessary to introduce couplings of a more general form. Due to the symmetry, the coupling functions $f_j(Q)$ are either even or odd in Q. The symmetrically coupled vibrations corresponding to the even functions f_j, such as $f_j = C_j Q^2$, are not reorganized in transition (i.e., their equilibrium positions do not shift) so that they do not contribute to the Franck-Condon factors. (In the language of Sethna, this means that the first term in the expansion of the exponent $\exp(-\omega_j|\tau|)$ after substitution to the formula for action gives zero.) On the other hand, they can strongly modulate the potential $V(Q)$ and promote tunneling. The antisymmetrically coupled vibrations (with odd functions f_j) lead to Franck-Condon factors in the usual way. For the situation represented in Figure 1.2, for instance, the normal lattice modes that shift the O atoms in the opposite directions are symmetrically coupled to the H coordinate, while those vibrations that move both O atoms as a whole are coupled antisymmetrically.

The symmetrical coupling case has been examined using Sethna's approximations for the kernel by Benderskii et al. [1990; 1991a]. For low-frequency bath oscillators the promoting effect appears in the second order of the expansion of the kernel in $\omega_j|\tau|$, and for a single bath oscillator in the model Hamiltonian (4.41) the instanton action has been found to be

$$S_{\text{sud}} = S_{\text{ins.1D}}(Q_t^3 + 2^{-3/2}3aQ_t^2) \qquad\qquad a = \frac{C^2}{4\Omega}$$

$$Q_t = -2^{-1/2}a(1-b)^{-1} + \left[1 + \frac{a^2(1-b)^{-2}}{2}\right]^{1/2} \qquad \Omega \ll (1-b)^{1/2}$$

$$(5.75)$$

This is the sudden approximation for symmetric potential. According to (5.75), the tunneling distance shortens from $2Q_0$ to $2Q_t$, where $Q_0 = 1$ in dimensionless units of (4.41). The corresponding tunneling trajectory in (Q, q) space is shown in Figure 2.15. In the opposite limit of high bath oscillator frequency the action is given by (4.46), and the trajectory is shown in the same figure. The exact instanton action value is compared in

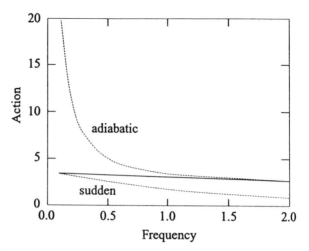

Figure 5.1. Dimensionless instanton action $S_{ins}\omega_0/V_0$ plotted against the q-vibration frequency Ω for PES (4.29) with $C = \Omega$. The solid line corresponds to the exact instanton solution; the dashed lines correspond to the sudden and adiabatic approximations.

Figure 5.1 with both the sudden and adiabatic approximations. For the purposes of demonstration, the adiabatic barrier height has been taken to be half the one-dimensional barrier $V_{\#} = V_0/2$, so that $b = \frac{1}{2}$, $C = \Omega$. One sees that the sudden approximation is realized only for fairly low vibrational frequencies, while the adiabatic approximation becomes excellent for $\Omega \geq 2$.

Suppose now that both types of vibrations are involved in transition. The symmetric modes shorten the effective tunneling distance to $2Q_t$, whereas the antisymmetric modes create the Franck-Condon factor in which the displacement $2Q_0$ now is to be replaced by the shorter tunneling distance $2Q_t$ [Benderskii et al., 1991a]:

$$\phi = 2 \sum_i Q_t^2 C_j^2 \omega_j^{-3} \tag{5.76}$$

Thus, the promoting vibrations reduce the Franck-Condon factor itself, which is not reflected in the spin-boson model of (5.56) and (5.68). As an illustration, three-dimensional trajectories for various interrelations between symmetric (ω_s) and antisymmetric (ω_a) vibration frequencies and ω_0 are shown in Figure 5.2 When both vibrations have high frequencies, $\omega_{a,s} \gg \omega_0$, the transition proceeds along the MEP (curve 1). In the opposite case of low frequencies, $\omega_{a,s} \ll \omega_0$, the tunneling occurs in the barrier, which is lowered and shortened by the symmetrically coupled vibration q_s, so that the position of the antisymmetrically coupled

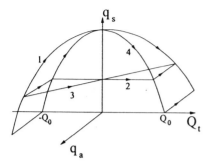

Figure 5.2. Three-dimensional instanton trajectories of a particle in a symmetric double well, interacting with symmetrically and antisymmetrically coupled vibrations with coordinates and frequencies q_s, ω_s and q_a, ω_a, respectively. 1, ω_a, $\omega_s \gg \omega_0$ (MEP); 2, ω_a, $\omega_s \ll \omega_0$ (sudden approximation); 3, $\omega_s \ll \omega_0$, $\omega_a \gg \omega_0$; 4, $\omega_s \gg \omega_0$, $\omega_a \ll \omega_0$.

oscillator q_s shifts through a shorter distance than that in the absence of coupling to q_s (curve 2). The cases $\omega_s \gg \omega_0$, $\omega_a \ll \omega_0$, and $\omega_s \ll \omega_0$, $\omega_a \gg \omega_0$, which are characterized by combined trajectories (sudden limit for one vibration and adiabatic for the other), are also illustrated in Figure 5.2.

5.4. DISSIPATIVE NONADIABATIC TUNNELING

The problem of nonadiabatic tunneling was formulated in Section 3.7, and in this section we study how dissipation affects the conclusions drawn there. The two-state Hamiltonian for the system coupled to a bath is conveniently rewritten via the Pauli matrices:

$$H = \left\{\frac{1}{2} p^2 + \frac{1}{2}[V_i(Q) + V_f(Q)]\right\}\mathbf{1} + \frac{1}{2}[V_f(Q) - V_i(Q)]\sigma_z + V_d(Q)\sigma_x$$

$$+ \frac{1}{2}\sum \left[p_j^2 + \omega_j^2\left(q_j + \frac{C_j Q}{\omega_j^2}\right)^2\right]\mathbf{1} \tag{5.77}$$

Here V_i and V_f are the diabatic terms of the initial and final states, and V_d is the diabatic coupling. We have explicitly added the counterterm $\sum C_j^2 Q^2/2\omega_j^2$ to cancel the adiabatic renormalization caused by vibrations. We shall consider the particular case of two harmonic diabatic terms:

$$H = \frac{1}{2} p^2 + \frac{1}{2}\omega_0^2\left(Q + \frac{F\sigma_z}{\omega_0^2}\right)^2 + V_d\sigma_x + \frac{1}{2}\epsilon\sigma_z$$

$$+ \frac{1}{2}\sum \left[p_j^2 + \omega_j^2\left(q_j + \frac{C_j Q}{\omega_j^2}\right)^2\right] \tag{5.78}$$

where the diabatic coupling V_d is assumed to be constant, F is half the difference of terms' slopes

$$F = \frac{1}{2} \left(\frac{dV_i}{dQ} - \frac{dV_f}{dQ} \right)_{Q=Q_c} \tag{5.79}$$

and the crossing point has the coordinate $Q_c = -\epsilon/2F$. The formal structure of (5.78) suggests that the reaction coordinate Q can be combined with the bath coordinates to form a new fictitious bath, so that the Hamiltonian takes the standard form of a dissipative TLS (5.56).

Suppose that the original spectrum of the bath is ohmic, with friction coefficient η. Then diagonalization of the total system $(Q, \{q_j\})$ gives a new effective spectral density [Garg et al., 1985]:

$$J_{eff}(\omega) = \omega_0^4 \chi''(\omega) = \frac{\eta \omega \omega_0^4}{(\omega_0^2 - \omega^2)^2 + \eta^2 \omega^2} \tag{5.80}$$

where χ'' is the imaginary part of susceptibility of the damped harmonic oscillator with frequency ω_0 and friction coefficient η. After formal substitution $Q_0 = F/\omega_0^2$, $\Delta_0/2 = V_d$ in (5.56), the Hamiltonian (5.78) becomes formally equivalent to (5.56) with the spectral density (5.80). We emphasize that, despite the formal similarity, the physical problems are quite different. Moreover, in general, the diabatic coupling V_d is not small, unlike the tunneling matrix element, and this circumstance does not allow one to apply the noninteracting blip approximation. So, even though it has been formulated in the standard spin-boson form, the problem remains rather sophisticated. In particular, it is difficult to explore the intermediate region between the nonadiabatic and adiabatic regions.

When V_d is small so that the transition is nonadiabatic, the usual golden rule analysis based on (5.67) can be performed to give [Garg et al., 1985; Wolynes, 1987]

$$k = V_d^2 \left(\frac{\pi \beta_{eff}}{E_0} \right)^{1/2} \exp\left(-\frac{Q_\#^2}{2\delta^2} \right) \tag{5.81}$$

where the distance from the minimum of the initial well to the barrier top is $Q_\# = Q_0 - \epsilon/2F$, and $E_0 = 2\omega_0^2 Q_0^2 = 2F^2/\omega_0^2$. The *effective temperature* is defined as

$$T_{eff} = \beta_{eff}^{-1} = \omega_0^2 \delta^2 (\eta, \beta) \tag{5.82}$$

where δ is defined in (5.51). At high temperatures $T_{eff} = T$, and (5.81) is nothing but the Marcus formula, irrespective of friction. As the temperature drops, (5.81) begins to incorporate quantum corrections. When friction increases, T_{eff} decreases and the prefactor in (5.81) increases. This means that reaction becomes more adiabatic. However, the increase of the prefactor is countered by a strong decrease in the leading exponent itself. The result (5.81) may be recast in a TST-like form as described below. If the transition were classical, the rate constant could be calculated as the average flux toward the product valley

$$k = \frac{1}{2} \int dv \, dQ \, B(v) v \rho_{cl}(v, Q) \delta(Q - Q^{\#})$$
(5.83)

where v is velocity, $B(v)$ the Landau-Zener prefactor (3.113) $B(v) = \pi V_d^2/vF$, and $\rho_{cl}(v, Q)$ the classical equilibrium distribution function in the initial well. In the quantum case, this function is replaced by its quantum counterpart, the Wigner function [Dakhnovskii and Ovchinnikov, 1985; Feynman, 1972; Garg et al., 1985] expressed via the density matrix as

$$W(v, Q) = (2\pi)^{-1} \int dQ' \exp(iQ'v) \rho \left(Q - \frac{Q'}{2}, Q + \frac{Q'}{2} \right)$$
(5.84)

Substitution of this for (5.83) gives the formula

$$k = \frac{\pi V_d^2 \rho(Q_{\#}, Q_{\#})}{2F}$$
(5.85)

identical to (5.81). The effect of friction is to slow down the motion, i.e., to decrease v, thereby increasing the Landau-Zener factor. Because the reaction flux is proportional to v while the factor B is inversely proportional to v, the high-temperature limit of Eq. (5.85) does not depend on friction coefficient, in contrast to the result (5.32) for adiabatic transitions.

In the deep tunneling regime, $T \to 0$, the velocity entering into the Landau-Zener factor is formally imaginary. For an asymmetric potential this limit can be studied with the usual Im F techniques. To explore the whole range of Landau-Zener parameters it is more expedient to deal with the original Hamiltonian (5.77). Further, the σ operators, as well as the q oscillators, can be integrated out of the problem by use of the quasienergy method, leading to the problem formulated in terms of the reaction coordinate alone. This approach is detailed in Appendix C; here we just write out the final result:

$$k = B k_{ad}$$
(5.86)

where k_{ad} is the rate of dissipative tunneling in the lower adiabatic term V_- (Figure 3.7), found according to the method of Section 5.2, and B is the prefactor described by (3.114), where the parameter δ is

$$\delta = \frac{V_d^2}{2Fv_{ins}} \tag{5.87}$$

and v_{ins} is the imaginary-time instanton velocity. In the nonadiabatic limit $(\delta \ll 1)$ $B = \pi V_d^2/v_{ins}F$, and at $\delta \gg 1$ the adiabatic result $k = k_{ad}$ holds. As shown in Section 5.2, with increasing η the instanton velocity decreases and the transition tends to be more adiabatic, as in the classical case. This conclusion is far from obvious, because one might expect that when losing energy, a particle should increase its upside-down barrier velocity. Instead, the energy losses are saturated to a finite η-independent value, and friction slows the tunneling motion down.

Generally speaking, the problem of nonadiabatic transitions presents an extremely rich dynamics resulting from the interplay of tunneling, resonances, interference, dephasing, and vibrational relaxation. Because of complexity of the total picture, studies of the problem were mostly restricted to sufficiently simple models revealing only some of the above aspects. For example, Shimshoni and Gefen [1991] considered the interplay of dissipation and interference in a simple model of two sequential nonadiabatic transitions caused by two crossing points. The resulting transition probability is dominated by interference effects, which become important when the separation between the crossing points is comparable with the characteristic size of the region where nonadiabatic effects dominate. Dissipation, however, may destroy this interference, leading to the regime where the two transitions are independent of each other. Various regimes of nonadiabatic transitions in the presence of dissipation may range from damped probability oscillations to rate dynamics or even localization [Makarov and Makri, 1993].

APPENDIX C

Dissipative Nonadiabatic Tunneling At $T = 0$

In this appendix we show how the quasienergy ideas developed in Section 5.1 can be applied to the problem of nonadiabatic tunneling. We use the Im F approach of Section 3.5 for the multidimensional system with

Hamiltonian

$$H = \left\{\frac{1}{2} p^2 + \frac{1}{2} [V_i(Q) + V_f(Q)]\right\}\hat{\mathbf{1}} + \frac{1}{2} [V_f(Q) - V_i(Q)]\sigma_z + V_d(Q)\sigma_x$$

$$+ \frac{1}{2} \sum_j \left[p_j^2 + \omega_j^2 \left(x_j + \frac{C_j Q}{\omega_j^2}\right)^2\right]\hat{\mathbf{1}} \tag{C.1}$$

where V_i and V_f are the initial and final potential energy terms, V_d is the diabatic coupling between them, $\mathbf{1}$ is the unit 2×2 matrix, and σ's are the Pauli matrices. Formally, this is a Hamiltonian of a single particle with potential $(V_i + V_f)/2$ (which may have no barrier at all) coupled to a two-level system and a bath of harmonic oscillators. Following the Im F method, we study the partition function of the system with the aim of presenting it as a path integral over only the $Q(\tau)$ paths. The oscillator degrees of freedom are traced out in a standard way producing the nonlocal action term, while integration over the TLS paths results in the quasienergy partition function $\tilde{Z} = 2 \cosh\{\beta\tilde{\epsilon}[Q(\tau)]\}$. After taking the $\beta \to \infty$ limit, only the lower quasienergy survives in $\tilde{Z}$, and one obtains

$$Z = \int D[Q(\tau)] \exp\left(- \int_0^\beta d\tau \left\{\frac{1}{2}\dot{Q}^2 + \frac{1}{2}[V_i(Q) + V_f(Q)]\right.\right.$$

$$- [V_d^2(Q) + \epsilon^2(Q)]^{1/2} + \int_0^\beta d\tau' K(\tau - \tau')Q(\tau)Q(\tau')\bigg\}\bigg)$$

$$\times \exp\left(\beta\tilde{\epsilon}[Q(\tau)] - \int_0^\beta d\tau\{V_d^2[Q(\tau)] + \epsilon^2[Q(\tau)]\}^{1/2}\right) \tag{C.2}$$

where

$$\epsilon(Q) = \frac{V_i(Q) - V_f(Q)}{2} \tag{C.3}$$

We have multiplied and divided the path integral in (C.2) by the same factor $\exp(\beta\tilde{\epsilon}_{ad})$ where

$$\tilde{\epsilon}_{ad}[Q(\tau)] = \beta^{-1} \int_0^\beta d\tau\{V_d^2[Q(\tau)] + \epsilon^2[Q(\tau)]\}^{1/2} \tag{C.4}$$

is the quasienergy in the adiabatic approximation for TLS. Without the last exponent, (C.2) would give the standard formula for dissipative adiabatic tunneling with the rate constant obtained in Section 5.2. Now we make the usual assumption in nonadiabatic transition theory that nonadiabaticity is important only in the vicinity of the crossing point Q_c where $\epsilon(Q_c) = 0$. Therefore, if the trajectory does not cross the dividing surface $Q = Q_c$, its contribution to the path integral is to a good accuracy described by adiabatic approximation, i.e., $\tilde{\epsilon} = \tilde{\epsilon}_{ad}$. Hence, the real part of partition function, Z_0, is the same as in adiabatic approximation. Then the rate constant may be written as

$$k = Bk_{ad} \tag{C.5}$$

where k_{ad} is the rate of adiabatic tunneling in the potential

$$V_-(Q) = \tfrac{1}{2}[V_i(Q) + V_f(Q)] - [V_d^2(Q) + \epsilon^2(Q)]^{1/2} \tag{C.6}$$

and the prefactor B equals

$$B = \langle B[Q(\tau)] \rangle_{S_{ad}} = \langle \exp\{\beta(\tilde{\epsilon}[Q(\tau)] - \tilde{\epsilon}_{ad})\} \rangle_{S_{ad}} \tag{C.7}$$

where the symbol $\langle \cdots \rangle_{S_{ad}}$ means averaging over the paths with the weight $\exp(-S_{ad})$

$$S_{ad}[Q(\tau)] = \int_0^\beta d\tau \left[\tfrac{1}{2}\dot{Q}^2 + V_-(Q) + \int_0^\beta d\tau' K(\tau - \tau')Q(\tau)Q(\tau') \right] \tag{C.8}$$

The equations for the quasienergies $\pm\tilde{\epsilon}$ are

$$\frac{dc_1}{d\tau} = -\epsilon[Q(\tau)]c_1 + V_d[Q(\tau)]c_2$$

$$\frac{dc_2}{d\tau} = \epsilon[Q(\tau)]c_2 + V_d[Q(\tau)]c_1 \tag{C.9}$$

with the condition

$$c_i(\tau + \beta) = \exp(\pm\beta\tilde{\epsilon})c_i(\tau) \tag{C.10}$$

Let $c_i(\tau)$ be an arbitrary solution to (C.9), which does not necessarily satisfy (C.10). Then it can be represented as a linear combination of exponentially increasing and decreasing linearly independent solutions (C.10). When $\beta \to \infty$, only the increasing solution survives after a large time, and one may write

$$\exp(\beta\tilde{\epsilon}) = \frac{c_i(\beta + \tau)}{c_i(\tau)} \tag{C.11}$$

where c_i is an arbitrary solution to (C.9) that is not subject to (C.10), and τ in (C.11) is sufficiently large. For this reason the boundary conditions (C.10) are of no concern. Let us define the phase ϕ as

$$\exp[\phi(\tau_2, \tau_1)] = \frac{c_1(\tau_2)}{c_1(\tau_1)} \tag{C.12}$$

Then $\beta\tilde{\epsilon}$ may be thought of as the phase accumulated by the function $c_1(\tau)$ during the period β. To find B in (C.7) we should compare the phase $\phi(\beta + \tau, \tau)$ to that calculated in the adiabatic approximation ϕ_{ad}. According to the standard arguments of Landau-Zener-Stueckelberg theory, this difference arises mostly from passing the point $Q(\tau^*) = Q_c$ where the adiabaticity is violated. In the vicinity of this point Eqs. (C.9) simplify to

$$\frac{dc_1}{d\tau} = -Fv(\tau - \tau^*)c_1 + V_d(Q_c)c_2$$

$$\frac{dc_2}{d\tau} = Fv(\tau - \tau^*)c_2 + V_d(Q_c)c_1 \tag{C.13}$$

where

$$F = \frac{1}{2}\left(\frac{dV_i}{dQ} - \frac{dV_f}{dQ}\right)_{Q=Q_c} \qquad v = \left(\frac{dQ}{d\tau}\right)_{\tau=\tau^*} \tag{C.14}$$

Taking up the dimensionless units $\tau - \tau^* = (2Fv)^{-1/2}\tau'$, $c_i = (2Fv)^{-1}c_i'$, $i = 1,2$, we rewrite (C.13) as

$$\frac{dc_1'}{d\tau'} = -\frac{\tau'c_1'}{2} + \delta^{1/2}c_2'$$

$$\frac{dc_2'}{d\tau'} = \frac{\tau'c_2'}{2}2 + \delta^{1/2}c_1' \tag{C.15}$$

with

$$\delta = \frac{V_d(Q_c)^2}{2Fv} \tag{C.16}$$

Equations (C.15) are exactly the same as those derived by Holstein [1978], and the following discussion draws on that paper. The pair of equations (C.15) may be represented as a single second-order differential equation:

$$-\frac{d^2 c_1}{d\tau^2} + \left(\frac{\tau^2}{4} + \delta - \frac{1}{2}\right)c_1 = 0 \tag{C.17}$$

where we have omitted the primes for simplicity. Its solution at $\tau \to \infty$ is an increasing parabolic cylinder function $D_{-\delta}(-\tau)$ with the following asymptotic forms:

$$c_1(\tau) \propto \tau^\delta \exp\left(-\frac{\tau^2}{4}\right) \qquad \tau < 0 \tag{C.18}$$

$$c_1(\tau) \propto \frac{(2\pi)^{1/2}\tau^{\delta-1}\exp(\tau^2/4)}{\Gamma(\delta)} \qquad \tau > 0 \tag{C.19}$$

where $\Gamma(\delta)$ is the gamma function. By using the definition (C.12) we obtain

$$\exp[\phi(\tau, -\tau)] = \frac{c_1(\tau)}{c_1(-\tau)} = \frac{(2\pi)^{1/2}\tau^{2\delta-1}\exp(-\tau^2/2)}{\Gamma(\delta)} \tag{C.20}$$

This should be compared with the adiabatic result. To do this we look for the solution c_i of (C.15) in the form

$$c_i(\tau) = u_i(\tau) \exp \int_{-\tau}^{\tau} \left(\delta^2 + \frac{\tau'^2}{4}\right)^{1/2} d\tau' \tag{C.21}$$

Substituting this for (C.15) and neglecting the derivatives one gets

$$u_1\left[\frac{\tau}{2} + \left(\frac{\tau^2}{4} + \delta\right)^{1/2}\right] = \delta^{1/2}u_2 \tag{C.22}$$

whence for $|\tau| \gg \delta^{1/2}$ the functions u_1 and u_2 are related by

$$u_1 \cong \frac{\delta^{1/2}u_2}{\tau} \qquad \tau > 0$$

$$u_2 \cong \frac{\delta^{1/2}u_1}{|\tau|} \qquad \tau < 0 \tag{C.23}$$

If we were to replace τ by $-\tau$ in (C.15), the solution u_1 would replace u_2 and vice versa. Therefore, one may write at $\tau > 0$

$$u_1(\tau) \cong \frac{\delta^{1/2}u_2(\tau)}{\tau} = \frac{\delta^{1/2}u_1(-\tau)}{\tau} \tag{C.24}$$

Thus, for the adiabatic phase ϕ_{ad} one obtains

$$\exp[\phi_{ad}(\tau, -\tau)] = \frac{c_1(\tau)}{c_1(-\tau)} = \tau^{-1}\delta^{1/2} \exp \int_{-\tau}^{\tau} \left(\delta^2 + \frac{\tau'^2}{4}\right)^{1/2} d\tau'$$

$$\cong e^{\delta}\delta^{-\delta}\tau^{2\delta-1} \exp\left(\frac{\tau^2}{2}\right) \tag{C.25}$$

Comparing (C.20) with (C.25) one finds

$$B[Q(\tau)] = \exp[\beta(\tilde{\epsilon} - \tilde{\epsilon}_{ad})] \cong \exp[n\phi(\tau, -\tau) - n\phi_{ad}(\tau, -\tau)]$$

$$= \left[\frac{(2\pi/\delta)^{1/2} e^{-\delta}\delta^{\delta}}{\Gamma(\delta)}\right]^n \tag{C.26}$$

where n is the number of times the trajectory crosses the point Q_c. The functional $B[Q(\tau)]$ actually depends only on the velocity $dQ/d\tau$ at the moment when the nonadiabatic region is crossed. If we take the path integral by the steepest descent method, considering that the prefactor $B[Q(\tau)]$ is much more weakly dependent on realization of the path than $S_{ad}[Q(\tau)]$, we obtain the instanton trajectory for the adiabatic potential V_{ad}, and $B[Q(\tau)]$ will have to be calculated for that trajectory. Since the instanton trajectory crosses the dividing surface twice, we finally have

$$B = \frac{2\pi\delta^{-1} e^{-2\delta}\delta^{2\delta}}{\Gamma^2(\delta)} \tag{C.27}$$

where the instanton velocity should be inserted into formula (C.16) for δ.

The ground-state tunneling splitting for two symmetrically placed diabatic terms can be found in the same manner, as described in Sections 3.6, and 4.2. Since the kink trajectory crosses the barrier once, we obtain

$$\Delta = \Delta_{ad}B^{1/2} \tag{C.28}$$

where Δ_{ad} is the tunneling splitting in the adiabatic potential and B is defined by (C.27). In the nonadiabatic regime, $\delta \ll 1$, the tunneling splitting is proportional to V_d and inversely proportional to the square root of the instanton velocity.

HYDROGEN TRANSFER

CONTENTS

In the early 1970s, low-temperature chemistry began to develop as a branch of solid-state chemistry. Most of the early experiments were directed toward studies of irreversible chemical reactions, particularly solid-state reactions of free radicals. The discovery of the crossover temperature and low-temperature limiting rate constant at $T < T_c$ was the most important achievement of that period. Experiments directed toward elucidating the spectroscopic manifestations of tunneling came later, after the development of high-resolution laser spectroscopic techniques. The primary subjects of these early studies were molecular solids and impurity molecules in noble gas matrices, in which tunneling is accompanied by reorganization of the environment. The effects of tunneling were often masked in these studies by inhomogeneous broadening resulting from intermolecular interactions.

The development of supersonic molecular beam techniques created new opportunities to study tunneling effects in gas-phase isolated molecules and dimers at ultralow translational and rotational temperatures. Modern low-temperature chemistry, therefore, includes the study of chemical dynamics of molecules in various states of aggregation.

Hydrogen tunneling occupies a special place in chemical dynamics for several reasons. First, owing to its small mass, the tunneling transfer of H or H^+ can be observed through higher and broader barriers than for heavier particles. The range of barrier heights is typically from 1 to 12–14 kcal/mol, and transfer distances are comparable with the lengths of hydrogen bonds. Second, the relative ease of measuring kinetic isotope effects for substitution of D for H permits the separation of two characteristic tunnel parameters: mass and tunneling distance. Third, the

151

frequencies of hydrogen stretching and bending vibrations are typically much higher than for nontunneling modes of the environment, so the participation of the latter in multidimensional tunneling is revealed more clearly than in the case of heavy-particle tunneling. The promotion and inhibition of hydrogen transfer by selected vibrations of different symmetry have been investigated in detail, as discussed in Chapter 2.

Intra- and intermolecular hydrogen transfer processes are important in a wide variety of chemical processes, ranging from free radical reactions (which make up the foundation of radiation chemistry) and tautomerization in the ground and excited states (a fundamental photochemical process) to bulk and surface diffusion (critical for heterogeneous catalytic processes). The exchange reaction $H_2 + H$ has always been the preeminent model for testing basic concepts of chemical dynamics theory because it is amenable to carrying out exact three-dimensional fully quantum mechanical calculations. This reaction is now studied in low-temperature solids as well.

6.1. SEMIEMPIRICAL TWO-DIMENSIONAL POTENTIAL OF A HYDROGEN BOND

A hydrogen bond can be loosely described as the interaction between two electronegative atoms and an intervening hydrogen atom, i.e., $-XH \cdots Y-$, where X and Y represent O, F, Cl, N, S, or C atoms. The potential surface that describes this interaction typically has two minima that correspond to formation of a strong XH or YH bonds. A one-dimensional potential that describes the motion of the central H atom between the two heavier atoms at a fixed distance R from each other can be represented by the empirical form

$$V(r) = \frac{\hbar\omega}{2}\left(v_2 r^2 + v_3 r^3 + v_4 r^4\right) \qquad (6.1)$$

where r is X–H distance. This potential is a double well if $v_2 < 0$ and $v_4 > 0$. Eigenfunctions of the corresponding Hamiltonian are expressed as linear combinations of harmonic oscillator functions:

$$\Phi_j(r) = \sum_n c_{nj} \phi_n(r) \qquad (6.2)$$

The expansion coefficients and energy eigenvalues are found from solutions of the secular equation $|H_{nn} - \delta_{nn}E| = 0$. The nonzero matrix elements H_{nn} can be expressed in terms of the coefficients of the potential

(6.1) using relations derived by Somorjai and Hornig (1962):

$$H_{nn} = (2n + 1)(1 + v_2) + \tfrac{3}{2}(2n^2 + 2n + 1)v_4$$

$$H_{n,n-1} = \tfrac{3}{2}n(2n)^{1/2}v_3$$

$$H_{n,n-2} = [n(n - 1)]^{1/2}(v_2 - 1) + (2n - 1)[n(n - 1)]^{1/2}v_4 \qquad (6.3)$$

$$H_{n,n-3} = [\tfrac{1}{2}n(n - 1)(n - 2)]^{1/2}v_3$$

$$H_{n,n-4} = [\tfrac{1}{2}n(n - 1)(n - 2)(n - 3)]^{1/2}v_4$$

This method of numerical quantization can be also used in the case of low barriers, for which the semiclassical approximation is no longer valid.

The empirical form of the potential in (6.1) implies that the co-efficients v_2, v_3, and v_4 depend parametrically on the distance R between the two heavy atoms X and Y. This dependence is quite strong, based on a large set of experimental data concerning the bond lengths and stretching vibrational frequencies XH of various compounds containing the same XH$\cdots$Y fragments. The equilibrium bond length and bond energy for XH changes significantly even for changes in R that approach the amplitude of zero-point vibrations. Therefore, a potential surface with at least two dimensions is required to correctly describe the tunneling dynamics.

Lippincott and Schroeder [1955, 1957] introduced a semiempirical two-dimensional PES and fitted their parameters from experimental data. Further studies in this direction were carried out by Savel'ev and Sokolov [1975] and Sokolov and Savel'ev [1977], Lautie and Novak [1980], Saitoh et al. [1981], and Emsley [1984]. These studies have shown that an adequate two-dimensional PES can be constructed from Morse functions of diatomic fragments XH and HY and repulsive functions representing the XY interaction. The values of r_{XH}^0 and ω_{XH} and isotope effects as a function of R are in agreement with the experimental ones for OH$\cdots$O, OH$\cdots$N, and NH$\cdots$N fragments. The dependencies $r_{XH}^0(R)$ and $\omega_{XH}(R)$ collected by Novak [1974] are shown in Figure 6.1. The method of Lippincott and Schroeder [1957] is one of the versions of the general semiempirical method of bond energy–bond order (BEBO) developed by Johnston and Parr [1963] to construct a two-dimensional PES.

Models of two-dimensional tunneling are described in Chapters 2 and 4. The first step in the analysis of experimental data is the determination of tunneling regime conditions, dependent on frequencies ω_{XH} and ω_{XY} and coupling parameters (see Sections 2.5 and 4.2). All of these parameters are readily estimated using empirical correlations similar to

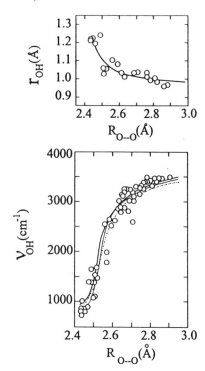

Figure 6.1. Correlation between (a) r_{OH} and R_{OO} and (b) ν_{OH} and R_{OO} in the OH$\cdots$O fragment. Open circles are the experimental values for different compounds. From Saitoh et al. [1911].)

the example shown in Figure 6.2 for the dependence of barrier height and transfer distance $(R - 2r_{XH}^0)$ on the distance between the two O atoms in the OH$\cdots$O fragment. The parameters of model PES (4.41) are

$$b = \frac{C^2}{2\Omega^2} = 1 - \frac{V^\#}{V_0} \qquad \Omega = 2\sqrt{2}\,\frac{\omega_{XY}}{\omega_{XH}} \qquad (6.4)$$

Therefore, all the information that is needed to construct PES is available. Of course, this two-dimensional PES of the hydrogen bond is a rather crude approximation, since the total set of skeleton motions is replaced by a single collinear XY stretching vibration with effective frequency ω_{XY} and coupling parameter C. Nevertheless, the approximation permits us to understand the behavior of tunneling trajectories for hydrogen transfer in typical cases by comparing the calculated values of C and Ω with domains shown in Figure 4.7.

For an intramolecular OH$\cdots$O fragment, a strong hydrogen bond corresponds to $R \le 2.55$ Å and $V^\# \le 8$ kcal/mol. In this case, C and Ω are close to unity and this corresponds to an intermediate case between the sudden and adiabatic regimes. Examples of such systems are malonaldehyde, tropolon and its derivatives, and the hydrogenoxalate anion.

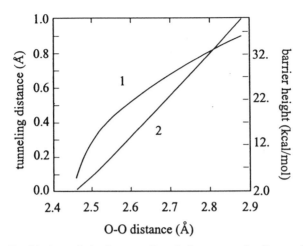

Figure 6.2. Empirical correlation between R_{OO}, hydrogen transfer distance (curve 1), and the barrier height (curve 2) in the OH $\cdots$ O fragment.

For intermolecular hydrogen bonds, where $b \ll 1$ and $\Omega \ll 1$, the sudden approximation is valid ($C \ll 1$, $\Omega \ll 1$). Good examples of this case are the carbonic acid dimers, as was demonstrated by Makri and Miller [1989b]. Owing to the large dipole moment of OH, NH, and SH bonds, hydrogen transfer reactions in polar media cause a significant dynamic polarization of the medium. This is characterized by a reorganization energy in the same manner as in electron transfer reactions (see Section 2.4). The effect is especially important in asymmetric potentials in which one of the equilibrium states is ionic (e.g., $XH \cdots Y \rightarrow X^- HY^+$). Reorganization of the medium changes the double well potential drastically. Hydrogen transfer in the OH $\cdots$ N fragment, which forms an ion pair $O^- \cdots HN^+$, has been treated for media having various degrees of polarity by Juanos i Timoneda and Hynes [1991].

6.2. HYDROGEN BOND AND TUNNELING SPLITTING

One of the most well-known examples of intramolecular tunneling is the isomerization of malonaldehyde consisting in hydrogen transfer in the fragment OH $\cdots$ O:

$$\tag{6.5}$$

Hydrogen exchange produces a tunneling splitting that is observed in the microwave spectrum of this molecule. In the ground vibrational state this splitting is equal to $21.6\,\text{cm}^{-1}$. It decreases to $3.0\,\text{cm}^{-1}$ when a D atom is substituted for H in the OH $\cdots$ O fragment [Baughcum et al., 1981, 1984; Turner et al., 1984]. A calculation of the one-dimensional tunneling probability was performed by de la Vega [1982]. The model system consisted of two equivalent equilibrium states and a rigid molecular skeleton. The values of the tunneling splitting Δ for H and D are about two orders of magnitude smaller than experimental ones. The origin of this disagreement, elucidated by Bicerano et al. [1983] and Carrington and Miller [1986], is the large displacement of heavy atoms accompanying the hydrogen transfer. The transition therefore cannot be reduced to a simple tunneling process through a static barrier. The skeletal reorganization is needed mainly due to the difference between single and double CO and CC bond lengths (1.33 and $1.23\,\text{Å}$ and 1.44 and $1.34\,\text{Å}$, respectively). A self-consistent calculation of the PES for multidimensional tunneling was made by Shida et al. [1989, 1991].

The transfer of hydrogen atoms is accompanied by a number of small-amplitude motions, including OH stretching, bending of the COH and OCC bonds, and stretching of CO. All of these vibrations contribute to the dynamical shortening of the tunneling distance. Experimental studies show that the vibrational modes at 318 and $1378\,\text{cm}^{-1}$ play a major role in assisting the tunneling. This is manifested by a severalfold increase of the tunneling splitting when these assisting vibrations are excited. Deformation of the skeleton in the transition state relative to initial state is shown in Figure 6.3. The tunneling distance for the H atom

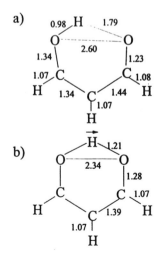

Figure 6.3. Geometries of (a) the reactants and (b) the transition state for the intramolecular proton transfer in malonaldehyde. The bond length is given in angstroms. From Bosch et al. [1991].)

in this configuration is only 0.47 Å, compared with 0.75 Å in the equilibrium configuration. The barrier height along the MEP is calculated by various quantum chemical methods to lie in the range 6.7–10.6 kcal/mol. Tunneling splittings close to observed ones (19 and 1.6 cm^{-1} for H and D transfer) are obtained when $V^{\#} = 6.3$ kcal/mol. From the temperature dependence of an NMR spectrum measured by Brown et al. [1979] it follows that the upper limit for the barrier height should be 6.1 kcal/mol. Calculations by Shida et al. [1989], Makri and Miller [1987a, 1989b], Bosch et al. [1990], and Rom et al. [1991] show that the major multidimensional features of this system can be represented by a two-dimensional PES (4.41). The parameters of this effective PES, taking into account the vibrationally adiabatic corrections for high-frequency degrees of freedom are

$$V_0 = 18.01 \text{ kcal/mol} \qquad C = 0.86 \qquad \Omega = 0.71 \qquad \omega_0 = 1.79 \times 10^{14} \text{ s}^{-1}$$

$$V^{\#} = V_0(1 - b) = 4.98 \text{ kcal/mol} \tag{6.6}$$

A contour plot of this surface is illustrated in Figure 6.4.

A number of empirical tunneling paths have been proposed in order to simplify the two-dimensional problem. Among those are the MEP [Kato et al., 1977], the sudden straight line [Makri and Miller, 1989b], and the so-called expectation value path [Shida et al., 1989]. The results of these papers are hard to compare because somewhat different PES's were used. As to the expectation value path, it was constructed as a parametric line $q(Q)$ on which the vibrational coordinate q takes its expectation value when Q is fixed. Clearly, for the PES at hand this path coincides with MEP, since q is the coordinate of a harmonic oscillator. The results for the tunneling splitting calculated with the use of some of the earlier proposed reaction paths for PES (4.41) are compiled by Bosch et al.

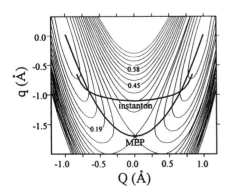

Figure 6.4. Contour plot, MEP, and instanton trajectory for hydrogen transfer in malonaldehyde. (From Benderskii et al. [1993].)

[1990]. All of them give values of Δ that are at least one order of magnitude less than the numerically exact value of $10.6\,cm^{-1}$, which is also given in that paper.

As we noted in the preceding section, the parameters C and Ω place this problem in the intermediate region between the sudden and adiabatic approximations, described in Sections 2.5 and 4.2, and neither of these approximations is quantitatively applicable to the problem. The tunneling trajectory has been numerically calculated by use of the instanton techniques of Section 4.2 [Benderskii et al., 1993] and is shown in Figure 6.4 together with the MEP. This extremal trajectory demonstrates that tunneling is essentially a two-dimensional transition that involves both degrees of freedom in a manner that is not described by the afore-mentioned approximate methods. The value of the prefactor obtained is $\tilde{B}_t = 110$, and the tunneling splitting, $13\,cm^{-1}$, compares well with both the experimental value and quantum calculations by Bosch et al. [1990]. The large value of the prefactor reflects the fact that the transverse vibrational frequency ω_t softens considerably from the initial state to the transition state. We note that this is a common feature of proton transfer reactions, and for this reason the simple semiclassical calculations done with exponential accuracy (i.e., without correctly calculating the value of the prefactor) underestimate the tunneling splitting. The barrier height along the optimum path is higher than along the MEP, but the barrier width is smaller due to the cutting of the corner. This trade-off results in a minimization of the action. As we noted earlier, the optimum path is temperature-dependent, and shifts toward the MEP with increasing temperature. For the PES having a single saddle point like (4.41), this path typically occupies an intermediate position between MEP and a straight line path corresponding to sudden approximation, so that only this portion of PES is significant for the transition probability calculation.

Unlike the gas-phase measurements, no tunneling has been detected in the IR spectra of malonaladehyde isolated in the rare gas matrices at 15–30 K [Firth et al., 1989]. This disappearance has been attributed to "detuning" of the potential as a result of weak asymmetric coupling to the environment.

Methyl-substituted malonaldehyde (α-methyl-β-hydroxyacrolein) provides an opportunity to study the role of asymmetry of the potential profile in the proton exchange. In the initial and final states, one of the C–H bonds of the methyl group is in the molecular plane and directed toward the proton position. The double well potential becomes symmetric only due to methyl group rotation over $\pi/6$, when the C–H bond lies in the plane perpendicular to the molecular one. As a result, proton tunneling occurs in combination with CH_3 hindered rotation and the splitting is about 10 times smaller than in malonaldehyde. This coupling of two tunneling motions has been suggested by Busch et al. [1980] from the analysis of the microwave spectrum of the molecule. A two-dimensional potential was proposed to analyze this problem:

$$V(Q, \theta) = V(Q) + \frac{V_3}{2}\left(1 - \frac{Q}{Q_0}\cos 3\theta\right) \qquad (6.7)$$

where Q_0 corresponds to proton at equilibrium. The motions of heavy atoms are not directly incorporated in this potential, and their role is assumed to produce an effective two-dimensional potential, i.e., to renormalize the parameters in Eq. (6.7). This renormalization can be done in two ways: One uses the adiabatic approximation based on the assumption that the heavy atoms adiabatically follow the system motion; this assumption leads to appreciable displacement of heavy atoms, especially oxygen. Another approach relies on the assumption that the positions of heavy atoms are fixed while tunneling transition occurs in potential (6.7). Moreover, these positions were assumed to take their average values, unlike the sudden approximation (2.79)–(2.81) and (4.33) in which these positions are optimized to give the largest tunneling probability. The tunneling splittings calculated for both two-dimensional potentials using a basis set method were in disagreement with experiment. The origin of this disagreement is clear from our discussion concerning proton transfer in malonaldehyde: The actual tunneling regime is neither adiabatic nor sudden, so neither form of potential adequately describes the tunneling dynamics. Since the torsional frequency is much smaller than those of all other modes, seemingly a better method would be to apply the sudden approximation with respect to torsion, averaging the tunneling probability over this coordinate. This example demonstrates the difficulties of quantum calculations in multidimensional systems, and the necessity to reduce the number of degrees of freedom.

In contrast with malonaldehyde, the tunneling splitting in the tropolone molecule is almost the same in the gas phase and in a neon matrix [Rossetti and Brus, 1980]. In both cases there is a similarity not only in Δ, but also in the equilibrium distances in the OH$\cdots$O fragments. Measurements of the tunneling splitting in various vibrational levels of the lowest excited A^1B_2 state of tropolone in supersonic jets have been carried out by Redington et al. [1988], Redington [1990], Sekiya et al. [1990a,b,c], and Fuke and Kaya [1989].

These studies provide direct experimental confirmation of the vibration-assisted tunneling model. The fluorescence excitation spectrum

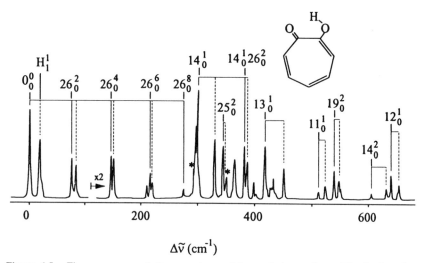

Figure 6.5. Fluorescence excitation spectrum of jet-cooled tropolone. The broken lines indicate transitions for the high-wave-numbered tunneling doublet components. The bands due to the hydrogen-bonded complex between tropolone and water are indicated by the asterisks. (From Sekiya et al. [1990b].)

obtained by Sekiya et al. [1990b] is shown in Figure 6.5. In the progressions of symmetrically coupled vibrations ν_{13} and ν_{14}, Δ is enhanced. These modes shorten the tunneling distance by deformation of the CCO and CCC bond angles (bending vibration ν_{13}) and by deformation of the out-of-plane axis-symmetrical seven-membered ring (ν_{14}). However, the tunneling splitting is suppressed by excitation of the antisymmetrically coupled vibrations ν_{26} and ν_{39}, as shown in Table 6.1. The skeletal motions of these vibrations are illustrated in Figure 6.6, and a detailed description of the tropolone vibrational modes is given by Redington and Bock [1991]. Out-of-plane bending vibrations ν_{25}, ν_{26}, and ν_{39} result in noncollinear displacements of the $OH \cdots O$ fragment, and thereby suppress the tunneling rate. The explanation of the vibrational selectivity and relevant formulae (2.87) and (2.90) for vibrations of different symmetry are given in Section 2.5.

The analysis of the IR spectrum of tropolone in a neon matrix [Redington, 1990], shown in Figure 6.7, is based on crystallographic data by Shimanouchi and Sasada [1973]. This analysis reveals that the displacements of the nontunneling heavy atoms (by ~ 0.07 Å) are comparable to or even greater than the amplitudes of zero-point vibrations. The hydrogen transfer is coupled strongly with in-plane C=O/C–O and C–C=C/C=C–C vibrations (q_1 and q_2 coordinates, respectively). The

TABLE 6.1
Tunneling Splitting of Vibrational Levels in the Excited A^1B_2 State of the Tropolone Molecule

Vibrational Band	Vibrational Frequency (cm^{-1})	Splitting (cm^{-1})	
		TROH	TROD
0_0^0	—	18.9	2.2
11_0^1	511	13	—
12_0^1	640	17	—
13_0^1	414	32^a	3
14_0^1	296	30.4^a	11
14_0^2	2×296	28	13
19_0^2	2×269	9	—
25_0^2	2×171	4.2^a	—
26_0^2	2×39	7.2^a	—
26_0^4	4×39	4.7	—
26_0^6	6×39	3.5	—
26_0^8	8×39	0.8	—

a Data from Redington et al. [1988]; other data from Sekiya et al. [1990b].

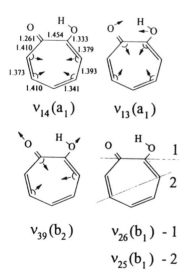

$$v_{14}(a_1) \qquad v_{13}(a_1)$$

$$v_{39}(b_2) \qquad v_{26}(b_1) - 1$$

$$v_{25}(b_1) - 2$$

Figure 6.6. Normal vibrations of tropolone molecule relevant for the tunneling tautomerization. The symmetry of each vibration is given in parentheses. The equilibrium bond length corresponds to the tropolone crystal. (From Redington et al. [1988].)

projection of this three-dimensional PES $V(r, q_1, q_2)$ onto the (q_1, q_2) plane is characterized by two saddle points. Multiple saddle points and, consequently, degenerate reaction paths, are typical for two-proton transfer and hydrogen transfer between interstitial sites in crystals. These

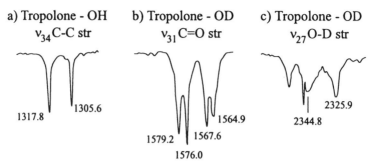

a) Tropolone - OH
ν_{34} C-C str

1317.8   1305.6

b) Tropolone - OD
ν_{31} C=O str

1579.2   1567.6
1576.0
1564.9

c) Tropolone - OD
ν_{27} O-D str

2344.8   2325.9

Figure 6.7. The ν_{27}, ν_{31}, and ν_{34} modes of tropolone isolated in neon matrix at 5 K. Frequencies are given in cm^{-1}. (From Redington [1990].)

cases are discussed in the next sections. We also note that this two-dimensional PES with several saddle points is characteristic of coupled tunneling rotation and interconversion in molecular complexes (see Chapter 8).

In the ground $\tilde{X}^1 A_1$ electronic state of tropolone the tunneling splitting (≤ 0.17 cm^{-1}) is smaller than in the excited state. Due to the aforementioned skeleton deformation that accompanies the hydrogen transfer, the barrier height is higher than that in molonaldehyde and equals 13.7 kcal/mol. The two conformers of tropolone that correspond to hydrogen atom localization in the left or right well both have C_s symmetry. The absence of symmetry restrictions leads to hybrid transitions for in-plane stretching vibrations CC (ν_{34}), CO (ν_{35}), and OH (ν_{27}). The fundamental vibrational band $n = 1 \leftarrow 0$ becomes a doublet. Tunneling splits each component of the doublet, yielding an overall quartet structure of the infrared spectra. The intradoublet separation is equal to a sum of splittings in the ground ($v = 0$) and excited ($v = 1$) states. Such a quartet structure of these bands has been observed by Redington and Redington [1979]. Their data are illustrated in Figure 6.7. The tunneling splitting sums range from 6.9 to 16.3 cm^{-1} depending on the mode that is excited. In the excited state Δ is greater than that in the ground state, due to enhancement of tunneling by these vibrations. The main contribution to the dispersional interaction of this impurity molecule with a rare gas matrix is caused by the strongly polar carbonyl group of tropolone. The energy of its interaction with neighboring neon atoms is about 28 cm^{-1} and causes an asymmetry of the potential (relative energy bias of wells, A) of about 2–3 cm^{-1}. In the ground state $A \gg \Delta$.

The effect of alkyl group substitution on the tunneling splitting in $A^2 B_1$ state of tropolone has been studied by Sekiya et al. [1990a] and Tsuji et al. [1991] using laser-induced fluorescence in supercooled jets. In isopropyltropolones substituted in positions 4 and 5, the splitting of the vibronic bands is missing, demonstrating that the barrier height increases in comparison with unsubstituted tropolone. In 3-isopropyltropolone the tunneling splitting is $58 \, \text{cm}^{-1}$ for 3-ITROH and $14 \, \text{cm}^{-1}$ for its deuterated form, 3-ITROD, respectively. The authors proposed that the substitution in this position increases the asymmetry of the double well potential. Since the energy splitting in the asymmetric potential is $(A^2 + \Delta^2)^{1/2}$, the greater A corresponds to greater band splitting. However, the larger isotope H/D effect is hardly consistent with this assumption. As shown in the second paper by Tsuji et al. [1991], substitution of a chlorine atom in the 3-position of tropolone does not change the tunneling splitting (21 and $3 \, \text{cm}^{-1}$ for H and D transfer, respectively). By contrast, the asymmetry for 3-bromotropolone is so large that no trace of proton tunneling was observed. The van der Waals radius of a chlorine atom $(1.8 \, \text{Å})$ is smaller than that of bromine atom $(1.95 \, \text{Å})$, so the steric repulsion, being the cause of potential asymmetry, is different for the above substituents.

A large tunneling splitting was observed by Rossetti et al. [1980], Bondybey et al. [1984], and Barbara et al. [1989] in the excited singlet (311 and $172 \, \text{cm}^{-1}$) and ground (69 and $12 \, \text{cm}^{-1}$) states of 9-hydroxy-phenalenone in noble gas matrices at 4.2 K for H and D transfer, respectively:

$$(6.8)$$

The fluorescence excitation spectrum of the deuterated compound, shown in Figure 6.8, exhibits a strong 0_+^+ band corresponding to the transition between the lowest tunneling states of the ground and excited potential surfaces. A weaker hot band, labeled 0_-^-, also appears in the spectrum. The temperature dependence of this band reveals the splitting in the ground state to be only $10 \, \text{cm}^{-1}$. The transition 0_+^- is weakly allowed due to the interaction of this impurity molecule with the neon host matrix. Its position reveals the tunneling splitting on the excited surface to be $170 \, \text{cm}^{-1}$, much larger than on the ground surface because the barrier in this case is lower than for the ground state. A symmetric one-dimensional double-well potential that is consistent with the observed splittings is also shown in Figure 6.8. A fluorescence study by Bondybey et al. [1984] confirmed these values for the deuterated compound, and showed that

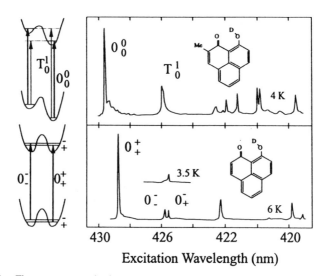

Excitation Wavelength (nm)

Figure 6.8. Fluorescence excitation spectra of matrix isolated 9-deuteroxyphenalenone (lower) and methyl-9-deuteroxyphenalenone (upper) as examples of nearly symmetric and asymmetric double well potentials for hydrogen transfer, shown on the left. The suppression of hot band 0^-_- is shown in a separate spectrum at 3.5 K. Due to asymmetry of the potential, the wave functions are linear combinations of the left and right well function with different amplitudes: $\psi_s = a\phi_1 + b\phi_2$, $\psi_a = -b\phi_1 + a\phi_2$, $b/a = 0.22$ and 0.80 in the ground and excited states of the methyl derivative. (From Barbara et al. [1989].)

the ground and excited state splittings for normal (hydrogenated) 9-hydroxyphenalenone are 69 and 311 cm^{-1}, respectively.

A quantum chemical calculation of the PES carried out by Kunze and de la Vega [1984] shows that the distance between oxygen atoms O_1 and O_2 in the transition state is 2.31 Å, which is shorter than in the equilibrium structure (2.52 Å). This leads to a decrease in the tunneling length to 0.42 Å. The dominant promoting role is played by bending vibrations COH and OCC. The splittings calculated in this work within the vibrationally adiabatic approximation (106 and 16 cm^{-1} for H and D transfer, respectively) are close to experimental values. The adiabatic barrier height on the excited surface is 5.2 kcal/mol, and the ground-state barrier height was found to be about 14 kcal/mol. According to the model of vibration-assisted tunneling, the static one-dimensional barrier is higher and narrower than that along the MEP. Moreover, the H/D isotope effect predicted by the one-dimensional model is twice as large as the observed value. As it was pointed out in Section 2.5, this discrepancy has a natural explanation in the frame of multidimensional models.

Substitution of a methyl group at the 2-position of 9-hydroxyphenalenone (α to the carbonyl group) causes an asymmetry of potential equal to 199 and 170 cm^{-1} for the H and D forms, respectively. Since the asymmetry is comparable to the tunneling splitting, fast hydrogen exchange is preserved [Rossetti et al., 1981]. A stronger asymmetry occurs in 6-hydroxybenzanthrone [Gillispie et al., 1986]:

(6.9)

The increased asymmetry allows the 0_+^- transition to be observed with intensity nearly equal to that of 0_+^+, so that the tunneling splitting on the excited surface was determined directly from the line positions in the fluorescence excitation spectrum. However, the same asymmetry reduces the population of the 0^- state to the point where the ground-state tunneling splitting could not be observed.

To study vibration-assisted tunneling, Sato and Iwata [1988] solved straightforwardly the Schrödinger equation in two dimensions for a collinear three-atom fragment ABA, the potential of which is modeled by two oppositely directed Morse curves for the AB interaction and a harmonic potential for AA interaction. They calculated the increase in tunneling probability as a function of quantum number (n) of the low-frequency AA vibration. For a barrier height of 1300 cm^{-1} (3.72 kcal/mol), assisting vibration frequency 450 cm^{-1}, displacement length 0.8 Å, and mass ratio $m_A/m_B = 100$, the tunneling splitting was found to be 111, 133, 151, 100, 193, and 203 cm^{-1} for $n = 0$–5, respectively, when $m_B = 1$ amu. The progression of energy levels for $m_B = 1$–3 amu is shown in Figure 6.9. The three lowest energy doublets for $m_A = 1$ and 2 amu correspond to vibrational states in the wells, while the energy of subsequent states exceeds the barrier height. The splitting decreases with increasing tunneling mass m_B, a fact that confirms the tunneling character of the states. The wave functions for this system (shown in Figure 6.10) clearly demonstrate the delocalization of the light particle. This result is in agreement with the semiclassical description set forth in Sections 2.5, 4.2, and 4.3. However, the simulation is beyond the framework of the semiclassical approximation, since Δ is comparable with ω_{AA}. In real systems, for higher barriers and more degrees of freedom, such simulations probably are not feasible using current computer technology. On the other hand, as tests show (see Section 4.1), semiclassical methods are

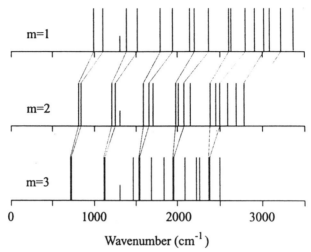

Figure 6.9. Energy level diagram for a model collinear system ABA. $m_A = 100\,\text{amu}$, $m_B = 1,2,3$ amu as indicated. The barrier height is $1300\,\text{cm}^{-1}$, and the frequencies of the AB and AA vibrations are 1800 and $450\,\text{cm}^{-1}$, respectively, $R_{AA} = 2.93\,\text{Å}$, $R_{AB} = 1.0\,\text{Å}$. (From Sato and Iwata [1988].)

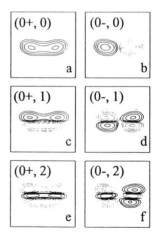

Figure 6.10. Wave functions for the first six states of model system, indicated in Figure 6.9. Dotted lines correspond to the negative part, $m_B = 1$. (From Sato and Iwata [1988].)

far less expensive and can be quite accurate even when the barrier is not very high.

A convenient system for simulating features of vibration-assisted tunneling is the hydrogenoxalate anion:

$$\begin{array}{c} H \cdots \\ O \quad\quad O^- \\ | \quad\quad\quad | \\ C - C \\ O \quad\quad\quad O \end{array} \tag{6.10}$$

In this case the adjustable parameters of the PES (4.41) are $V_0 = 18.52 \, \text{kcal/mol}$, $V^{\#} = 5.62 \, \text{kcal/mol}$, $C = 1.07$, $\Omega = 0.91$, and $\omega_0 = 1.50 \times 10^{14} \, \text{s}^{-1}$ [Bosch et al., 1990]. As in the case of malonaldehyde, the PES parameters place this system between the sudden and adiabatic regimes. The PES contour map and the instanton trajectory for this case are shown in Figure 6.11. Benderskii et al. [1993] have utilized the instanton analysis to obtain the prefactor $\tilde{B}_t = 54$ and the tunneling splitting $1.4 \, \text{cm}^{-1}$, which is in excellent agreement with the value $1.30 \, \text{cm}^{-1}$ obtained by Bosch et al. [1990] from a quantum mechanical calculation.

Two proton exchange in pairs of $OH \cdots O$ fragments of various carbonic acids dimers, e.g.,

$$-C\begin{array}{c} O \text{------} HO \\ \\ OH \text{------} O \end{array}C- \tag{6.11}$$

is well studied with NMR relaxation measurements (T_1-NMR) [Idziak and Pislewski, 1987; Meir et al., 1982; Nagaoka et al., 1983], incoherent inelastic neutron scattering (IINS) [Horsewill and Aibout, 1989a], and impurity fluorescence at high spectral resolution [Rambaud et al., 1989, 1990]. In these systems, as well as in many other cases of translational

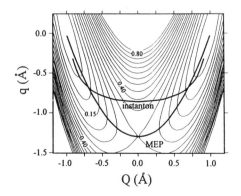

Figure 6.11. Contour plot, MEP and instanton trajectory for isomerization of hydrogenoxalate ion (6.9). (From Benderskii et al. [1993].)

tunneling, the energy bias caused by the asymmetry of crystalline fields precludes proton delocalization. In carbonic acid crystals the asymmetry A is usually about $60\ cm^{-1}$, which is about two orders of magnitude larger than Δ. For this reason the tunneling splitting is difficult to observe. A unique example of tunneling splitting observation is given by Oppenlander et al. [1989]. Upon replacing the host benzoic acid dimer by the thioindigo molecule of nearly the same size, the resulting bias accidentally turns out to be small, of the same order as Δ. The 4×4 Hamiltonian of a complex consisting of two dimers and the guest molecule is

$$
H =
\begin{array}{c c c c c}
 & \alpha\alpha & \alpha\beta & \beta\alpha & \beta\beta \\
\begin{vmatrix} \\ \\ \\ \\ \end{vmatrix}
& \begin{matrix} -A \\ \Delta/2 \\ \Delta/2 \\ 0 \end{matrix}
& \begin{matrix} \Delta/2 \\ B \\ 0 \\ \Delta/2 \end{matrix}
& \begin{matrix} \Delta/2 \\ 0 \\ B \\ \Delta/2 \end{matrix}
& \begin{matrix} 0 \\ \Delta/2 \\ \Delta/2 \\ \Delta \end{matrix}
\begin{vmatrix} \\ \\ \\ \\ \end{vmatrix}
\end{array}
\qquad (6.12)
$$

where α and β denote two possible states of two-proton transfer in a single dimer. The asymmetry A is twice the bias of a bare dimer and B is the shift in energy due to the guest molecule. Although B, $A \gg \Delta$, $|B - A| \sim \Delta$, and tunneling splitting becomes observable. The *hole-burning* spectra measured by Oppenlander et al. [1989] are reproduced in Figure 6.12. In these experiments, a narrow-band hole-burning laser is tuned to a particular frequency within an inhomogeneously broadened band. This saturates one of the lines of the tunneling multiplet associated with a particular crystalline environment. The probe laser is then tuned over the bands to detect the changes in spectra that are induced by proton exchange in the subset of molecules in the selected environment. In this way, the effects of inhomogeneous broadening are overcome, and tunneling splittings are revealed spectroscopically. The eigenfunctions of the three lowest levels of the Hamiltonian (6.12) are

$$0.82|\alpha\alpha\rangle - 0.41(|\alpha\beta\rangle + |\beta\alpha\rangle) + 0.05|\beta\beta\rangle$$

$$0.71(|\alpha\beta\rangle - |\beta\alpha\rangle) \qquad (6.13)$$

$$0.58|\alpha\alpha\rangle + 0.57(|\alpha\beta\rangle + |\beta\alpha\rangle) - 0.10|\beta\beta\rangle$$

The coefficients reveal the extent of delocalization of the wave function in this case. The tunneling splitting is found to be $8.4 \pm 0.1\ GHz$, and the net energy bias $|A - B| = 4.8 \pm 0.3\ GHz$. The other two methods, T_1-NMR and IINS, allow measuring the thermal hopping rates of proton transfer, and are described in Section 6.4.

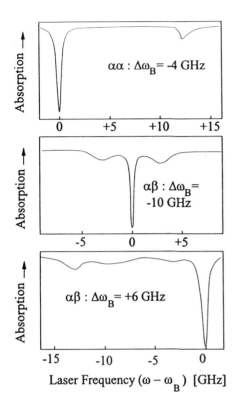

Figure 6.12. Hole-burning spectra of thioindigo in benzene acid crystal at 1.35 K. The scanning laser frequency ω is measured with respect to the burning laser frequency ω_B; $\Delta\omega_B$ is the detuning of the burning laser relative to the center of unresolved absorption band. The curves are fits of experimental points (not shown). (From Oppenlander et al. [1989].)

6.3. TAUTOMERIZATION IN EXCITED ELECTRONIC STATES

Coherent tunneling of hydrogen atoms in excited electronic states occurs in vibronic spectra only in the case of the symmetric and slightly asymmetric double well potentials. When the asymmetry is large enough, these transitions become irreversible due to vibrational relaxation, and rate constants of incoherent transfer can be measured using time-resolved spectroscopy. A typical scheme of vibronic transitions in the case of asymmetric terms of the ground and excited states is shown in Figure 6.13. The X form is stable in the ground state, but the energy of this form in the excited state is greater than that in Y form. Proton transfer relating to the radiationless transition $X^* \to Y^*$ competes with X^* deactivation ($X^* \to X$). When the rates of these processes are comparable, the fluorescence spectrum consists of the ordinary bands of $X^* \to X$ transitions as well as additional bands of $Y^* \to Y$ transitions. Due to energetic considerations, the latter exhibit a larger Stokes shift. When the barrier to

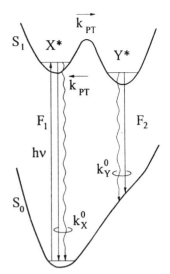

Figure 6.13. Kinetic scheme of the population and depopulation of the primary excited (X^*) and the tautomer (Y^*) forms. Wavy arrows show the radiationless transitions.

$Y \rightarrow X$ relaxation on the lower surface is small or absent, the entire photochemical cycle is reversible.

Dual fluorescence spectra of many species have been studied in detail (see a review by Barbara et al. [1989]) and allow the rate constants for hydrogen and deuterium transfer to be measured as a function of temperature. As a typical example, the spectrum of 2,5-bis(2-benzoxazonile)-4-methoxyphenone

$$(6.14)$$

was measured by Mordzinski and Kühnle [1986]. Also, the spectrum of 7-azoindole dimer

$$(6.15)$$

was obtained by Tokumura et al. [1986]. These spectra are illustrated in Figures 6.14 and 6.15. The latter demonstrates that the ratio of the dual fluorescence band intensity changes drastically upon deuteration of the

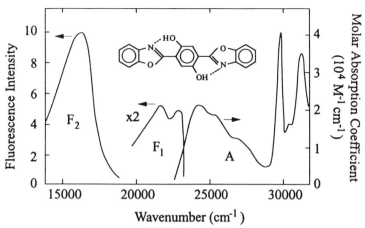

Figure 6.14. Absorption (*a*) and dual fluorescence spectra of compound (6.14). (From Mordzinski and Kuhle [1986].)

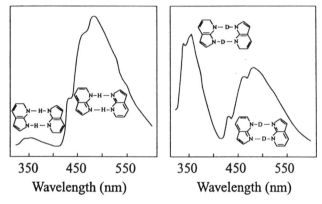

Figure 6.15. Steady-state dual fluorescence spectra of 7-azoindole-h_2 and 7-azoindole-d_2 dimer in 3-methylpentane at 132 K. (From Tokumura et al. [1986].)

molecule, due to the H/D kinetic isotope effect on the tautomerization rate constant $X^* \to Y^*$.

Fuke and Kaya [1989] studied the tautomerization of dimer (6.15) formed in a supersonic jet to investigate the vibrational selectivity of concerted two-proton transfer in excited electronic state. Tunneling in the NH $\cdots$ N fragments leads to broadening of certain vibrational bands in the fluorescence excitation spectrum (Figure 6.16). Tunneling is promoted by the symmetric intermolecular vibration with frequency $120\ cm^{-1}$. The widths of bands with $n = 0$, 1, and 2 are 5, 10, and

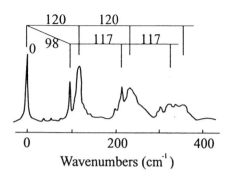

Figure 6.16. Fluorescence excitation spectrum of jet-cooled 7-azoindole dimer. (From Fuke and Kaya [1989].)

30 cm^{-1}, respectively. In contrast, the bending intermolecular vibration at 98 cm^{-1} reduces the transition probability. As expected, the broadening is absent in the deuterated isotopomer due to the overall slower tunneling rate.

This example demonstrates the spectroscopic manifestations of incoherent tunneling which give rise to spectral line broadening rather than splitting. The problem is reminiscent of the spin exchange problem known in magnetic resonance spectroscopy. In the spin exchange problem there are two Zeeman frequencies, ω_1 and ω_2, corresponding to two types of spins, and the frequency of double flip-flop exchange between these spins, ω_{ex}. The latter is analogous to the rate constant of the double proton exchange reaction above. To describe the incoherent exchange the pulse model was used by Lynden-Bell [1964], Johnson [1967], and Stunzas and Benderskii [1971]. An effective exchange term, whose real part is proportional to ω_{ex}, is introduced in the von Neuman equation:

$$\frac{\partial \rho}{\partial t} = -i\hbar^{-1}[H, \rho] + \left(\frac{\partial \rho}{\partial t}\right)_{ex} \qquad (6.16)$$

The exchange leads to a spectrum consisting of two Lorentzian lines with positions

$$\tilde{\omega}_{1,2} = \tfrac{1}{2}(\omega_1 + \omega_2) \pm \tfrac{1}{4}[(\omega_1 - \omega_2)^2 - \omega_{ex}^2]^{1/2} \qquad (6.17)$$

and widths

$$\Gamma = \Gamma_0 + \omega_{ex} \qquad (6.18)$$

where Γ_0 is the intrinsic width. Therefore, due to incoherent exchange the lines tend to merge and to widen, in contrast with coherent case where the lines move apart without broadening (see (3.107)).

In contrast to the scheme represented in Figure 6.13, a barrier exists for the reverse hydrogen transfer in the ground state of 7-azoindole dimer. The temperature dependence of the transfer rate constant measured by Tokumura et al. [1986] transforms from an Arrhenius form to the low-temperature plateau at about 130 K, where $k_c = 4 \times 10^3\,s^{-1}$ (see Figure 1.1, curve 7). The H/D isotope effect is 3.2 and 14 for 273 and 172 K, respectively. The transfer rate in the excited S_1 state is much greater than in the ground state. Arrhenius dependence occurs at $T >$ 100 K with an apparent activation energy 1.1 kcal/mol. The crossover temperature is 50 K and $k_c \sim 10^{10}\,s^{-1}$.

In the 1-azocarbazole dimer

$$(6.19)$$

the rate constant of the two-proton transfer is $\sim 10^9\,s^{-1}$, which is less than in the previous example because of the asymmetry of the potential. Fuke and Kaya [1989] found that the vibrational selectivity of transfer manifests itself as changes in the intensity of tautomer Y fluorescence, depending on which vibronic bands are excited in the excitation spectrum. The intermolecular stretching vibration at 109 cm^{-1} promotes the transition, but intermolecular bending vibrations at 55 and 67 cm^{-1} decrease the rate constant.

The intensity of vibronic bands in the tautomer fluorescence excitation spectrum is determined by the ratio of rate constants for hydrogen transfer and deactivation of corresponding vibronic level of X*. The latter is a sum of radiative (k_{s0}) and radiationless (k_{s1}) rate constants. The k_{s1} values are, generally speaking, different for various modes, but k_{s0} is generally constant over a wide range of the vibronic spectrum. For this reason, the vibrational selectivity can be observed only when $k_{s0} > k_{s1}$. This condition is valid for low-frequency vibrations of isolated molecules or dimers in supercooled jets, because the vibrational spectrum does not contain low-frequency lattice modes that are always present in solid-state studies. In solids, k_{s1} is determined by vibrational relaxation because the rate of energy transfer in the lattice, as a rule, exceeds k_{s0}. The rapid vibrational relaxation in excited electronic states is most commonly revealed as an absence of fluorescence from vibrationally excited levels of S_1 (see Section 2.4). This phenomenon is typical for organic solids and prevents this type of spectroscopic observation of vibrational selectivity for proton transfer in solids.

Analogous hydrogen transfer in the lowest triplet state is detected by nonstationary absorption of ^3X and ^3Y forms and time-resolved phosphorescence after nanosecond laser flash photolysis. Since the lifetime of the triplet states is typically 3–5 orders of magnitude greater than that in singlet states, hydrogen transfer in the ^3T states can be studied over a much wider range of time intervals than in the ^1S state. In the ground state of 2-(2-hydroxyphenyl)benzoxazole (R = H)

$$(6.20)$$

the enol form $^1E_0$ is more stable than the keto form $^1K_0$ by about 14 kcal/mol. Upon excitation by the laser, rapid hydrogen transfer $^1K_1 \leftarrow {}^1E_1$ takes place in the excited singlet state, so the intersystem crossing $^3E_1 \leftarrow {}^1E_1$ is practically insignificant. The keto form $^1K_1$ converts to $^3K_1$ on a nanosecond time scale [Grellmann et al., 1989; Mordzinski and Grellmann, 1986]. Subsequent slower hydrogen transfer turns $^3K_1$ into $^3E_1$ and a dual phosphorescence spectrum is formed due to radiative transitions from both triplet states $^1K_0 \leftarrow {}^3K_1$ and $^1E_0 \leftarrow {}^3E_1$. The difference between $^3K_1$ and $^3E_1$ state energies does not exceed $\pm 15\,\text{cm}^{-1}$; therefore, the observed rate constant for establishing the tautomeric equilibrium $^3K_1 \leftrightarrow {}^3E_1$ is a sum of forward and reverse reaction rate constants:

$$k(T) = k_f(T) + k_r(T) \qquad (6.21)$$

The temperature dependence of this rate constant was measured by Al-Soufi et al. [1991], and is shown in Figure 6.17. It exhibits a low-temperature limit of rate constant $k_c = 8 \times 10^5\,\text{s}^{-1}$ and a crossover temperature $T_c \simeq 80$ K. In accordance with the discussion in Section 2.5, the crossover temperature is approximately the same for hydrogen and deuterium transfer, showing that the low-temperature limit appears when the low-frequency vibrations, whose masses are independent of tunneling mass, become quantal at $T < T_c$. The kinetic H/D isotope effect increases with decreasing temperature in the Arrhenius region by about two orders of magnitude and approaches a constant value $k_H/k_D = 1.5 \times 10^3$ at $T < T_c$.

Al-Soufi et al. [1991] measured the temperature-dependent rate constants (6.21) in nonpolar media with very different temperature-

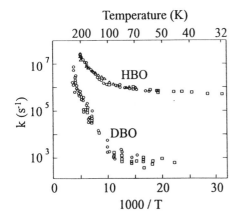

Temperature (K)

Figure 6.17. Arrhenius plot of the rate constant for intramolecular hydrogen (upper trace) and deuterium (lower trace) transfer in the lowest triplet state of compound (6.20) in three different solvents: 3-methylpentane (Δ), its $1:1$ mixture with isopentane ($\bigcirc$), and a $2:1:1$ mixture of both with methylcyclopentane ($\square$). (From Al-Soufi et al. [1991].)

dependent viscosities. One of them, 3-methylpentane, forms a rigid glass in the range 88–77 K, whereas two others are still liquids. From Figure 6.17 it is clear that the rate constant is unaffected by this drastic variation of viscosity. Therefore, the low-frequency modes participating in tunneling are assigned to intramolecular vibrations and are not associated with vibrations of the medium, since otherwise one would expect tunneling to be very sensitive to changes in the character of the solvent. The apparent activation energy and prefactor in the temperature range 140–220 K are 3.83 kcal/mol and $2.5 \times 10^9 \, s^{-1}$, respectively. As shown in Chapter 2, the low prefactor and apparent activation energy are typical of the intermediate regime between the Arrhenius dependence and the rate constant quantum limit.

In the methyl-substituted compound (6.20), $R = CH_3$, the rate constants k_H^T and k_D^T are approximately the same as for $R = H$, though the energy of $^3K_1$ level is at least 0.7 kcal/mol greater than that of $^3E_1$ level. Since the transition $^3E_1 \rightarrow \, ^3K_1$ is endoergic, $k_{XY} \gg k_{YX}$ [Eisenberger et al., 1991]. We emphasized earlier that the tunneling frequencies of coherent transitions are very sensitive to any asymmetry of the potential. However, in this case the rate constant for the *incoherent* transition is unaffected by the difference in energies of the initial and final states.

The keto–enol tautomerization in the excited triplet state of 2-methylacetophenone is associated with hydrogen transfer in the $CH \cdots O$ fragment:

$$ (6.22) $$

Grellmann et al. [1983] determined the tautomerization rate constant by measuring the transient absorption of the enol form generated by flash photolysis. The measured rate constants $k_H(T)$ and $k_D(T)$ exhibit an Arrhenius region ($E_a = 9.6\,\text{kcal/mol}$ and prefactor $\sim 10^{15}\,\text{s}^{-1}$) and a low-temperature plateau below $110\,\text{K}$, where $k_{CH} = 10^5\,\text{s}^{-1}$, $k_{CD} = 2 \times 10^2\,\text{s}^{-1}$. The promotion of tunneling is due to torsional vibrations of OH and CH_2 groups as well as bending vibration of the oxy group. In the *trans*-enol form the transition does not occur because of the large tunneling distance for the hydrogen atom ($2.8\,\text{Å}$). In the *cis* form this length shortens to $\sim 1\,\text{Å}$, but the transition to this form is unfavorable energetically ($\Delta H \approx 4.8\,\text{kcal/mol}$). In the transition state the OH bond is located out of the plane of the ring. The analysis of all these promoting motions has not been done. Siebrand et al. [1984] shows only that the experimental curves $k_H(T)$ and $k_D(T)$ can be described within the framework of a two-dimensional model in the sudden approximation, where the promoting collinear vibration has a frequency of $120\,\text{cm}^{-1}$ and reduced mass of 15 amu. The tunneling distance is estimated to be $1.8\,\text{Å}$, so the transition state differs from the *cis* form. Note that the rate constant for reaction (6.22) is measured over the range greater than 5 orders of magnitude. Similar to reaction (6.20), the crossover temperatures for k_H and k_D almost coincide, which completely contradicts the predictions of one-dimensional tunneling models.

Grellman et al. [1982] and Bartelt et al. [1985] found that the photoconversion of enamine (I) in its T_1 state into hexahydrocarbazole (III) proceeds through the intermediate formation of an intensely colored zwitterion (II), the decay rate of which is measured through the variation in the optical density of the sample:

$$(6.23)$$

I   II   III

The rate constant for hydrogen atom transfer (conversion II into III) spans six orders of magnitude in the range 290–$80\,\text{K}$. The quantum limit of the rate constant and crossover temperature are $5 \times 10^{-3}\,\text{s}^{-1}$ and $100\,\text{K}$, respectively. The ratio k_H/k_D increases from 10 to 5×10^3 as the temperature falls from 290 to $100\,\text{K}$. It is the H atom in position a that is transferred, since the substitution of deuterium atom at position b ($R = H$) does not change the rate constant.

Kensy et al. [1993] showed that the zwitterion formation and subsequent cyclization due to hydrogen transfer take place in the lowest triplet excited state of diphenylamine and its methyl substituents. The crossover temperature is about 100 K, and the values of $k_c(\mathrm{H})$ are 10^{-2}–$10^{-4}\,\mathrm{s}^{-1}$ for various substituents. The H/D kinetic isotope effect at $T < T_c$ is about 10^2.

The photoconversion of N-ethyl diphenylamine results in the two-proton intermolecular transfer and leads to the formation of N-ethyl carbazole and N-ethyltetrahydrocarbazole. In this reaction $T_c \simeq 130\,\mathrm{K}$, $k_c(\mathrm{H}) = 10^{-3}\,\mathrm{s}^{-1}$, $k_c(\mathrm{D}) \simeq 10^{-5}\,\mathrm{s}^{-1}$. Like the reactions described above, k_H and k_D do not change in media with different viscosities, demonstrating that the medium does not affect intermolecular hydrogen transfers having very small rate constants. We emphasize that in these reactions the values of k_c span some 15 orders of magnitude, whereas the crossover temperature is about constant. The change of k_c is due to variations of the tunneling length; the relatively constant values of T_c, on the other hand, are due to the participation of low-frequency intramolecular vibrations, which have similar values in these materials.

We now consider hydrogen transfer reactions between the excited impurity molecules and the neighboring host molecules in crystals. Prass et al. [1988, 1989] and Steidl et al. [1988] studied the abstraction of an hydrogen atom from fluorene by an impurity acridine molecule in its lowest triplet state. The fluorene molecule is oriented in a favorable position for the transfer (Figure 6.18). The radical pair thus formed is deactivated by the reverse transition. H atom abstraction by acridine molecules competes with the radiative deactivation (phosphorescence) of the ^3T state, and the temperature dependence of transfer rate constant is inferred from the kinetic measurements in the range 33–143 K. Below 72 K, $k(T)$ is described by Eq. (2.30) with $n = 1$, while at $T > 70\,\mathrm{K}$ the Arrhenius law holds with the apparent activation energy of 0.33 kcal/mol (120 cm^{-1}). The value of E_a corresponds to the thermal excitation of the symmetric vibration that is observed in the Raman spectrum of the host crystal. The shift in its frequency after deuteration shows that this is a libration; i.e., the tunneling is enhanced by hindered molecular rotation in crystal.

The H(D) atom abstraction rate constants in durene crystals by the impurity molecules quinoline, isoquinoline, quinoxaline, and quinozaline in their excited triplet state were measured by Hoshi et al. [1990] using the phosphorescence method described above. The transfer occurs in the fragment CH $\cdots$ N formed by a methyl group of durene and a nitrogen atom of the impurity molecule. In the interval 300–100 K the activation energy drops from 3.5 kcal/mol to 1.6 kcal/mol. Deuteration reduces the

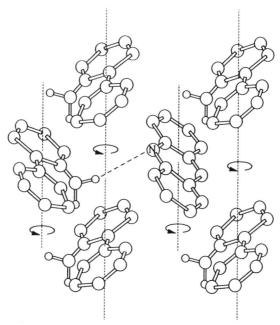

Figure 6.18. Position of a acrydine molecule (substitutional impurity) in fluorene crystal. Dashed lines show the direction of hydrogen transfer. The arrows indicate the libration that assists the tunneling transition. (From Prass et al. [1989].)

rate constant by approximately one order of magnitude at 200 K, whereas the activation energy increases to 4.5 kcal/mol (within the interval 220–300 K). Comparisons of the four systems reveal that an increase in the transfer length (as measured from the crystallographic data) correlates with a decrease of the rate constant. The temperature dependences of k_H and k_D are in qualitative agreement with the predictions of the vibration-assisted tunneling model and are interpreted as an effect of impurity molecule libration with a frequency of 230 cm^{-1}.

Hydrogen atom transfer from anthracene, excited into its lowest excited singlet state, to anthraquinone impurity molecules creates a radical pair that strongly quenches the fluorescence from anthracene crystals. The reverse transfer rate constant, found from measurements of fluorescence intensity and its characteristic lifetime at different moments after the creation of the radical pair, varies from 10^6 to 10^5 s^{-1} in the range 110–65 K, $k_c = 4 \times 10^4$ s^{-1}, $T_c = 60$ K. The k_c values drops to $\sim 10^2$ s^{-1} in the deuteroanthracene crystal [Lavrushko and Benderskii, 1978].

The proton transfer rate constants in matrix-isolated complexes of α- and β-napthol with ammonia were measured by Brucker and Kelley [1989] with the picosecond fluorescence method. The values of k_α and k_β are $4.5 \times 10^{10} \text{ s}^{-1}$ and $4.1 \times 10^{10} \text{ s}^{-1}$, respectively, and are temperature-independent over the range $T = 10\text{--}24$ K. The proton displacement in the fragment $OH \cdots N$ is ~ 0.75 Å. The height of the one-dimensional parabolic barrier corresponding to these values of rate constant and tunneling length is 3.6 kcal/mol. This is a case in which proton tunneling is accompanied by charge transfer.

6.4. TWO-PROTON TRANSFER

Two-proton transfer in crystals of carboxylic acids has been studied thoroughly by the T_1-NMR and IINS methods. The proton spin-lattice relaxation time, measured by T_1-NMR, is associated with the potential asymmetry A, induced by the crystalline field. The rate constant of thermally activated hopping between the acid monomers can be found from T_1 using the theory of spin exchange [Look and Lowe, 1966]:

$$T_1^{-1} = C \frac{\exp(-\beta A)}{[1 + \exp(-\beta A)]^2} \left(\frac{\tau_c}{1 + \tau_c^2 \omega_Z^2} + \frac{4\tau_c}{1 + 4\tau_c^2 \omega_Z^2} \right) \qquad (6.24)$$

where τ_c^{-1} is the sum of rate constants for hopping between two wells 1 and 2,

$$\tau_c^{-1} = k_{12} + k_{21} \qquad (6.25)$$

ω_Z is Zeeman frequency, and C is the lattice sum for dipole–dipole interaction. The transfer parameters A and τ_c were determined from the temperature dependence of T_1 by Horsewill and Aibout [1989a] (see also Nagaoka et al. [1983]). These are listed in Table 6.1. In the Arrhenius region, the rate constants for direct and reverse transfer are

$$k_{12} = k_0 \exp(-\beta V_0)$$
$$k_{21} = k_0 \exp[-\beta(V_0 - A)] \qquad (6.26)$$

For the various acids, the rate constants for hopping range from 10^8 to $2 \times 10^{10} \text{ s}^{-1}$ in the temperature range 40–100 K with apparent activation energy 1–2 kcal/mol. The distance R_{00} in the dimers and the change of CO bond lengths that accompany proton transfer are listed in Table 6.2. For the benzoic acid crystal, where both tunneling splitting and thermal hopping were observed, the transition between these two regimes occurs

TABLE 6.2
Two-Proton Transfer in Carboxylic Acid Dimer Crystals

Acid	V_0 (kcal/mol)	A (kcal/mol)	k_0 (s^{-1})	k_c (s^{-1})	R_{O-O} (Å)	r_{C-O} (Å)	$r_{C=O}$ (Å)
Diglycolic	1.33	0.6	2.5×10^{11}	4.8×10^8	2.67	1.300	1.226
Suberic	1.59	0.58	2.5×10^{11}	4.0×10^8	2.65	1.309	1.227
Maloric	1.49	0.64	2.0×10^{10}	5.0×10^8	2.68	1.31	1.22
Benzoic	0.99	0.15	1.7×10^{11}	2.8×10^8	2.63	1.268	1.258
Terephthalic	1.19	0.26	1.0×10^{11}	2.1×10^9	2.61	1.272	1.262

at ~ 40 K. Formally, this was the crossover temperature introduced in Section 2.1 on the basis of the spectral criterion.

As argued in Section 2.3, when asymmetry A far exceeds the tunneling matrix element Δ, phonons should easily destroy the coherence, and relaxation should persist even in the tunneling regime. This incoherent tunneling, which is characterized by a rate constant, requires a change in the quantum numbers of vibrations coupled with the reaction coordinate. In Section 2.3 we derived the expression for the intradoublet relaxation rate with the assumption that only the one-phonon processes are relevant. Working within the same weak coupling approximation, it takes little effort to produce the expression for the rate constant in the asymmetric case by simply replacing Δ in (2.44)–(2.45) by the energy bias A. This expression was obtained by Skinner and Trommsdorf [1988]. When $\beta A > 1$, the rate constants k_{12} and k_{21} are proportional to $\bar{n}(A)$ and $\bar{n}(A) + 1$, respectively, where $\bar{n}(A)$ is the equilibrium number of phonons with energy A given by (2.55). The relaxation rate is

$$\tau_c^{-1} = k_c \coth\left(\frac{\beta A}{2}\right) \qquad (6.27)$$

and k_c is determined by the coupling coefficient (see Section 2.3). The crossover temperature predicted by (6.27) is associated with the asymmetry of the potential:

$$T_c = \frac{A}{2k_B} \qquad (6.28)$$

This model is based on the weak coupling approximation and it takes into account neither reorganization nor vibration assistance in the sense of Section 2.5. Although the term "vibration-assisted tunneling" is also applied to the Skinner-Trommsdorf model, this assistance signifies only that vibrations supply the energy needed to provide a resonance.

The role of two-phonon processes in the relaxation of tunneling systems was analyzed by Silbey and Trommsdorf [1990]. Unlike the model of a two-level system coupled linearly to a harmonic bath (2.41), bilinear coupling to phonons of the form $C_{ij}q_iq_j\sigma_z$ was considered in this model. In the deformation potential approximation the coupling constant C_{ij} is proportional to $\omega_i\omega_j$. There are two leading two-phonon processes with different dependence of the relaxation rate on temperature and energy gap $\tilde{\Delta} = (A^2 + \Delta^2)^{1/2}$. Two-phonon emission prevails at low temperatures. Since its probability is proportional to $[\bar{n}(\omega_j) + 1][\bar{n}(\omega_j) + 1]$ and the energy of two emitted phonons is equal to $\tilde{\Delta}$, two-phonon emission is temperature independent and its rate is proportional to $\tilde{\Delta}^5$ for the Debye model $(\rho(\omega) \sim \omega^3)$ when $\beta\tilde{\Delta} \gg 1$ and $\tilde{\Delta} \le \hbar\omega_D$ (ω_D is the Debye frequency).

In the opposite case, $\beta\tilde{\Delta} < 1$, the Raman process is dominant, the probability of which is proportional to $\bar{n}(\omega_i)[\bar{n}(\omega_j) + 1)]$ under the condition $\hbar(\omega_i - \omega_j) = \tilde{\Delta}$. At $\beta\tilde{\Delta} \ll 1$ the rate of this process is proportional to $\tilde{\Delta}T^4$. The temperature dependences of both processes for various $\tilde{\Delta}$ are shown in Figure 6.19. The nonmonotonic dependence of τ_c^{-1} on $\tilde{\Delta}$ predicted by this model was not observed experimentally because this effect is pertinent to inaccessibly low temperatures. However, the T^4 dependence was verified experimentally for benzonic acid crystals by Oppenlander et al. [1989].

Quantum-chemical calculations of the formic acid dimer structure were carried out by Graf et al. [1981] and Meier et al. [1982]. The bond lengths and angles in the ground and transition states are listed in Table 6.3. Hydrogen transfer in two OH·····O fragments occurs when the distance

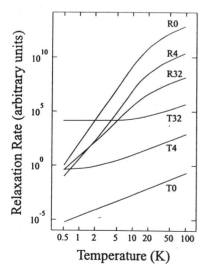

Figure 6.19. Relaxation rate for Raman transitions (R) and two-photon emission (T) in a two-level system as a function of temperature. The value of asymmetry is 0, 4, and 32 cm^{-1} as indicated. The tunneling matrix element is 0.25 cm^{-1} and the Debye frequency is 80 cm^{-1}. (From Silbey and Trommsdorf [1990].)

TABLE 6.3
The structure of Formic Acid Dimer in Equilibrium and Transition States

Bond Length Bond Angle	Equilibrium Structure		Transition State Calculated
	Experimental	Calculated	
C=O (Å)	1.217	1.25 (1.221)[a]	1.28 (1.268)
C–O (Å)	1.300	1.36 (1.324)	1.28 (1.268)
C···C (Å)	3.820	4.00 (4.112)	3.70 (3.650)
O–H (Å)	1.033	— (0.971)	— (1.192)
OH···O (Å)	2.696	2.73 (2.722)	2.30 (2.381)
∡ COH (deg)	126.2	121.0 (125.0)	114 (125.1)
∡ COH (deg)	108.5	111.6 (113.5)	120 (120.0)

a The results of a calculation by Shida et al. [1991a,b] are given in parentheses.

between oxygen atoms decreases by ~0.4 Å, shortening the distance for hydrogen atom tunneling from 0.7 to ~0.3 Å. At fixed heavy atom coordinates, the barrier is greater than 30 kcal/mol, but decreases to 12–17 kcal/mol as a result of the displacement. In the subsequent calculations of Hayashi et al. [1984] and Chang et al. [1987] the energy difference between transition- and ground-state configurations was found to be 12.3 and 16.5 kcal/mol. It is noteworthy that the barier height predicted by the empirical Lippincott-Schroeder correlation for $R_{00} =$ 2.69–2.72 Å (see Figure 6.1) is significantly greater than the experimental value. The origin of this discrepancy is that the tunneling length is shortened not only due to the oxygen atom displacements, but also due to the dimer stretching motion, and the latter is not taken into account by this correlation. Meier et al. [1982] assumed that the variation of bond lengths and angles along the MEP can be parametrized as a function of two coordinates: the concerted proton displacement Q and the skeleton deformation q. Since $U(Q, q)$ is an even function of q and $q \neq 0$ in the transition state, there are two MEPs situated symmetrically relative to straight line $q = 0$.

A calculation of the tunneling splitting in formic acid dimer was undertaken by Makri and Miller [1989b] for a model two-dimensional polynomial potential with antisymmetric coupling. The semiclassical analysis, which exploited a version of the sudden approximation, gave $\Delta = 0.9 \, cm^{-1}$, whereas the numerically exact result is $1.8 \, cm^{-1}$. An analogous calculation by Shida et al. [1991a,b] showed that in this case the sudden approximation is better than the various versions of the adiabatic one. Although the barrier height along the straight line path of sudden passage is 22.5 kcal/mol and along MEP it is only 9.5 kcal/mol, the tunneling splitting is 25 times greater in the former than the latter.

Some possible causes for underestimation of the tunneling splitting by the adiabatic approximation were discussed in Section 6.2. Since the comparison of various approximations was the main goal pursued by these model calculations, the asymmetry caused by the crystalline environment was not taken into account.

Generally speaking, the PES of two-proton transfer in two coupled $XH \cdots Y$ fragments is at least four dimensional. As shown by Shida et al. [1991a,b], it is possible to choose the coordinates

$$\rho_1 = r_1 + r_2 - r_3 - r_4$$
$$\rho_2 = R_1 + R_2$$
$$\rho_3 = r_1 - r_2 - r_3 + r_4 \qquad (6.29)$$
$$\rho_4 = R_1 - R_2$$

where r_1, r_2 and r_3, r_4 are the intra- and intermolecular OH distances, respectively, and R_1, R_2 are the O–O distances. The coordinates ρ_1 and ρ_3 describe the synchronous and asynchronous displacements of the two protons, whereas ρ_2 is associated with decreasing of both the XY distances due to a promoting stretching vibration. The coordinate ρ_4 relates to the wagging skeleton vibration with antisymmetric coupling with proton motion. It is possible to reduce the problem by including this low-frequency ρ_4 vibration in the set of the bath modes (see Section 1.2). The projection of the three-dimensional PES $U(\rho_1, \rho_2, \rho_3)$ onto the (ρ_1, ρ_2) plane at $\rho_3 = 0$ gives a two-dimensional contour map shown in Figure 6.20a. This two-dimensional PES $U(\rho_1, \rho_2)$ describes a synchronous two-proton transfer, e.g., the motion of two rigidly coupled protons promoted by a heavy particle displacement. The difference between this PES and the previously examined two-dimensional PES is that the potential $U(\rho_1^{\#}, \rho_2^{\#}, \rho_3)$ at dividing line $\rho_1 = \rho_1^{\#}$, $\rho_2 = \rho_2^{\#}$ can have one minimum at $\rho_3 = 0$ (points A and B) or two minima at $\rho_3 \neq 0$ (points C and D). In other words, the PES is capable of describing the bifurcation of the path, which breaks up to two equivalent paths in three dimensions corresponding to asynchronous motion of the two protons. This is shown schematically on the contour map in Figure 6.20b, which corresponds to the projection of $U(\rho_1, \rho_2, \rho_3)$ onto the proton coordinate plane at a fixed configuration of the heavy atoms. When the maximum of the potential corresponds to $\rho_3 = 0$ and it is high enough, the MEPs going across two saddle points with $\rho_3 \neq 0$ relate to stepwise proton transfer.

The contour maps shown in Figure 6.20 reflect two different approximations in the reduction of the problem to two dimensions. The first of these is realized for 7-azoinodole and carbonic acid dimers, where the

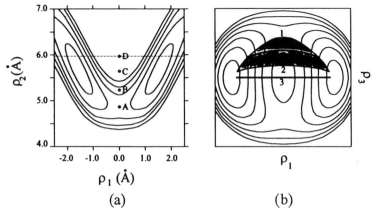

(a)   (b)

Figure 6.20. (a) Projection of a three-dimensional PES $V(\rho_1, \rho_2, \rho_3)$ for two-proton transfer in formic acid dimer onto the (ρ_1, ρ_2) and (ρ_1, ρ_3) planes. In contrast with points A and B, in points C and D the potential along the ρ_3 coordinate is a double well resulting in bifurcation of the reaction path [from Shida et al., 1991b]. (b) The contour lines correspond to equilibrium value of ρ_3 and potential (6.37) when $V(Q) = V_0(Q^4 - 2Q^2)$, $V_0 = 21$ kcal/mol, $C = 5.09V_0$, $A = 5.35V_0$, $Q_0 = 0.5$. When $Q_0 > Q_c$, two-dimensional tunneling trajectories exist in the shaded region between curves 1 and 2. Curve 3 corresponds to synchronous transfer.

trajectories with $\rho_3 \neq 0$ give a negligible contribution in the synchronous transfer rate and the appropriate coordinates are ρ_1 and ρ_2. The opposite case of asynchronous transfer, described in coordinates ρ_1 and ρ_2, is discussed below.

Tautomerization of metal-free porphyrin (H_2P) and pthalocyanine (H_2Pc) involves the transfer of two hydrogen atoms inside a planar 16-member heterocycle (Figure 6.21). In the stable *trans* form the H atoms lie along the diagonal of the square, formed by four equivalent nitrogen atoms. The energy of *cis* form, in which both the H atoms are positioned at one of the sides, is 3–7 kcal/mol higher than that of the

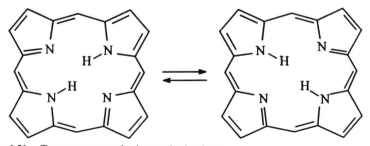

Figure 6.21. Two-proton transfer in porphyrine base.

trans form [Merz and Reynolds, 1988; Smedarchina et al., 1989]. The transition state energies for *trans–cis* and *trans–trans* isomerization are calculated in these papers using semiempirical quantum chemical methods. The values are $V_0 = 35$–42 kcal/mol and $V'_0 = 60$–66 kcal/mol, respectively. Although the barrier height appears to be overestimated, the calculations show that, in contrast with carbonic acid dimers, synchronous two-proton transfer in these compounds is energetically less favorable than stepwise transfer. Dewar [1984] pointed out that the transformations associated with reorganization of several chemical bonds are generally stepwise rather than synchronous.

The temperature dependence of the rate constants for the two-proton exchange in H_2P, HDP and D_2P were measured with NMR methods in the range of 200–320 K, where $k(T)$ grows from 10^1 to 10^5 s^{-1} [Crossley et al., 1987; Frydman et al., 1988; Hennig and Limbach, 1979; Schlabach et al., 1986]. In the range 95–110 K the rate constants of H and D transfer were measured by Butenhoff and Moore [1988] by the method of spectral hole burning to be 10^{-4}–10^{-2} s^{-1}. Comparison of the data from the two methods shows that $k_H(T)$ and $k_D(T)$ do not obey the Arrhenius law. The apparent activation energy E_a^H varies from 10.4 kcal/mol at $T = 200$–320 K to 6.4 kcal/mol at 95–110 K, while the activation energy for the D atom exchange is 3.3 kcal/mol greater than for H. The isotope effect is ~ 25 and ~ 250 at 250 and 111 K, respectively. Schlabach et al. [1986] concluded that at $T \geq 200$ K tautomerization corresponds to stepwise transfer because the rate constants for HDP and D_2P are similar, i.e., in both isotopomers the same stage of D transfer is rate-limiting.

The PES determined by Smedarchina et al. [1989] has two *cis*-form local minima separated by four saddle points from the global *trans*-form minima. Owing to the endoergicity of the first state, stepwise transfer (*trans–cis*, *cis–trans*) exhibits Arrhenius behavior even at $T < T_c$. The concerted transfer of two hydrogen atoms was thought to become prevalent at sufficiently low temperature. However, because of the high barrier for the concerted *trans–trans* transition, this region was experimentally inaccessible due to unmeasurably small values of the rate constant. According to the calculations of Sarai [1982], hydrogen transfer is accompanied by a strong deformation of the heterocycle. The total reorganization energy is 7.4 kcal/mol. Owing to the low-frequency skeleton vibrations, the tunneling distance at the energy E_{CT} of the local minimum of the *cis* form is shortened to 0.6 Å, while the geometric length of transfer between global minima is 1.7 Å. Butenhoff and Moore [1988] explained the persistence of Arrhenius behavior with a small activation energy together with a large isotope effect at low temperature by supposing that tunneling occurs via a thermally activated state with energy E_{CT}. Following the discussion in Section 2.1, for endoergic reactions there are two apparent activation energies: V_0 and E_{CT} at $T > T_c$ and $T < T_c$, respectively. Using the simple one-dimensional formula (2.6) they obtained $V_0 = 15.6\,\text{kcal/mol}$ and $E_{CT} = 5.4\,\text{kcal/mol}$. At $T < T_c$ the rate constant is described by the approximate relation

$$k = \frac{\omega_0}{2\pi} \exp\left[-\frac{2\pi}{\hbar\omega^{\#}} (V_0 - E_{CT}) - \beta E_{CT} \right] \tag{6.30}$$

where V_0 is the barrier height for the *trans–cis* transition. The stepwise transition corresponds to a formal kinetic scheme

$$T \underset{k_{TC}}{\overset{k_{CT}}{\rightleftharpoons}} C \xrightarrow{k'_{CT}} T \tag{6.31}$$

which takes into account the reverse transition from the intermediate *cis* form (C) into the initial *trans* form (T). The rate constant of the overall transfer is

$$k = \frac{k_{TC} k'_{CT}}{k_{CT} + k'_{CT}} \tag{6.32}$$

Since $k_{TC} \ll k_{CT}$, the *cis*-state population is in quasiequilibrium (steady-state concentration). In H_2P and D_2P $k_{CT} = k'_{CT}$, but in HDP the D atom transfer is a rate-limiting step and the two isotope effects are

$$\frac{k_{HH}}{k_{HD}} = \frac{k^H_{CT}}{k^D_{CT}} \gg 1$$

$$\frac{k_{HD}}{k_{DD}} = 1 \tag{6.33}$$

These relations explain the difference between isotope effects on successive deuteration.

Two-proton transfer in azophenyl

$$\tag{6.34}$$

is also thought to be stepwise [Rumpel et al., 1989]. The structure of the PES is similar to that described above for H_2P [Holloway et al., 1989].

Two-proton transfer in cyclic diarylamidine dimers

$$(6.35)$$

was studied by Meschede et al. [1988], Meschede and Limbach [1991] using ^1H and ^{19}F NMR. When R is p-fluorophenyl, the Arrhenius dependence of k_{HH}, k_{HD}, and k_{DD} is characterized by activation energies 4.52, 6.39, and 7.99 kcal/mol, respectively, in the range 164–261 K. In contrast with H_2P, where $k_{HH}/k_{HD} \gg 1$ and $k_{HD}/k_{DD} = 1$, in these species $k_{HH}/k_{HD} = 23 \pm 4$ and $k_{HD}/k_{DD} = 10 \pm 2$. The fact that both isotope effects are large indicates that both hydrogen atoms participate in the rate-limiting step, i.e., the exchange is synchronous. The ratio $k_{HD}/k_{DD} \gg 1$ also occurs in acetic acid dimers [Limbach et al., 1982].

Two-proton transfer in naphthazarin crystals

$$(6.36)$$

was observed with ^{13}C-NMR [Shian et al., 1980; Bratan and Strohbusch, 1980]. There are two equivalent paths for the step-wise transfer and, therefore, the transition state and MEP are two-fold, if the stepwise transfer is energetically preferable. On the other hand, the pathway for concerted transfer is unique, and lies between the saddle points. Based on this reasoning, de la Vega et al. [1982] found that the barrier for stepwise transfer (~25 kcal/mol) is 3.1 kcal/mol lower than that for the concerted transfer. These authors proposed a model two-dimensional PES:

$$V(Q, q) = V(Q) + \tfrac{1}{2}Cq^2(Q^2 - Q_0^2) + \tfrac{1}{4}Aq^4 \qquad (6.37)$$

where Q and q are the two-proton coordinates similar to ρ_1 and ρ_3 introduced in (6.29). One-dimensional motion along Q corresponds to concerted transfer $(q = 0)$, whereas the two MEPs associated with stepwise transfer are

$$q = \pm\left(\frac{C}{A}\right)^{1/2}(Q_0^2 - Q^2)^{1/2} \qquad (6.38)$$

The energy difference between the saddle point and the maximum of the potential at $Q = q = 0$ is

$$\Delta V = \frac{C^2}{4A}Q_0^4 \qquad (6.39)$$

The potential (6.37) corresponds with the previously discussed projection of the three-dimensional PES $V(\rho_1, \rho_2, \rho_3)$ onto the proton coordinate plane (ρ_1, ρ_3), shown in Figure 6.20b. As pointed out by Miller [1983], the bifurcation of reaction path and resulting existence of more than one transition state is a rather common event. This implies that at least one transverse vibration, q in the case at hand, turns into a double-well potential. The instanton analysis of the PES (6.37) was carried out by Benderskii et al. [1991b]. The existence of the one-dimensional optimum trajectory with $q \equiv 0$, corresponding to the concerted transfer, is evident. On the other hand, it is clear that in the classical regime, $T > T_{c1}$ (T_{c1} is the crossover temperature for stepwise transfer), the transition should be stepwise and occur through one of the saddle points. Therefore, there may exist another characteristic temperature, T_{c2}, above which there exists two other two-dimensional tunneling paths with smaller action than that of the one-dimensional instanton. It is these trajectories that collapse to the saddle points at $T = T_{c1}$. The existence of the second crossover temperature T_{c2} for two-proton transfer was noted by Dakhnovskii and Semenov [1989].

Earlier, a similar instanton analysis for a PES with two transition states was performed by Ivlev and Ovchinnikov [1987], in connection with tunneling in Josephson junctions. In the language of stability parameters introduced in Section 4.1, the appearance of two-dimensional tunneling paths is signaled by vanishing of the stability parameter. As follows from (4.24), the one-dimensional tunneling path formally becomes infinitely wide, i.e., it loses its stability,[1] and inclusion of the higher order terms in q results in splitting of the unstable instanton path.

Formally, a small stability parameter $\lambda \ll 1$ is equivalent to the existence of a classical transverse degree of freedom, and this mode contributes an Arrhenius dependence (6.30) with a small activation energy. The bifurcation diagram (Figure 6.22) shows how the (Q_0, β) plane of parameters breaks up into domains exhibiting different behaviors of the instanton. In the Arrhenius region at $T > T_{c1}$, classical transitions take place through both the saddle points. When $T < T_{c2}$, the optimum path is a one-dimensional instanton that crosses the maximum barrier point $Q = q = 0$. Domains i and iii in Figure 6.22 are divided by domain ii, where quantum two-dimensional motion occurs. The crossover temperatures T_{c1} and T_{c2} depend on Q_0 (i.e., on ΔV). When $Q_0^2 \ll Q_c^2 = [(1 + 2C)^{1/2} - 1]/C$, domain ii is narrow $(T_{c1} \approx T_{c2})$, so that in the classical regime the transfer is stepwise, while the quantum motion is a two-proton concerted transfer. This is the case in which the tunneling path differs from the classical one. When $Q_0 \lesssim Q_c$, the intermediate region $T_{c2} < T < T_{c1}$ widens and all three types of trajectories are present (see Figure 6.20). At $T \le T_{c1}$, small amplitude tunneling trajectories (curve 1) passing near the saddle point exist. As the temperature lowers, the trajectory gradually moves away from the saddle point and gets closer to trajectory 3. In the intermediate temperature region the tunneling trajectory can be ascribed neither to synchronous nor to stepwise transfer. A small apparent activation energy, which decreases with lowering the temperature, characterizes the $k(T)$ behavior in the intermediate region between T_{c1} and T_{c2} (Figure 6.23). When $Q_0 > Q_c$, the one-dimensional trajectory cannot be achieved and the proton transfer is stepwise at all temperatures. This consideration can be extended to the aforementioned three-dimensional problem, taking into account the dependence of Q_0 and C on the low-frequency skeleton coordinate ρ_2 and

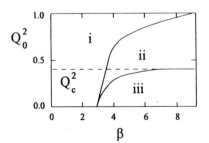

Figure 6.22. Bifurcation diagram for potential (6.37) with $C = 16V_0$, $A = 90V_0$, $V(Q) = V_0(Q^4 - 2Q^2)$ in (Q^2, β) plane. Domains i, ii, and iii correspond to Arrhenius dependence (stepwise thermally activated transfer), two-dimensional, and one-dimensional instantons (concerted transfer), respectively. The value $Q = Q_c$ divides domains ii and iii at $\beta \to \infty$. (From Benderskii et al. [1991b].)

[1] By stability we mean here the property to minimize the tunneling action, i.e., the barrier free energy (see Section 3.5).

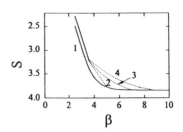

Figure 6.23. Action (in units V_0/ω_0) versus β for potential (6.37) at $Q_0 = 0$, 0.33, 0.346, and 0.35 for the curves 1–4, respectively; $C = 16V_0$, $A = 90V_0$. Dashed lines correspond to the two-dimensional instanton region. (From Benderskii et al. [1991b].)

using the sudden approximation for the rate constant calculation. More complex PES's having a number of equivalent transition states, such as those of porphyrin, can be studied in the framework of a similar approach. As we will see further in Chapters 7 and 8, PES's with several saddle points are typical of systems containing at least two coupled wide-amplitude motions.

Protonated acetylene, $C_2H_3^+$, has an unusual internal-rotation tunneling degree of freedom, in which three protons rotate around the C_2 core with the molecule remaining planar. Proton transfer has been studied by Crofton et al. [1989] using high-resolution infrared spectroscopy. The observed doublet structure of the vibration–rotation pattern with the asymmetric C–H stretching are definitely assigned to the nonclassical (bridged) structure I. The arrangement corresponding to tunneling exchange of the apex proton and end protons occurs via the classical transition state structure II:

$$H-C\equiv C-H \quad \xrightarrow{\quad} \quad \overset{H}{\underset{H}{>}}C=\overset{+}{C}-H \quad \xrightarrow{\quad} \quad H-C\equiv C-H \qquad (6.40)$$

$$\text{I} \qquad\qquad\qquad \text{II} \qquad\qquad\qquad \text{I}$$

The energy of the classical structure is 6.0 and 4.0 kcal/mol greater than that of **I** in the ground and vibrationally excited states; the corresponding tunneling splittings are $\leq 0.005 \text{ cm}^{-1}$ and 0.06 cm^{-1}, respectively. Formerly, Hougen [1987] proposed a model calculation in which he assumed that three protons are fixed in an equilateral triangle and rotate around the C_2 frame, thus covering six equivalent classical and nonclassical structures. In this model, only one variable is necessary to describe the strongly coupled transfer of all three protons. The generalized angular coordinate was introduced by Hougen using the permutation-inversion group analysis, which we discuss in Chapter 8. A similar sixfold one-dimensional potential for concerted proton transfer was proposed by Escribano et al. [1988].

The high-level quantum chemical calculations performed by Lee and

Schaefer [1986], Lindh et al. [1987], and Pople [1987] have demonstrated that proton transfer is accompanied by considerable stretching of the CC and CH bonds as well as alternating of vibrational frequencies. For example, the CH bond of the apex atom in the bridged structure is 1.296 Å, in comparison with 1.116 Å in the classical structure. The tunneling splitting of several rotation–vibration bands, calculated by Gomez and Bunker [1990] within a one-dimensional model, are in disagreement with experimental values, emphasizing the significant role of nonreactive modes.

6.5. FREE RADICAL CONVERSION IN SOLIDS

The abstraction of an H atom from crystalline and glass-like matrices of saturated organic compounds RH

$$CH_3 + RH \rightarrow CH_4 + R \tag{6.41}$$

is one of the most thoroughly studied examples of radical reactions in solids. This reaction in the γ-irradiated aliphatic alcohols was studied by Sprague and Williams [1971] and Campion and Williams [1972] by EPR spectroscopy to measure the decrease in intensity of methyl radical lines and the simultaneous increase of the lines of matrix radicals (CH_2OH and CH_3CHOH in methanol and ethanol, respectively). Toriyama and Iwasaki [1979] observed the conversion of CH_3 to CH_2OH with a characteristic time 3×10^3 s in a methanol glass at 4.2 K. These authors showed that radiolysis results in the creation of a primary radical pair $\dot{C}H_3$–$\dot{C}H_2OH$ by elimination of a water molecule. From the analysis of the EPR spectrum of this pair it follows that the distance between the carbon atoms of the two radicals is somewhat enlarged in this process (5.0 Å in comparison with the 4.75 Å intermolecular C–C distance in the crystal prior to irradiation). The displacement of CH_3 toward the neighboring methanol molecule (Figure 6.24) shortens the distance for hydrogen transfer. However, even after this displacement the C–C distance in the CH $\cdots$ C fragment where the reaction (6.41) takes place exceeds 3.4–3.5 Å, so that the hydrogen transfer distance is much larger than in the species considered previously. Reaction (6.41) is characterized by a low apparent activation energy (≤ 2 kcal/mol) in combination with a large isotope H/D effect ($\geq 10^2$–10^3) and the existence of a low-temperature limiting rate constant, all of which implicate tunneling as the mechanism of H atom transfer at $T < T_c$. Some values of k_c and T_c obtained by Le Roy et al. [1980] and Doba et al. [1984] are given in Table 6.4. A typical Arrhenius plot is depicted in Figure 1.1 (curve 11).

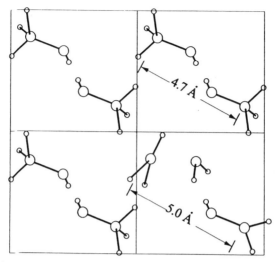

Figure 6.24. Primary radical pair [CH_2O–CH_3] in methanol crystal. The hydrogen atom is abstracted by the CH_3 radical from the adjacent CH_3 molecule. (From Toriyama and Iwasaki [1979].)

TABLE 6.4

X–Y Distances and Intermolecular Vibration Frequencies ω_{XY} Fitting the Temperature Dependence of the Rate Constant for Reactions (6.4.1), $X + HY \rightarrow XH + T$, in the Sudden Approximation

X	Y	R_{XY} (Å)	$\omega_{XY} \times 10^{-13}$ (s^{-1})
DMG[a]	DMG	3.17	2.9
CH_3	CH_3OH	3.66	3.6
CH_3	CH_3CH_2OH	3.64	3.5
CH_3	Acetonitrile I[b]	3.82	3.3
CH_3	Acetonitrile II	3.89	3.1
CH_3	Methylisocyanide	3.86	2.9

[a] Dimethylglyoxime.
[b] Two different crystallographic modifications.

Yakimchenko and Lebedev [1971], Toriyama et al. [1977] studied the isomerization of the primary radical pair in a γ-irradiated dimethylglyoxime crystal. The rate constant for hydrogen transfer in the intermolecular fragment OH $\cdots$ O obeys the Arrhenius law with activation energy 9.9 kcal/mol in the temperature range 150–200 K. The low-temperature plateau, where $k_c = 10^{-5}$ s^{-1}, is observed from 50 K down to 4.2 K. The variation of the rate constant in this wide temperature range does not exceed a factor of two (see curve 12 in Figure 1.1). According to crystallographic data, the O–O distance in the dimethylglyoxime crystal is

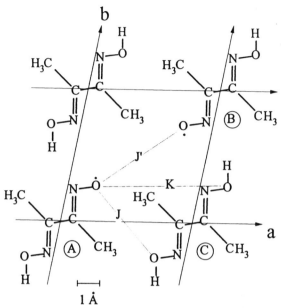

Figure 6.25. Spatial arrangement of reactants for long-range hydrogen transfer in a γ-irradiated dimethylglyoxime crystal. Radical pair *J* transforms into radical pair *J* as a result of hydrogen transfer from molecule C to radical B. (From Toriyama et al. [1977].)

3.2 Å (Figure 6.25). Therefore, this, too, is a case of long-range hydrogen transfer reaction. In the first papers devoted to the analysis of low-temperature solid-state reaction (6.41) the barrier height was assumed to be the same as in the gas-phase reaction. The transfer distance determining the one-dimensional barrier width was used as an adjustable parameter to fit the $k(T)$ dependencies calculated from (2.1) to experimental ones. The values of the effective barrier width d derived from these calculations by Le Roy et al. [1980] are presented in Table 6.4. Although such a comparison reproduces the experimental curves satisfactorily, the values of d are much shorter than those determined from spectroscopic and crystallographic data. The origin of this disagreement becomes clear if one compares the reaction (6.41) with the well-known gas-phase exchange reaction:

$$CH_3 + CH_4 \rightarrow CH_4 + CH_4 \tag{6.42}$$

The structure of the transition state calculated by Sana et al. [1984] corresponds to a linear $C \cdots H \cdots C$ fragment with two equivalent C–H bonds stretched to 1.25–1.28 Å. Using the two-dimensional PES of this fragment, Hipes and Kupperman [1986] showed that the C–C distance in the gas-phase reaction is 2.7–2.8 Å . Tunneling cuts corners by bypassing a saddle point, and the hydrogen transfer distance does not exceed 0.54–0.59 Å. It is the barrier along the MEP that determines the activation energy of the gas-phase reaction (Figure 6.26). However, the crossover temperature for this one-dimensional barrier calculated by relation (1.7) would exceed 500 K, while the observed values are ~50 K.

If one started from the d values calculated from the experimental data, the barrier height would go up to 30–40 kcal/mol, making any reaction impossible. The disparity between V_0 and d in the one-dimensional model is illustrated in Figure 6.26, where the PES cross section for the typical value $d = 1.0$ Å is compared with the one-dimensional barrier fitting the experiment. The barrier height predicted by the Lippincott-Schroeder correlation (see Section 6.1) for hydrogen transfer distance in dimethylglyoxime also exceeds 40 kcal/mol. For these reasons, the solid-state tunneling reactions of free radicals were the first to reveal the need to go beyond the one-dimensional consideration of H atom tunneling.

In view of the correlation between V_0 and d, it is important to remember that the van der Waals distances between the reactants in a lattice are much longer than interreactant approach distances in gas-phase reaction complexes. For instance, C–C distance corresponding to typical intermolecular contact is 3.7 Å, so that the tunneling distance in the C–H $\cdots$ C fragment should exceed 1.5 Å (1.3 Å if the zero-point energy is taken into account). Therefore, the challenge is not only in explaining the $k(T)$ dependence from interrelated V_0 and d, but also in reconciling these values with the packing of reactant molecules in the lattice. Obviously, these requirements cannot be met in the framework of a

Figure 6.26. PES cross section for reaction (6.41), $C^{(1)}H_3 + C^{(2)}H_3OH \rightarrow C^{(1)}H_4 + C^{(2)}H_2OH$. The C–C distance corresponds to the hydrogen transfer distance 1.3 Å. 1,1′, The diabatic Morse terms for CH bonds in CH_3OH and CH_4, respectively; 2, the adiabatic potential; 3, the one-dimensional parabolic barrier fitting the experimental $k(T)$ dependence; 4, the potential along the reaction path for the exchange gas-phase reaction $CH_3 + CH_4 \rightarrow CH_4 + CH_3$.

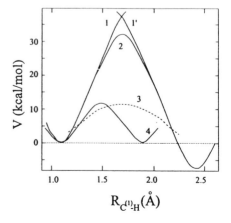

one-dimensional model. As explained in Section 2.5, these difficulties are naturally circumvented in the vibration-assisted tunneling model.

Ovchinnikova (1979), Trakhtenberg et al. [1982], and Doba et al. [1984] demonstrated that a satisfactory description of a solid-state reaction (6.41) can be achieved using the two-dimensional model with one effective low-frequency oscillator promoting tunneling (see Section 2.5). The frequency of this collinear oscillator and the interreactant distance R_0 were found from the temperature dependencies of the rate constant and isotope H/D effect within the sudden approximation. The results of Trakhtenberg et al. [1982] are given in Table 6.4. As one may expect for the vibrations that bring neighboring reactants together, the frequency values correspond to the Debye temperature in typical molecular solids. The R_0 values are close to the sum of van der Waals radii. Although these calculations are not based on knowledge of the real structure and dynamics of the crystal, the correspondence can be regarded as credible testimony for the vibration-assisted tunneling model. The dynamics of the transition is supposed to be similar to that shown in Figure 2.16. Owing to the promoting intermolecular vibration, the corner-cutting trajectory corresponds to a C–C distance that is just 0.2–0.3 Å larger than this distance in the transition state of the gas-phase reaction.

Some other solid-state low-temperature reactions having non-Arrhenius behavior and large H/D isotope effect have been interpreted to involve tunneling hydrogen transfer. Dubinskaya [1990] observed the quantum limit of the rate constant for hydrogen transfer from a matrix of malonic acid to the polyvinylacetate macroradical, $k_c = 2 \times 10^{-4}\,s^{-1}$, $T_c \simeq$ 110 K. The apparent activation energy in the range 200–240 K (~9.1 kcal/mol), as well as the value of k_c corresponds to approximately the same barrier parameters as in the reactions of methyl radical. Senthilnathan and Platz [1982] and Platz et al. [1982] observed low apparent activation energies and prefactors in the reaction of triplet diphenylcarbene with various organic solids, in which radical pairs are created:

$$(6.43)$$

At $T = 77$–150 K, E_a is 2–4 kcal/mol and the prefactor is 10^2–$10^4\,s^{-1}$.

The barrier height is about 15 kcal/mol by analogy to the similar gas-phase reaction $H_2C: +CH_4 \rightarrow 2CH_3$, which was calculated by Bausch-licher et al. [1976]. The C–C distance calculated in the same work is found to be equal to 2.65 Å. Like reaction (6.41) any attempt to reconcile the experimental dependence $k(T)$ with a model of one-dimensional tunneling in the barrier of indicated height leads to a tunneling distance d that is far shorter than could be reasonably rationalized based on this C–C distance.

There have been many studies of cryochemical solid-phase reactions of hydrogen atoms with organic molecules and polymers (see the review by Dubinskaya [1990]). Evidence of tunneling in these reactions comes from the decrease of the apparent activation energy to 1.5–2.5 kcal/mol in the range of 100–150 K from its high-temperature value ($E \equiv 6$ kcal/mol at 200–250 K). Both activation energies are significantly lower than the barrier height (10–12 kcal/mol). The low-temperature limit $k_c = (0.5$–$1.0) \times 10^{-2} s^{-1}$ has been observed at $T < 110$ K for the reaction of H atom with malonic acid, for which the activation energy above 250 K is 6.2 kcal/mol.

In gases composed of both methane and ethane, H atoms react with ethane at 10–20 K, forming C_2H_5 radicals, according to Toriyama et al. [1979, 1980]. These authors observed similar reactions $H + HR \rightarrow H_2 + R$ with saturated hydrocarbons, beginning with ethane, along with addition of H to ethylene at 30–50 K in inert gas matrices. The rates of these reactions are limited by hydrogen diffusion. The evaluation of the diffusion coefficient of H atoms in xenon matrices ($\sim 10^{-14}$ cm^2 s^{-1} at 50 K) leads to the conclusion that the transitions occur between neigh-boring interstitial positions of the fcc lattice (i.e., the diffusion jump length is ~ 4.4 Å). The sites are separated by low, wide barriers, the fluctuations of which are caused by low-frequency vibrations of the crystal. Within the framework of the vibration-assisted tunneling model, if the Debye temperature of xenon ($\theta_D = \hbar \omega_D / k_B$) is 55 K, it could be assumed that T_c for hydrogen diffusion in this crystal should be ~ 20 K.

Non-Arrhenius behavior of the isomerization rate constant of sterically hindered aryl radicals 2,4,6-triterbutylphenyl

(6.44)

and 2,4,6-tri-(1'-adamantyl)phenyl was observed by Brunton et al.

[1976]. In the range 245–123 K the H/D kinetic isotope effect increases from 40 to $\sim 10^4$. The difference between the apparent activation energies for H and D transfer (6.2 and 9.4 kcal/mol, respectively, at 200 K) significantly exceeds the difference of zero-point energies. Therefore, the isotope effect cannot be explained simply in terms of classical kinetics. The Arrhenius plot is shown in Figure 1.1 (curve 10), where $T_c \sim 100$ K and $k_c \sim 10^2 \, s^{-1}$. The hydrogen transfer distance in this case is 1.34 Å.

According to Bolshakov and Tolkachev [1976] and Zaskulnikov et al. [1981], the kinetics of radical conversion in glassy matrices consists of an initial part determined by the rate constant $k(T)$ described above and a subsequent conversion of the remaining molecules which obeys the empirical Kolrausch law:

$$R(t) = R_0 \exp\left[-\left(\frac{t}{\tau}\right)^\alpha \right] \qquad (6.45)$$

where $R(t)$ describes the time-dependent decay of the CH_3 radical concentration. The parameter α in (6.45) increases with temperature and tends to unity above the glass-transition temperature T_g (103 K for methanol), where the kinetic curves become exponential. The Kolrausch law is a characteristic of inhomogeneous species exhibiting a distribution of different reactant configurations (see, for example, Klaffer and Schlesinger [1986]). Formally, the set is characterized by a distribution of rate constants, and the mean reactant concentration is the average over the distribution of configurations with various decay rates. In the case of low-temperature reactions the existence of the set of reactant configurations inherent to glass-like matrices can be phenomenologically accounted for by introduction of a distribution of barrier heights. The introduction of an equilibrium reactant distance distribution (for instance, the C–C distance in reactions (6.41) and (6.43)) into the model of vibration-assisted tunneling enables us to quantitatively describe the empirical relation (6.44) and to explain the temperature dependence of α. Since the rate constants of the tunneling reactions depend exponentially on both of the barrier parameters, their distribution is more sensitive to variations in the environment than that for usual reactions with Arrhenius behavior [Goldanskii et al., 1989a,b].

The concept of tunneling has recently been invoked to explain the mechanism of photodissociation of matrix-isolated molecules. Previously, photodissociation was customarily accounted for by the fact that translationally hot photofragments escape from the cage and stabilize in separate matrix sites, thereby avoiding recombination. Using the time-dependent self-consistent field approximation for molecular dynamics simulations,

Gerber and Alimi [1990] showed that H atoms, created by photodissociation of HI in a xenon matrix at 2 K, lose their excess energy of ~35 kcal/mol as a result of collisions with surrounding heavy atoms much faster (within 2.5×10^{-13} s) than they leave the cage. This fast relaxation reduces their energy to the zero-point level (3.5 kcal/mol), which is located below the top of the barrier leading to the neighboring interstitial site. Under these conditions the classical probability of cage escape is zero, and the dissociation is solely due to the tuneling decay of the metastable state. The maximum probability of leaving the cage, in competition with the geminate recombination, is achieved for the times of ~5×10^{-12} s and equals ~8×10^{-4}. It seems plausible that tunneling effects play a major role in numerous primary processes in radiolysis and photolysis of low-temperature solids (cf., Section 9.2).

Since tunneling conversion of radicals is usually a strongly exoergic process, it can be well described using the theory for decay of metastable state discussed in Section 4.1. These reactions present examples in which the vibrationally adiabatic approximation can be strongly violated even though the tunneling path is to a good accuracy described by the MEP. This breakdown of the vibrationally adiabatic approximation is a noticeable effect when the barrier is low enough, as was demonstrated by Carrington et al. [1984] for the vinylidene–acetylene rearrangement. This rearrangement proceeds via a transition state being a planar three-member cycle:

$$
\begin{array}{ccccc}
\text{H} \diagdown & & \text{H} \diagdown & & \\
\quad \text{C} = \text{C} \colon & \longrightarrow & \quad \text{C} - \text{C} & \longrightarrow & \text{H} - \text{C} \equiv \text{C} - \text{H} \\
\text{H} \diagup & & \text{H} \diagup & &
\end{array} \qquad (6.46)
$$

The geometry of the vinylidene molecule is such that the MEP coincides with the direction of the CH_2 rocking vibration. The two hydrogen atoms are initially equivalent, so there are two equivalent paths of decay, and the potential along MEP corresponding to this picture can be modeled by a polynomial:

$$
V(s) = As^2 + Bs^4 + Cs^6 \qquad (6.47)
$$

The barrier height is assumed to be 2–4 kcal/mol, which agrees with a recent quantum chemical calculation [Gallo et al., 1990]. The vibrationally adiabatic approximation describes tunneling decay when the quantum numbers of all transverse vibrations are conserved along the tunneling path. However, when coupling between the reaction coordinate and a transverse mode is strong enough, an initially excited transverse vibration (say, with $n = 1$) may transfer its energy to the reaction coordinate, changing its quantum number to $n = 0$. This mechanism of vibrational predissociation leads to tunneling in an effectively lower barrier with greater probability. For vinylidene, the vibration that is strongly coupled to the reaction coordinate is the CH_2 scissors mode. Excitation of this mode results in an increase of the rearrangement rate.

6.6. QUANTUM DIFFUSION OF HYDROGEN

Diffusion coefficients of hydrogen atoms in metals are 10^{-10}–10^{-4} cm^2/s in the range of 100–500 K, which is by 10–15 orders of magnitude larger than those for heavier impurities (e.g., for oxygen and nitrogen atoms). They are characterized by a strong isotope (H/D) effect and low activation energies (≤ 6 kcal/mol) (see, for example, the review by Volnl and Alefeld [1978]). The anomalous mobility of hydrogen, deuterium, and tritium atoms and muonium (with the mass of $0.11m_H$) is due to vigorous jumps of these light impurities between the interstitial sites of the fcc and bcc lattices [Fukai and Sugimoto, 1985; Richter, 1986]. Insertion into tetrahedral and octahedral sites in these lattices causes the displacement of the surrounding heavy atoms. The potential energy surface for hopping into equivalent neighboring sites is therefore asymmetric because this motion requires reorganization of environment. This process of autolocalization of the impurity H(D) atom is similar to the formation of Landau-Pekar small-radius polarons in ion crystals [Pekar, 1954]. Linear coupling of the particle motion to the vibrations of the initial well ensures the asymmetry of the potential.

The impurity energies in deformed lattices of this type have been calculated by Puska and Nielmien [1984]. It has been shown that in the bcc lattices of V, Nb, Ta, Cr, and Fe the hydrogen atoms can be localized in both octahedral and tetrahedral sites, whereas localization occurs only in octahedral sites of fcc lattices in Al, Ni, Cu, and Pd. In the former case the hops occur along the lines connecting the O and T positions, the typical barrier heights being 1.2–1.3 kcal/mol. In fcc lattices the barrier heights are 5–10 kcal/mol. The relative change of the lattice parameters in the first coordination sphere is 0.04–0.11, while in the second sphere it does not exceed 0.01–0.02.

When the displacements of the nuclei are considered in terms of the phonon modes of the crystal, the reorganization energy can be expressed in terms of the relative shifts of the equilibrium positions of acoustic vibrations of acoustic vibrations. Using the Debye model and assuming that the friction is independent of frequency, the reorganization energy is proportional to the friction coefficient, $E_r = 4/3\pi\eta\omega_D^2 Q_0^2$ (see Section 2.3).

Since the frequencies of vibrations of H and D impurities ($\sim 10^3$ cm^{-1}) exceed the Debye frequency ω_D by almost an order of magnitude, early models were concerned primarily with jumps of a light impurity at a frozen lattice configuration [Holstein, 1959]. The transition takes place via tunneling when suitable heavy atom displacement equalizes the energy levels of the initial and final states. This model corresponds to the sudden approximation (nonadiabatic regime in Holstein's terminology). The transition probability is given by golden rule formula (1.14). Due to the temporary degeneracy (termed a "coincidence event") the light particle jumps between sites with finite probability. Thus, the transition rate is determined by a product of two terms: the thermally averaged probability of generating a "coincidence" and the rate of particle tunneling during the resonance. The first term is simply the Frank-Condon factor (i.e., the square of the overlap integral for the shifting mode) and depends on the thermally averaged values of phonon occupation numbers (2.56). The second term is proportional to the square of the tunneling matrix element. To calculate the hopping rate, one normally uses a number of additional simplifications. Within the Condon approximation, the tunneling matrix element $J_{p,p'}$ for the interstitial transition from site p to neighboring site p' is independent of the phonon coordinates, and the tunneling and phonon motions are separated from each other. The set of particle degrees of freedom is reduced to only one coordinate coupled linearly to the thermal bath, so that tunneling at the coincidence is treated as one dimensional. The additional resonances between the particle's levels with different quantum numbers are ignored.

There are two different temperature regimes of diffusive behavior; they are analogous to those described by Holstein [1959] for polaron motion. At the lowest temperatures, coherent motion takes place in which the lattice oscillations are not excited; transitions in which the phonon occupation numbers are not changed are dominant. The Frank-Condon factor is described by (2.51), and for the resonant case one has in the Debye model:

$$\Phi_0 = \frac{1}{2} \sum_{\omega < \omega_D} \frac{\Delta q_\omega^2}{\delta_\omega^2} = \frac{5E_r}{4\hbar\omega_D} \qquad (6.48)$$

where Δq_ω are phonon displacements. Equilibrium positions of phonons change in the transition, but the phonon system is otherwise unaffected. In these *diagonal* transitions, the initial and final states are related by translation operators. The crystal momentum of the impurity is conserved, and a band picture is appropriate. The bandwidth is of the order $J_{p,p'}$, and the characteristic time for the impurity to travel through the nearest-neighbor distance is $\sim\hbar/J_{p,p'}$. However, this coherent motion takes place over a very limited range of temperatures near absolute zero.

At higher temperatures, the transfer rate of the impurity molecule between sites is enhanced by phonon emission or absorption; this can be viewed as a "phonon-dressed" state transfer. When these off-diagonal processes are included in the analysis, the transition probability can be broken up into phonon processes of different order. One-phonon processes do not contribute, because momentum is not conserved in these events. Among the two-phonon processes, only Raman events contribute. These lead to a T^7 dependence of the transition probability (see, for example, Flynn and Stoneham [1970]). This regime takes place when $k_B T \ll \frac{1}{2}\hbar\omega_D$.

Finally, when $k_B T > \frac{1}{2}\hbar\omega_D$, the transition rate is described by a straight analogue of Marcus formula (2.66) in which the tunneling matrix element J takes the place of the electronic coupling V. This Arrhenius dependence is characterized by an activation energy $E_a = \frac{1}{4}E_r$, which corresponds to the thermal fluctuation that drives the system through the saddle point. The model is very similar to that of electron transfer in polar media (see Section 2.4). Here, the electron motion is assumed to occur at fixed nuclear coordinates (due to the large disparity between their characteristic frequencies and the characteristic time of electron motion) and the slow motions of the nuclei (which are quantum when $T < \hbar\omega/k_B$) serve to create configurations that are favorable for the transition.

Flynn and Stoneham [1970] were the first to suggest that diffusion of the light impurity atom is enhanced by symmetrically coupled vibrations. To incorporate this effect, which is called *fluctuational barrier preparation*, the authors have proposed to take into account the dependence of the tunneling matrix element on displacements of the heavy nuclei. This approach goes beyond the familiar Condon approximation. In this version of phonon-assisted tunneling, the phonon-dressed incoherent transitions are also induced by a suitable reduction in the barrier height via emission or absorption of phonons.

This model of barrier modulation was further developed by Emin et al. [1979] and Dhawan and Prakash [1984] and Teichler and Klamt [1985]. Trakhtenberg et al. [1982] expressed this dependence of the tunneling matrix element on the displacement of the heavy nuclei as an expansion of the action $S(E)$ in (2.3):

$$S(E, R) = S(E, R_0) + (R - R_0)\left(\frac{\partial S}{\partial R}\right)_{R=R_0} + \frac{1}{2}(R - R_0)^2\left(\frac{\partial^2 S}{\partial R^2}\right)_{R=R_0}$$

$$(6.49)$$

It is clear from the discussion in Section 2.5 that use of this expression implies the sudden approximation (see also Goldanskii [1989b]). To obtain the criteria of validity for this approach, note that the action of a particle tunneling through a barrier of height V_0 and width d during time τ_0 is roughly $S \approx md^2/2\tau_0 + \tau_0 V_0/2$. Minimizing this action with respect to τ_0, we obtain characteristic jump time $\tau_0 \cong (md^2/V_0)^{1/2}$. If $\hbar/\tau_0 \gg \hbar\omega_D$, the jump is instantaneous on the time scale of characteristic phonon vibration periods, which corresponds to condition (2.86). As noted in Section 2.5, the potential along the optimum path in this approximation differs from the fixed-nuclei barrier because $R < R_0$ for the corner-cutting trajectory (see Figure 2.16). Figure 6.27 illustrates the barrier profiles for the jump of a hydrogen atom between tetrahedral sites of a niobium crystal, as calculated by Chakraborty et al. [1988]. The barrier $V(Q)$ of height ~15 kcal/mol corresponds to the equilibrium nuclear configuration in its autolocalized state. The effective barrier for the sudden transition V_s is 9 kcal/mol. Finally, for the case of a vibrationally adiabatic transition in which the minimum energy configuration is attained at each point along the reaction path, the barrier V_{vad} is only 3.5 kcal/mol.

For muonium and hydrogen atom diffusion, the transition is believed to be sudden, but for deuterium and tritium this approximation is not valid. Proceeding to the adiabatic transition for heavier particles means an effective lowering of the barrier (from V_s to V_{vad}) because of the reorganization. For the same reason, $\ln k$ decreases more slowly than $m^{1/2}$ as m is increased, as pointed out in Section 2.5 (see Figure 2.18 and the corresponding discussion in the text). The tunneling matrix elements and Franck-Condon factors for diffusion of light impurities in some metals are given in Table 6.5, which is adapted from the review by Fukai and Sagimoto [1985]. The same table lists parameters for transitions of H and D between the equilibrium sites located near oxygen atom included in the lattice.

Until recently, hydrogen diffusion on metal surfaces was considered to

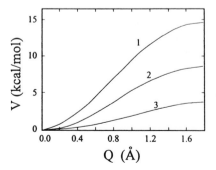

Figure 6.27. The potential curves for sudden (2) and adiabatic (3) hydrogen transfer between the tetrahedral interstitial positions in niobium crystal. Curve 1 is the potential in the unrelaxed lattice. (From Chakraborty et al. [1988].)

TABLE 6.5
Tunneling Matrix Elements J_0 and Franck-Condon Factors for
Diffusion of Light Impurities in Metals

Particle	Metal	Δ_0 (s^{-1})	FC
μ	Cu	9×10^{13}	6.1
μ	Al	1.5×10^{14}	3.2
H	Nb	9×10^{11}	3.3
D	Nb	5×10^{10}	2.9
H	Nb(O$_x$H$_y$)	1.4×10^{12}	3.4
D	Nb(O$_x$D$_y$)	9×10^{10}	3.1

be the result of jumps on a fixed lattice (see, for example, Tully [1981]). Difoggio and Gomer [1982] and Wang and Gomer [1985] observed tunneling diffusion of H and D atoms on the [110] surface of tungsten, and have drawn attention to the effect of lattice vibrations. A sharp appearance of the low-temperature limit ($D = D_c$) occurs at 130–140 K, as shown in Figure 6.28. The values of D_c depend only slightly on the mass of the tunneling particle: D_c for D and T atoms is only 10 and 15 times smaller than that for H, respectively. The value $D_c \cong 2 \times 10^{-13}$ cm^2 s^{-1} corresponds to a transition probability of $\sim 10^3$ s^{-1} ($W_c = 4D_c/d^2$, where $d \sim 2.7$ Å is the distance between equilibrium positions). The characteristic activation energy of $D(T)$ above T_c grows from 4.0 to 5.8 kcal/mol as the surface coverage is increased. Moreover, it does not exhibit the expected behavior for the kinetic isotope effect. It follows from the electron energy loss spectrum (EELS) of Blanchet et al. [1982] that a hydrogen atom on the [110] surface of a tungsten crystal partici-pates in the vibrations of frequencies 1310, 820, and, probably, 660 cm^{-1}. The latter two vibrations are the motions along the face and, therefore, can modulate the barrier thereby promoting tunneling. However, the

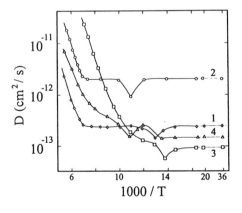

Figure 6.28. Temperature dependence of the surface diffusion coefficient for H (curves 1 and 2) and D (3 and 4) on the [110] face of the tungsten crystal at coverage 0.1 (1 and 3) and 0.5 (2 and 4). (From Wang and Gomer [1985].)

Debye frequency of the surface vibrations corresponds better to the observed value of T_c. The results for $D(T)$ leave no doubt that surface diffusion of hydrogen isotopes is quantal. However, the weak dependence of D_c on mass cannot be rationalized within the framework of any one-dimensional model for the tunneling.

Using a model PES for chemisorption proposed by McGreery and Wolken [1975], Jaguet and Miller [1985] studied transitions of an H atom between neighboring equilibrium positions on the [100] surface of tungsten. They noted that the transition probability increases as coupling to the lattice vibrations is increased. The most favorable situation is for low lattice frequencies. In this calculation, which was performed for fairly high temperatures $(T > T_c)$, the flux–flux formalism was used along with the vibrationally adiabatic approximation (see Section 3.6). Muttalib and Sethna [1985] showed that the weak dependence of D on the mass of the tunneling particle can be explained by using the vibrationally adiabatic approximation. As explained in Section 5.3, this approximation leads to the concept of effective tunneling mass, which is a sum of the bare mass of the particle and contributions from the vibrations that adiabatically follow the particle.

The analysis of the experimental data within the framework of this model has shown that the effective H atom mass M^* ranges from $10m_H$ to $6m_H$, depending on the change in transverse vibrational frequencies along the trajectory. In the case of sufficiently strong coupling $(E_s \gg \omega_D)$ M^* contains only a small contribution from the bare mass of the H atom, and according to Eq. (5.74), $S \sim M^{*1/2}$. Therefore, the tunneling probability depends only weakly on the mass of the particle being transferred.

The nonmonotonic dependence of the diffusion coefficient D on the degree of coverage θ has been explained by a change in the adsorption potential. When some of the neighboring nodes are occupied by adsorbate atoms and thereby shifted from the equilibrium positions, there is an additional reorganization energy involved in the transition. The increase in $D(\theta)$ for $\theta < 0.5$ is due to a decrease in the vibrationally adiabatic barrier height. For higher coverages $(0.5 < \theta < 1)$ the process can be viewed as the diffusion of vacancies in the two-dimensional adsorption layer; this diminishes D. More detailed calculations within the vibrationally adiabatic model have been performed using a more realistic PES by Lauderdale and Truhlar [1985] and by Truong and Truhlar [1987].

Although our consideration of hydrogen diffusion in and on metals has been brief, we conclude that this type of process is characterized by overcoming barriers that are wide and low. Depending on the coupling symmetry, coupling to lattice phonons either slows tunneling down (increasing effective tunneling mass) or enhances the transition probability by effectively lowering the barrier. Because of the above peculiarity of barriers, the diffusion jumps occur in the vibrationally adiabatic regime rather than in the sudden regime, despite the fact that the frequencies of the bath vibrations are lower by far than typical frequencies associated with the equilibrium vibrations of the impurity itself.

6.7. EXCHANGE REACTIONS IN HYDROGEN CRYSTALS

The diffusion of H and D atoms in crystals of molecular hydrogen has been explored in detail with the EPR method. The atoms were generated by γ-irradiation of the crystals or by photolysis of H atom precursors doped into the crystals. Miyazaki et al. [1984] and Itskowski et al. [1986] observed that the half-lifetime of H atoms prepared at an initial concentration of 4×10^{-8} mol/cm^3 in H$_2$ crystals is $\sim 10^4$ s at both 4.2 and 1.9 K. The bimolecular recombination rate constant, $k_H = 82$ cm^3 mol^{-1} s^{-1}, is diffusion limited at these temperatures. Due to the low concentration of H atoms, each encounter of recombination partners requires 10^5–10^6 hops, each of which has a length of approximately the lattice period. If we assume that the critical diameter for recombination of H + H is $R_H = 2$ Å, then the diffusion coefficient $D = k_H / 4\pi R_H$ is calculated to be 2.7×10^{-16} cm^2 s^{-1}. The fact that the diffusion coefficient is practically independent of temperature strongly indicates that the mechanism is not thermally activated, because even a small value of E_a would give rise to an enormous change in diffusion rates at these low temperatures.

The diffusion coefficient of D atoms in a D$_2$ crystal (measured in the same way as described above) is also temperature-independent in the range 1.9 to 4.2 K, though it is ~ 4 orders of magnitude smaller than for H in H$_2$ [Lee et al., 1987; Miyazaki et al., 1989]. In mixed crystals of D$_2$ and HD (from 20:1 to 7:1) a decrease in the D-atom concentration coincides with an increase in that of H, so that the total concentration remains constant. This means that the conversion is due to the diffusion-limited exchange reaction

$$HD + D \rightarrow H + D_2 \qquad (6.50)$$

At $T \leq 11$ K, the self-diffusion of H$_2$ in an H$_2$ crystal is due to tunneling of a molecule from a lattice node to a vacancy, the formation of which requires 0.22 kcal/mol [Weinhaus and Meyer, 1973]. If diffusion of H and D were caused by motion of H$_2$ or D$_2$ through the lattice, then the process would be expected to be thermally activated, contrary to the experimental results. This discrepancy strongly suggests that the mechanism of the quantum diffusion of H and D atoms is not a molecular one, but is caused by the exchange reactions

$$H + H_2 \rightarrow H_2 + H$$
$$D + D_2 \rightarrow D_2 + D \tag{6.51}$$

as a result of which the atom moves to the nearest equilibrium position. The rate constants of reactions (6.51) are related to the diffusion coefficients by a simple formula ($k = 6D/d^2V_0$, where d is the jump length (4.6 Å for the H_2 crystal), and V_0 is the crystal molar volume (20.0, 20.6, and 22.2 cm^3/mol for H_2, HD, and D_2, respectively) [Silvera, 1980]. The rate constants of reactions (6.50) and (6.51) are given in Table 6.6.

The spatial localization of H atoms in molecular H_2 and HD crystals has been inferred by Miyazaki et al. [1991] from analysis of the hyperfine structure in the EPR spectrum caused by the interaction of the unpaired electron with the matrix protons. The mean distance between an H atom and protons of the nearest molecules, inferred from the ratio of line intensities for the allowed ($\Delta m_i = 0$) and forbidden ($\Delta m_i = \pm 1$) transitions, is 3.6–4.0 Å and ~2.3 Å for H_2 and HD, respectively. It follows from the comparison of these distances with parameters of the hcp lattice of H_2 that the H atoms in the H_2 crystal occupy substitutional sites, whereas in the HD crystal they occupy octahedral interstitial sites.

The intermolecular distance in the H_2 crystal (3.79 Å) is almost 5 times larger than the H–H equilibrium bond length, but is close to the equilibrium distance in the linear H_3 van der Waals complex (3.5 Å). As a substitutional impurity, a hydrogen atom moves almost freely in the cavity with the radius of ~0.6 Å, which suggests that a gas-phase model might be appropriate for calculating the rate coefficients of reactions (6.50) and (6.51). This approach has been taken by Takayanagi et al. [1987] and by Hancock et al. [1989]. The binding energy of the linear H_3 complex is 0.055 kcal/mol, and its "intermolecular" vibrational frequency is less than 50 cm^{-1}. The vibrationally adiabatic barrier height in the

TABLE 6
Rate Constant for Exchange Reactions of H and D atoms at
1.9 and 4.2 K

Reaction[a]	k_c (cm^3 (mol)$^{-1}$ s)	k_c/V_0 (s^{-1})
$H + H_2$	1.8	0.9
$D + DH$	2.3×10^{-3}	1.1×10^{-4}
$D + D_2$	1.8×10^{-3}	7.8×10^{-5}

[a] Reactions $H + D_2$ and $H + HD$ are endoergic due to zero-point energy differences.

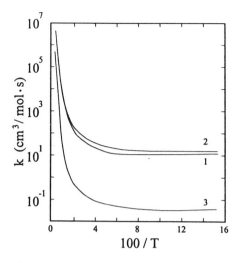

Figure 6.29. Temperature dependences of the rate constants for the exchange reactions $H_2 + H$ (1), $H_2 + D$ (2), and $HD + D$ (3) calculated by Hancock et al. [1989] within the semiclassical small curvature approximation. (From Hancock et al. [1989].)

transition state is $\sim 9.4\ \text{kcal/mol}$, and the corresponding H–H distance at this point is $0.82\ \text{Å}$. Takayanagi et al. [1987] modeled the barrier by an Eckart potential with width of 1.5–$1.8\ \text{Å}$. The rate constant has been calculated from Eqs. (2.1)–(2.4) using the barrier height as an adjustable parameter. This yields a value of V_0 similar to that of the gas-phase $H + H_2$ reaction.

Hancock et al. [1989] used a version of the small curvature semiclassical adiabatic approach introduced by Truhlar et al. [1982] to calculate the temperature dependence of the rate constant, as shown in Figure 6.29. Variations in $k(T)$ below the crossover point (25–30 K) are due to changes in the prefactor due to zero-point vibrations of the H atom in the crystal. Obviously, the gas-phase model does not take these into account. The absolute values of the rate constant differ by 1–2 orders of magnitude from the experimental ones for the same reason.

Interestingly, Miyazaki et al. [1991b] showed that the rate constant depends on the initial rotational quantum state of H_2; i.e., reaction (6.50) is three times faster in *para*-H_2 than in *ortho*-H_2. Although the exchange reactions considered here constitute a fundamental reaction system that has been studied by quantum theory for more than half a century, further investigation is needed to understand chemical dynamics in real crystals.

TUNNELING ROTATION

CONTENTS

The fine structure of torsion vibration spectra of small symmetric molecules and groups such as CH_3, CH_4, NH_3, and NH_4^+ is one of the best examples of the manifestations of tunneling. Inelastic neutron scattering (INS) and nuclear magnetic resonance (NMR) in low magnetic fields, as well as some less universal methods, have permitted the study of tunneling rotation in a wide variety of molecular solids over a wide range of tunneling frequencies from 10^5 to 10^{11} Hz. Since these frequencies are exponentially dependent on the barrier height, they are very sensitive to the changes in potential resulting from small variations of chemical structure and molecular environment. As a rule, the tunneling frequencies in solids are smaller than those in isolated molecules in gas phase, so comparison of these data allows us to obtain unique information about weak intermolecular interactions in organic solids. These problems are detailed in several reviews and books (see, e.g., Press [1981] and Heidemann et al. [1987]).

The dynamics of tunneling rotation of hindered rotors interacting with intra- and intermolecular vibrations has received much less attention than structural studies. Such interactions shift and broaden tunneling spectral lines and, when temperature is raised, lead to transitions from coherent tunneling to thermally activated hopping.

In this chapter, we do not provide a detailed description of numerous experimental results obtained in this field. Our discussion is restricted to basic theoretical ideas and selected examples. The main goal pursued in this chapter is to demonstrate that the general approach to tunneling dynamics discussed earlier is totally applicable to this phenomenon.

7.1. SPLITTING OF TORSION VIBRATIONAL LEVELS

Any CH_3 group can rotate about the bond that connects it to the rest of the molecule. The potential $V(\phi)$ for this hindered rotation is typically threefold. In the extreme case when the barrier is absent, the group rotates freely, while in opposite limit of an infinite barrier it oscillates in one of the three minima. How the energy levels change when moving from free rotation to torsional vibration is shown in Figure 7.1. Tunneling partially lifts the triple degeneracy of each vibrational level creating a singlet A and doublet E (E_a, E_b), in accordance with the irreducible representation of the C_{3v} point symmetry group, which is isomorphous to the permutation group G_6. Figure 7.1 shows that the A and E levels alternate in the progression of torsional multiplets $n = 0, 1, \ldots$, and the sign of the tunneling splitting is $(-1)^n$. The potential $V(\phi)$ may be expanded in a Fourier series, and usually the first harmonic suffices:

$$V(\phi) = \frac{V_3}{2}(1 - \cos 3\phi) \qquad (7.1)$$

where ϕ is the dihedral angle of rotation of the CH_3 group as a whole relative to the fixed framework. The frequency ω_0 of small amplitude torsional vibrations is related to the barrier height by

$$V_3 = \tfrac{2}{9}I\omega_0^2 \qquad (7.2)$$

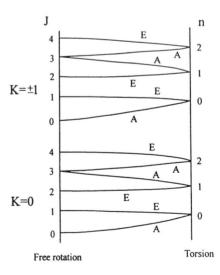

Figure 7.1. Correlation between the energy levels of free rotation of a symmetric top and those of torsion vibration in the threefold potential. Quantum numbers J and K are rotational levels; n represents vibrational levels. Relative positions of A and E levels in multiplets $n = 0, 1, \ldots$, are shown on the right.

where

$$\omega_0 = \left(\frac{k_r}{I}\right)^{1/2} \tag{7.3}$$

I is the reduced moment of inertia, $I^{-1} = I_1^{-1} + I_2^{-1}$, and I_1 and I_2 are the moments of inertia of CH_3 group and the rest of the molecule. For the potential of a general form including the harmonics $3m\phi$, $m = 2, 3, \ldots$, with amplitudes V_{3m}, ω_0 and barrier height are independent parameters. Magnetic interactions of neutrons with protons or deuterons of the CH_3 (CD_3) group lead to the appearance of peaks of quasielastic and inelastic scattering corresponding to AE and $E_a E_b$ transitions, respectively. At liquid helium temperature, the energy separation between the elastic central peak and the sidebands of inelastic scattering is equal to the tunneling splitting. The shape of a spectral line is described by Eq. (2.53). A typical example of an INS spectrum is given in Figure 7.2, which is adapted from the paper by Horsewill et al. [1987]. The spectrum dramatically changes when the system makes the transition from tunnel-

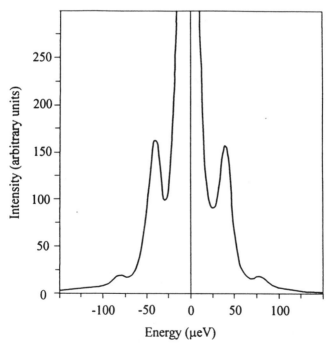

Figure 7.2. INS spectrum of acetylacetone at 4.5 K. (From Horsewill et al. [1987].)

ing to thermally activated rotation (hopping). The sidebands disappear and the quasielastic peak becomes Lorentzian with half-width $\Gamma = 3/2\tau_h^{-1}$, where τ_h is the characteristic hopping time, i.e., the inverse rate constant $\tau_h^{-1} = k$ (see, for example, Scold [1968]). In spin-relaxation theory [Zweers and Brown, 1977], τ_h is equal to the correlation time of the two level Zeeman system, τ_c.

The states A and E have total spins of protons $\frac{3}{2}$ and $\frac{1}{2}$, respectively. The diagram of Zeeman splitting of the lowest tunneling octet AE with $n = 0$ is shown in Figure 7.3. Since the spin wave functions belong to the same symmetry group as that of hindered rotation, the spin and rotational states are fully correlated. The total wave functions belong to the A representation and are the products $|\psi_A \chi_A\rangle$, $|\psi_{Ea} \chi_{Eb}\rangle$, and $|\psi_{Eb} \chi_{Ea}\rangle$ of rotational ($|\psi_i\rangle$) and spin ($|\chi_i\rangle$) functions. The transitions observed in NMR spectra are $\Delta m_I = \pm 1$ and $\Delta m_I = \pm 2$ and include, aside from the Zeeman frequencies, sidebands shifted by Δ. The special technique of dipole–dipole-driven low-field NMR in the time and frequency domain [Weitekamp et al., 1983; Clough et al., 1985] allowed the detection of these sidebands directly. Using the same method, Horsewill and Aibout [1989b] also observed the transitions with $\Delta m_I = 0$. These transitions with conserved spin projection are due to coupling of the tunneling and spin reservoirs in the vicinity of crossing of Zeeman levels belonging to the different tunneling states $(A, \frac{3}{2})$, $(E, -\frac{1}{2})$ and $(A, \frac{1}{2})$, $(E, -\frac{1}{2})$. The NMR spectrum of thioamisole, measured by Horsewill and Aibout [1989b], is shown in Figure 7., demonstrating the change of the resonance frequencies with magnetic field.

A more traditional technique, measuring the temperature dependence

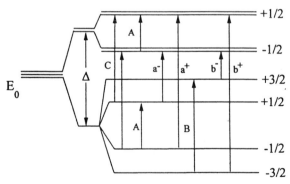

Figure 7.3. Zeeman splitting of the lowest AE octet of CH_3 group. The levels of the E state with $m = \pm\frac{1}{2}$ are doubly degenerate.

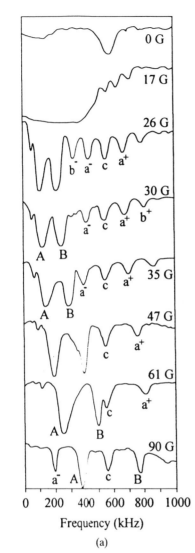

Figure 7.4. (a) The frequency-swept dipole–dipole-driven NMR spectra of thioanisole recorded at a variety of magnetic field intensities. (b) The NMR and sideband transitions observed in thioanisole. Data of Figure 7.4a are presented as a plot of magnetic field versus transition frequency. (From Horsewill and Aibout [1989b].)

of longitudinal spin–lattice relaxation time T_1 of methyl group protons, is also widely used for determining the tunneling frequencies $\Delta/\hbar$ and τ_h. The interrelation between T_1 and τ_h is due to modulation of the dipole–dipole interactions by random proton motions (see, for example, Abraham [1961] and Goldman [1970]). Similar to (6.24), this interrelation is

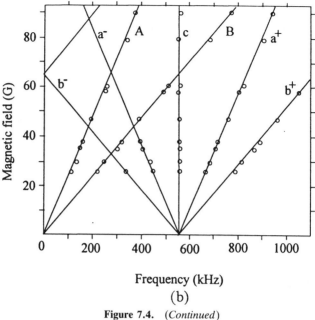

Frequency (kHz)

(b)

Figure 7.4. (*Continued*)

described by [Clough et al., 1982]

$$T_1^{-1} = C_1 \left(\frac{\tau_h}{1 + \omega_Z^2 \tau_h^2} + \frac{4\tau_h}{1 + 4\omega_Z^2 \tau_h^2} \right)$$

$$+ C_2 \left[\frac{\tau_h}{1 + (\omega_Z + \Delta)^2 \tau_h^2} + \frac{\tau_h}{1 + (\omega_Z - \Delta)^2 \tau_h^2} \right.$$

$$\left. + \frac{4\tau_h}{1 + (2\omega_Z + \Delta)^2 \tau_h^2} + \frac{4\tau_h}{1 + (2\omega_Z - \Delta)^2 \tau_h^2} \right] \qquad (7.4)$$

where C_1 and C_2 stand for lattice sums of matrix elements of dipole–dipole interactions. The terms proportional to C_1 and C_2 correspond to EE and AE transitions, respectively. The Arrhenius behavior of τ_h leads to the appearance of minima in T_1 when the temperature is varied. When $\Delta \ll \omega_Z$, T_1 has one minimum for the each type of rotor. Several minima are frequently observed for $T_1(T)$, implying that there are several inequivalent rotors with different τ_h. For example, the temperature dependence of T_1 for tiglic acid [Horsewill et al., 1989] is shown in Figure

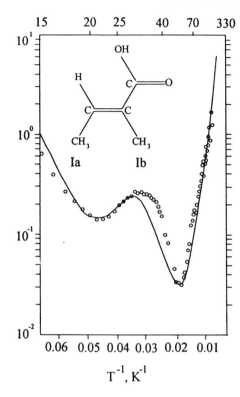

Figure 7.5. The temperature dependence of T_1 in tiglic acid (7.27, II). The solid curve represents the fit of experimental points. (From Horsewill et al. [1989].)

7.5. The low-temperature minimum of this dependence associated with more rapid rotation corresponds to the CH_3 group neighboring the carbonyl group.

Clough et al. [1981, 1982] found a universal correlation between the temperature at which T_1 has a minimum, T_{min}, and Δ, when the measurements are performed at the same Zeeman frequency. This correlation, demonstrated in Figure 7.6, holds for all molecular solids studied thus far, even though Δ covers a range of about four orders of magnitude. The deviations are within factor of 2. At first sight, the existence of a universal dependence $\Delta(T_{min})$ is surprising because it implies a correlation between two dynamical parameters relating to quite different processes: coherent tunneling and incoherent thermally activated hopping. In actuality, this correlation is merely a consequence of the fact that the barrier shape is approximately the same for various species. The barrier is overcome either by hopping (with the rate constant (2.14)) or tunneling (with frequency (2.20)). Based on Sections 2.1 and

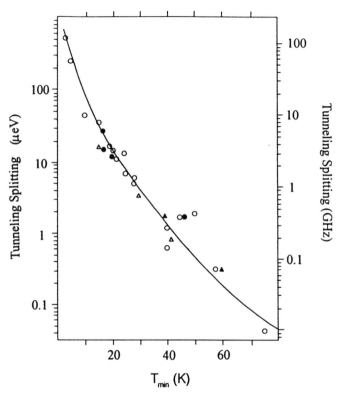

Figure 7.6. Universal correlation between T_{min} and tunneling splitting. The Zeeman frequency is 21 MHz. The points correspond to different chemical species. (From Clough et al. [1982].)

3.5, formula (2.20) for one-dimensional tunneling can be written as

$$\Delta \sim \exp\left(-r\frac{V_3}{\hbar\omega_0}\right) \tag{7.5}$$

where the factor r is constant due to the above-mentioned reason. Where $\omega_Z \gg \Delta$, formula (7.4) determines V_3 as a function of T_{min} and (7.5) turns into universal dependence:

$$\Delta = \Delta_0 \exp(-\gamma T_{min}) \tag{7.6}$$

where γ is a parameter depending on ω_Z, ω_0, and r. Note that Δ is not measured directly but is calculated from the barrier height, which is extracted from the Arrhenius dependence of τ_h^{-1}. The commonly used

Zeeman frequencies 10.5 and 21 MHz and T_{min} located in the range 20–150 K, correspond to rotation barrier heights from 0.65 to 3.4 kcal/mol.

Although the above explanation relied on a crude semiclassical estimate (with exponential accuracy), it can easily be refined either by exactly solving the Schrödinger equation for the one-dimensional potential (7.1) (see, for example, Press [1981]) or, for sufficiently high barriers ($V_0/\hbar\omega_0 \gtrsim 2$), by employing the WKB approximation. The eigenfunctions of stationary states A and E

$$\psi_A = \frac{1}{\sqrt{3}}\left[\psi_0(\phi) + \psi_0\left(\phi - \frac{2\pi}{3}\right) + \psi_0\left(\phi + \frac{2\pi}{3}\right)\right]$$

$$\psi_{E_a} = \frac{1}{\sqrt{3}}\left[\psi_0(\phi) + \exp\left(-i\frac{2\pi}{3}\right)\psi_0\left(\phi - \frac{2\pi}{3}\right) + \exp\left(i\frac{2\pi}{3}\right)\psi_0\left(\phi + \frac{2\pi}{3}\right)\right]$$

$$\psi_{E_b} = \psi_{E_a}^* \tag{7.7}$$

are linear combinations of the site functions describing the intrawell motion. Following the lines of Section 3.4 (see Appendix A), one derives the tunneling-induced shift of the ground vibrational state $n = 0$:

$$E_A - E_0 = -\frac{2\hbar^2}{I}\,\psi_0(\phi)\psi_0'(\phi)\big|_{\phi = \pi/3}$$

$$E_E - E_0 = \frac{1}{2}(E_0 - E_A) \tag{7.8}$$

where $E_0 = \frac{1}{2}\hbar\omega_0$. Using the WKB asymptotes of the site wave functions, one finally arrives at the WKB formula for the tunneling splitting:

$$\Delta = E_E - E_A = \frac{3}{2\pi}\hbar\omega_0 \exp\left[-\frac{1}{\hbar}\int_{\phi_*}^{2\pi/3 - \phi_*} d\phi\left[2I\left(V(\phi) - \frac{\hbar\omega_0}{2}\right)\right]^{1/2}\right] \tag{7.9}$$

where ϕ_* and $2\pi/3 - \phi_*$ are the turning points, $\phi_* = \frac{1}{3}\cos^{-1}(1 - \hbar\omega_0/V_0)$. Peternelj et al. [1987] showed that in the Gaussian packet approximation Δ is described by a formula that can be derived from (7.9) using the steepest descent method:

$$\Delta = \frac{3}{2}\hbar\omega_0 \exp\left(-\frac{\pi^2}{9}\frac{I\omega_0}{\hbar}\right)\left[\frac{I\omega_0}{\hbar}\left(\frac{\pi^2}{9} - \frac{4}{9}\right) + 1 + \cdots\right] \tag{7.10}$$

The accuracy of WKB formulas (7.9) and (7.10) is illustrated in Table 7.1

TABLE 7.1
Tunneling Splitting in Potential (7.1)

	$\Delta/\hbar\omega_0$		
$V_3/\hbar\omega_0$	Exact Calculation	WKB Approximation (7.9)	Gaussian Packet Approximation (7.10)
1.234	2.91×10^{-2}	3.79×10^{-2}	1.58×10^{-2}
1.744	5.04×10^{-3}	5.76×10^{-3}	1.68×10^{-3}
2.467	3.53×10^{-4}	3.77×10^{-4}	6.39×10^{-5}
3.020	4.36×10^{-5}	4.54×10^{-5}	4.99×10^{-6}
3.429	7.26×10^{-6}	7.46×10^{-6}	5.62×10^{-7}

taken from Peternelj and Jencic [1989]. As seen from this table, the WKB approximation is reasonably accurate, even for very shallow potentials.

At $T = 0$ the hindered rotation is a coherent tunneling process like that studied in Section 2.3 for the double well. If, for instance, the system is initially prepared in "pure" state ψ_i localized in one of the wells, then the density matrix in the coordinate representation is given by

$$\rho(\phi, \phi_1 | t) = \sum_{n,m} \exp\left[-\frac{i}{\hbar} (E_n - E_m)t \right] \langle \psi_m | \psi_i \rangle \langle \psi_i | \psi_n \rangle \psi_m(\phi)\psi_n^*(\phi_1)$$

(7.11)

where Ψ_m and Ψ_n are the eigenfunctions of stationary states A_1, E_a, and E_b with eigenvalues E_m and E_n, respectively. The amplitude of being in the final state Ψ_f at time t is

$$P[\psi_f(t)] = \int_{-\pi}^{\pi} d\phi \, d\phi_1 \psi_f^*(\phi)\rho(\phi, \phi_1 | t)\psi_f(\phi_1)$$ (7.12)

Choosing initial and final states

$$\psi_i = \psi_0(\phi) \qquad \psi_f = \psi_0\left(\phi \pm \frac{2\pi}{3}\right)$$ (7.13)

and taking into account only the intramultiplet transitions, we obtain the expressions for the probability of finding the system in one of the other wells $P(\pm 2\pi/3, t)$ and the survival probability $P(0, t)$:

$$P(0, t) = 1 - \frac{8}{9} \sin^2\left(\frac{\Delta}{2\hbar} t\right) \qquad P\left(\pm \frac{2\pi}{3}, t\right) = \frac{4}{9} \sin^2\left(\frac{\Delta}{2\hbar} t\right)$$ (7.14)

The transition amplitude $A(t)$, defined as

$$P\left(\pm\frac{2\pi}{3},t\right) = |A(t)|^2 \tag{7.15}$$

is related to the tunneling frequency by

$$\lim_{t\to 0}\left(\frac{1}{t}|A(t)|\right) = \frac{\Delta}{3\hbar} \tag{7.16}$$

The discussion of Section 2.3, which is concerned with the destruction of coherence by interaction with the environment, applies to this case. As in the case of tunneling in a double well, $|A(t)|^2$ can be represented as a product of the attempt frequency of hitting the turning point and the barrier transparency (see Section 3.4). The tunneling splitting is determined by the same parameters and contains only an additional prefactor $3/2$ because of the symmetry.

Let us consider in more detail the connection between the tunneling and hopping frequencies. In view of the symmetry of the rotation Hamiltonian, each eigenfunction can be expanded in a series of eigenfunctions of angular momentum $|m\rangle = (2\pi)^{-1/2}\exp(-im\phi)$, where m is an integer. In this representation, the kinetic energy operator corresponding to free rotation is diagonal and its matrix elements are equal to $(\hbar^2/2I)m^2$. The potential energy operator (7.1) has nonzero off-diagonal elements connecting the states of the free rotor with quantum numbers m and $m \pm 3$. The Hamiltonian matrix breaks up into three blocks corresponding to states A, E_a, and E_b, with m equal to 0, ± 3, $\pm 6, \ldots$; 1, 1 ± 3, $1 \pm 6, \ldots$; and -1, -1 ± 3, $-1 \pm 6, \ldots$, respectively. The torsion-tunneling eigenfunctions are

$$\psi_n(\phi) = (2\pi)^{-1/2}\sum_m A_{mn}\exp(im\phi) \tag{7.17}$$

Clough et al. [1982] found the coefficients A_{mn} by numerically solving the Schrödinger equation. As $V_3 \to 0$, all these coefficients except A_{nn} vanish. Each eigenstate may be characterized by its eigenvalue E_n and the expectation value of angular momentum is defined as

$$\hbar\langle m_n\rangle = -i\hbar\int_{-\pi}^{\pi}d\phi\,\psi_n^*\nabla\psi_n = \hbar\sum_{m>0}mA_{mn}^2 \tag{7.18}$$

For A states, $\langle m_n\rangle = 0$, while for states E_a and E_b, $\langle m_n\rangle$ have equal magnitudes and opposite signs. A plot of E_n as a function of $\langle m_n\rangle$ for E states is given in Figure 7.7 for several values of V_3. From this diagram,

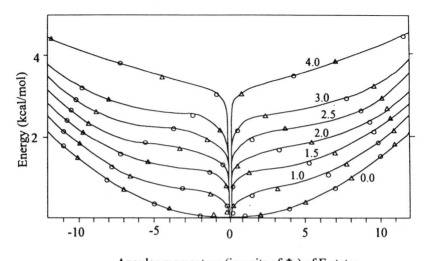

Angular momentum (in units of $\hbar$) of E states

Figure 7.7. A plot of the energy of E eigenstates of the rotational Hamiltonian as a function of mean angular momentum $\langle m_E \rangle$ for different barrier heights, indicated in kcal/mol. (From Clough et al. [1982].)

one clearly distinguishes the cases of being in the well and near the top of the barrier. For torsional levels in the well, $\langle m_n \rangle$ are small:

$$\langle m_n \rangle \ll \left(\frac{2I}{\hbar^2} E_n \right)^{1/2} \tag{7.19}$$

The number of levels for which condition (7.19) is valid increases when V_3 is increased. The energy of a free rotor is proportional to square of angular momentum, so for levels outside the well

$$E_n \simeq \frac{\hbar^2}{2I} \langle m_n \rangle^2 \tag{7.20}$$

When $\langle m_n \rangle \gg 1$, expansion (7.18) contains a large number of waves with different m. Interference of many such waves creates a wave packet, whose group velocity is proportional to $\langle m_n \rangle$. Rotation through the angle $\pm 2\pi/3$ then can be related to the hopping time

$$\tau_n^{-1} = \frac{3}{2\pi} \frac{\hbar}{I} \langle m_n \rangle \tag{7.21}$$

or, after statistically averaging,

$$\tau_h^{-1} = \langle \tau_n^{-1} \rangle_T = \frac{3}{2\pi} \frac{\hbar}{I} Z_0^{-1} \sum_n \langle m_n \rangle \exp(-\beta E_n) \qquad (7.22)$$

The structure of Eq. (7.22) is the same as that of Eq. (2.1). It determines the statistically averaged probability of incoherent transition at $T > T_c$. Note that it is the interference of many waves that renders the transition incoherent. Formally below the crossover temperature this formula should describe the rate of the incoherent transitions from the ground state. However, the transition under these conditions should be coherent. This means that the incoherent hopping rate should vanish at $T = 0$. To take this fact into account Clough et al. [1982] calculated τ_h^{-1} statistically averaging $\langle m_n \rangle - \langle m_0 \rangle$ rather than $\langle m_n \rangle$. Their results show that this correction is significant only for low barriers and $T \sim T_c$.

The semiclassical estimates for tunneling splitting and for incoherent tunneling rates are (see Eqs. (2.18), (2.20), and (7.9))

$$\Delta = \frac{3\hbar\omega_0}{2\pi} \exp\left(-\frac{1}{\hbar} S\right) \qquad k_c = \frac{\omega_0}{2\pi} \exp\left(-\frac{2}{\hbar} S\right) \qquad (7.23)$$

where S is the action between the turning points. It is evident from Eq. (7.23) that $\Delta \gg k_c \sim \tau_h^{-1}$. Based on this reasoning we can formulate a general conclusion: Since the exponent in k_c is twice that in Δ, the transition from coherent tunneling to incoherent thermally activated hopping occurs at higher temperature than the usual T_c defined for the rate constant (see the discussion of (2.20) in Section 2.1). This is the reason why tunneling spectra can be observed at rather high temperatures and their temperature dependence is not due to contribution of thermally activated hopping, but to the interaction of rotation with lattice vibrations, according to the mechanism described in Section 2.3. Using Eq. (7.23) with typical prefactor $\omega_0/2\pi = 10^{13} \text{ s}^{-1}$, the range of Δ shown in Figure 7.6 corresponds to the range of k_c from 10^7 to 10^{-1} s^{-1}. In other words, spectroscopic manifestations of tunneling rotation are associated with approximately the same barrier transparencies as low-temperature quantum chemical reactions.

Rotation of an end methyl group of a large molecule ($I \sim I_1$) implies transfer of the reduced mass $m = 3m_H m_C/(3m_H + m_C) = 2.4$ amu (for CD_3, $m = 4.0$ amu) through the distance 1.74–1.79 Å. This tunneling distance is much greater than typical distances of hydrogen transfer and, therefore, the barrier should be lower to observe the tunneling rotation. In the harmonic approximation for the well, the zero-point amplitude entering into (7.9) is

$$\phi_*^2 = \langle \phi^2 \rangle = \frac{\hbar}{I\omega_0} = \frac{2}{9} \frac{\hbar\omega_0}{V_3} \qquad (7.24)$$

Even for barriers as high as 3.5 kcal/mol, when $\omega_0 \simeq 200\,\text{cm}^{-1}$, $\langle \phi^2 \rangle \simeq$ 0.2 rad, so that zero-point linear displacements of H atoms are 0.20–0.22 Å. Thus, the torsion vibrations, unlike stretching modes, are really motions with wide amplitudes in the full sense of these words. The temperature dependence of $\langle \phi^2 \rangle$ can be found in the harmonic approximation using (2.82); experimentally, it can be extracted from the temperature dependence of the width of INS peaks assigned to torsion vibrations. As a typical example of this dependence, the results of Trevino et al. [1980] for deuterated nitromethane crystal are represented in Figure 7.8.

7.2. POTENTIAL FOR HINDERED ROTATION

Torsional frequencies ω_0 can be extracted from INS, Raman, or IR spectra. A set of values Δ, $\tau_h(T)$, and ω_0 permits the determination of the parameters of the rotational potential of a more general form than (7.1). Namely, the V_3, V_6 potential

$$V(\phi) = \frac{V_3}{2}(1 - \cos 3\phi) + \frac{V_6}{2}[1 - \cos 6(\phi + \delta_0)] \qquad (7.25)$$

is customarily used to incorporate the influence of intra- and intermolecular interactions with the atoms that are not bonded chemically with the CH_3 group. Since in molecular solids intermolecular interactions are much weaker than intramolecular ones and they almost do not disturb the intramolecular geometry, the potential $V(\phi)$ is a sum of intra- and intermolecular contributions. The first is the potential of hindered rotation in the gas phase, which is known from IR and microwave spectra. Comparison of potentials (7.25) for gas phase with those for solids shows that the rotation is only weakly hindered by intermolecular interactions (see for example, Abed and Clough [1987] and Prager et al. [1990]). Gajdos and Bleha [1983] showed by model calculations that the small intermolecular part of $V(\phi)$ has nothing to do with the weakness of van der Waals interaction, the energy of which is comparable or even greater than $V_3 + V_6$, but is due to the weak orientational dependence of distant molecular group interactions. This fact was recognized by Flory [1969] in his well-known studies of long hydrocarbon chain molecules in solids, where the interaction of CH_2 fragments of neighboring molecules

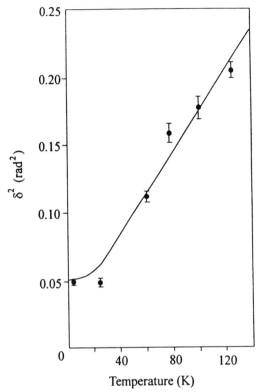

Figure 7.8. Mean square amplitude of methyl group libration in CD_3NO_2 as a function of temperature. The solid line corresponds to $\langle \phi^2 \rangle$ for harmonic oscillator with frequency $42.7\,cm^{-1}$, calculated from (2.82). (From Trevino et al. [1980].)

is determined by only the distance between carbon atoms (see also Vacatello et al. [1980].

Rotation around an ordinary C–C bond is characterized by the highest barrier. In n-alkane crystals $V_1 > 0$, the molecules lie in the zigzag-like *trans* configuration and form layers with parallel alignment of long axes. The end CH_3 groups are situated in parallel planes separated by 3.62–3.65 Å [Kitaigorodskii, 1961; Mnyukh, 1963], depending on intermolecular contacts [Mathiensen et al., 1967]. Because of these features of n-alkane crystals, the rotational potential is independent of the chain length and depends on the interaction of CH_3 with adjacent CH_2 groups. The barrier height in crystals of C_2H_{2n+2}, $n \geq 6$ is the same as in the isolated n-propane molecule (3.33 kcal/mol). The effect of intermolecular interactions shows up in the difference of tunneling frequencies in crystals with odd and even n [Abed and Clough, 1987]. For odd n, there are two different directions of rotation axes, and the tunneling frequencies of two inequivalent groups equal 140 and 300 kHz. The difference of these frequencies corresponds to a difference in barrier heights no larger than 2%. For crystals with even n from 6 to 18 and odd n from 11 to 17, there is only one tunneling frequency, equal to 300 and 340 kHz, respectively.

In the isolated toluene molecule, rotation is almost free. The crystal of α-toluene has two inequivalent molecules in the elementary cell belonging to crystallographic group $P_{21/c}$. Cavagnat et al. [1986] observed two pairs of well-resolved sidebands in the INS spectrum (Figure 7.9) and assigned them to two inequivalent rotors with different barrier heights. According to the crystallographic data of Anderson et al. [1977], the molecules of type I have a closely packed environment with mean intermolecular distance $\sim$4.42 Å, while for molecules of type II this distance is $\sim$5.2 Å. Cavagnat et al. [1986] calculated the potential (7.25)

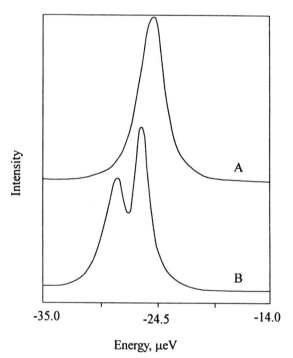

Figure 7.9. Tunneling lines in INS spectra of (A) α-$C_6D_5CH_3$ and (B) α-$C_6H_5CH_3$ at 5 K. (From Cavagnat et al. [1986].)

for molecules of both the types using the semiempirical atom–atom potential method (AAP). Numerous calculations of geometry and energy of crystalline lattices for various organic solids demonstrate that this method gives, as a rule, satisfactory results (see, for example, [Pertsin and Kitaigorodskii, 1987]). The potential of hindered rotation is calculated as a ϕ-dependent component of crystal energy taken as a sum of electrostatic and dispersion interactions. The former is determined by Coulomb interactions of partial atomic charges found from semiempirical quantum chemical calculations. Repulsive and dispersion interactions of chemically nonbonded atoms m and n are described by the 6-exp potential proposed by Buckingham [1967]:

$$V_{nm}(r_{nm}) = K_m K_n \left[\frac{A_{mm}}{r_{nm}^6} + \beta_{nm} \exp(-\alpha_{nm} r_{nm}) \right] \qquad (7.26)$$

where r_{nm} is the interatomic distance, and the constants K_n, K_m, A_{nm}, β_{nm}, α_{nm} are adjusted to give the best fit of calculated and experimentally found geometric parameters and lattice energies for a large number of species containing n and m atoms. The global minimum of the intermolecular potential, which is associated with the stable crystal configuration, is found by solving the variational problem for a lattice of a given crystallographs group using the Euler angles and intermolecular distances as variables. The AAP method can also be used to calculate frequencies of lattice vibrations. Comparison with the known vibration spectrum offers an additional test of AAP. One of the AAP versions is described by Caillet and Claverie [1975]. The AAP parameters were selected by Dashevskii and Kitaigorodskii [1970]. The calculated hindered rotation potentials in α-toluene crystal are presented in Figure 7.10. The potential for type I molecules is asymmetric because of the large values of $V_6(|V_6|/V_3| \sim 0.5)$. A more symmetric potential characterizes the rotation of type II molecules.

The barrier height in molecules and crystals of o-substituted toluenes is determined by van der Waals contact of the substituent with the CH_3 group. In contrast, the lower barriers in m- and p-substituted species are due to intermolecular interactions. Similar low barriers (0.5–1.0 kcal/mol) are typical of the rotation in tetra-, penta-, and hexamethylbenzenes [Clough et al., 1981a,b; Takeda and Chihara, 1983; Takeda et al., 1980]. In these species, the barrier is due to interactions between CH_3 groups, which result not in their coupled rotation but rather in inequivalent arrangements of rotors in the crystal.

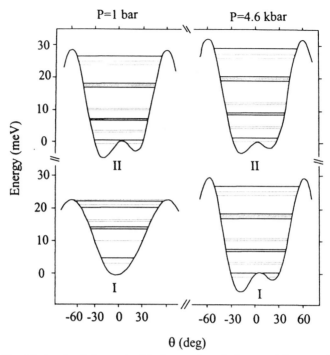

Figure 7.10. Hindered rotation potentials for two inequivalent methyl groups I and II in α-toluene calculated by the atom–atom potential method. The values of V_3, V_6, and δ_0 are equal to 29.8 meV, −14.9 meV, and 17.2° (I) and 23.0 meV, −1.8 meV, and 35° (II). Solid and dashed lines indicate the levels of torsional vibrations for CH_3 and CD_3 groups. The same potentials under pressure 4.6 kbar are shown in the right. (From Cavagnat et al. [1986].)

The barrier for methyl group rotation in the isolated acetone molecule measured with IR and microwave spectroscopy is 0.78–0.81 kcal/mol [Forel and Fokassier, 1967; Nelson and Pierce, 1965]. In the crystal, this barrier increases to 1.41 kcal/mol [Clough et al., 1984a]. In crystals of ketones $CH_3-(CH_2)_m-CO-(CH_2)_n-CH_3$ the barrier height depends on the number of CH_2 fragments between the CH_3 and CO groups. For $m = 0$, $V_3 = 0.8$–1.0 kcal/mol, which is close to isolated acetone molecule, while for m, $n > 4$, V_3 is the same as in crystals of n-alkanes [Abed et al., 1988]. In the acetophenone crystal, studied by Aslanoosi et al. [1989], the barrier height is 2.36 kcal/mol. According to their AAP calculation, the increase in V_3 as compared with the aforementioned ketones is caused by the influence of phenyl ring, as shown in Figure 7.11. The intermolecular interaction of C_3 symmetry is only 0.18 kcal/mol.

Horsewill et al. [1989] studied tunneling in 2-methyl-2-butene and tiglic

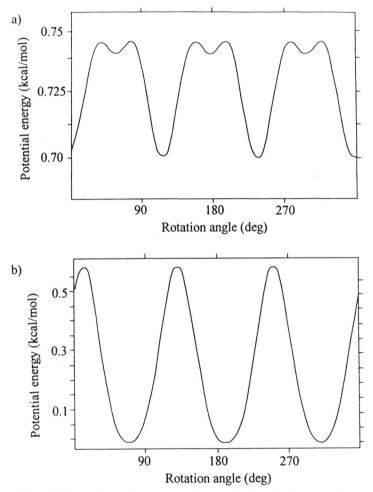

Figure 7.11. (a) The calculated intermolecular contribution to the potential barrier in acetophenone. (b) The calculated potential due to interaction between atoms of methyl groups and the nearest H atom of phenyl group. (From Aslanoosi et al. [1989].)

acid:

$$
\begin{array}{ccc}
\text{site a} \quad \text{site b} & \qquad & \text{site a} \quad \text{site b} \\
\text{I} & & \text{II}
\end{array}
\tag{7.27}
$$

The temperature dependence of T_1 in II is shown in Figure 7.5. It demonstrates the existence of two inequivalent rotors, for which V_3 is equal to 0.69 and 1.36 kcal/mol. Similarly, in I, $V_3 = 1.95$ and 1.22 kcal/mol. The lower of the two barriers in I is assigned to rotation of the CH_3 group at site a. The corresponding group in tiglic acid (IIa) has a similar barrier (1.36 kcal/mol). The higher barrier in I is associated with coupled rotation of the two other groups (site b and *trans* to site a). The rotation of the IIb group is characterized by the lowest barrier. The interaction between Ia and Ib groups calculated by AAP method does not exceed 0.1 kcal/mol.

In the isolated nitromethane molecule, the CH_3 group rotates almost freely [Townes and Shawlow, 1955], whereas the barrier in the crystal is created by intermolecular interactions. The parameters of the potential (7.25) calculated by Cavagnat and Pesquer [1986] are in excellent agreement with experimental data by Trevino and Rymes [1980], Alefeld et al. [1982]. Dispersion interactions make the major contribution to the barrier, while electrostatic interactions are 2–3 times weaker and almost independent of ϕ. The H–H interactions are dominant because they exhibit the strongest orientational dependence. This calculation assumed that the environment relaxes to equilibrium at each dihedral angle ϕ; i.e., the resulting potential $V(\phi)$ is that along the MEP. Although little information is presently available about coupling ϕ with environment displacements, the good agreement of experimental data with the calculation provides evidence that the MEP is a good tunneling path for rotation. This suggestion is in line with our general reasoning of Sections 2.5 and 4.2, where we have motivated the idea that the optimum path for heavy particle transfer should be close to MEP.

A direct way to affect the tunneling frequency that can be realized in the case of tunneling rotations is to apply pressure to the system. Because pressure will change intermolecular distances, thereby changing the barrier height, and because the tunneling probability is very sensitive to the barrier height, one expects a noticeable dependence of the tunneling frequency on pressure. Indeed, Cavagnat et al. [1986] observed that the higher and lower tunneling frequencies shift in opposite directions in the α-toluene crystal (from 6.90 to 7.67 GHz and from 6.29 to 4.04 GHz, respectively) when the applied pressure is varied from 10^{-3} to 4.6 kbar The different behavior of inequivalent CH_3 groups means that the barrier height increases for type I molecules in close-packed environment and decreases for type II molecules having free volume (see Figure 7.10). McDonald et al. [1989] found a nonmonotonic pressure dependence of tunneling frequencies in 4-methyl-2,6-ditertiary butylphenol. The barrier grows at $P < 1.6$ kbar and decreases in the range 1.6–2.4 kbar. This type of dependence is caused by different changes of attractive and repulsive terms in (7.26).

It is known that perdeuteration of hydrogen-containing organic solids reduces the specific volume by ~0.5% [Kooner and van Hook, 1988], which is equivalent to the compression under external pressure of ~0.5 kbar. Therefore, deuteration of the matrix effectively increases the barrier height [Cavagnat et al., 1989]. The shift of tunneling peaks under partial deuteration is accompanied by inhomogeneous broadening resulting from clustering of various isotopomers and creation of internal strains in isotopically inhomogeneous solids.

In all of the previous examples, the role of the methyl group environment was played by relatively small amplitude vibrations around a single stable configuration. From a theoretical point of view, the system–bath Hamiltonians like Eqs. (3.29) and (3.30) are nicely suited to describe this situation. Tunneling rotations present a number of cases when tunneling is coupled to wide amplitude motions, for which the assumption about a single stable configuration breaks down. A theoretical description of these cases involves considering multiple tunneling paths and degenerate saddle points, as described in the previous chapter. Among these examples are the species with intramolecular hydrogen bonds. In methyl-substituted malonaldehyde, rotation of the CH_3 group is strongly coupled to hydrogen transfer, and the frequencies of both the tunneling motions are comparable (see Section 6.2).

Another example is the acetylacetone molecule. In the gas phase, it may be in keto and enol forms, the latter being more stable:

$$CH_3-\overset{\overset{O}{\|}}{C}-CH_2-\overset{\overset{O}{\|}}{C}-CH_3 \quad \longleftrightarrow \quad CH_3-\overset{\overset{O^{\diagup H\cdots}O}{}}{C}=CH-\overset{\overset{O}{\|}}{C}-CH_3 \qquad (7.28)$$

keto   enol

Its INS spectrum, measured by Horsewill et al. [1987], contains two tunneling peaks (Figure 7.2) that are assigned to two inequivalent methyl groups associated with barriers of 0.45 and 1.18 kcal/mol. This is possible when the proton tunneling frequency in the OH $\cdots$ O fragment is smaller than the frequency of tunneling rotation. In the opposite case, both rotors should be equivalent. A quantum chemical calculation undertaken by Riso and Rodriguez [1990] for this case gives correct values of the rotational barrier height for both rotors, but the calculated frequency for proton transfer $(2 \times 10^{11}\,\text{s}^{-1})$ is too high to explain the INS spectrum.

The presence of an intramolecular hydrogen bond reduces the barrier in oxyacetone I compared to pyruvic acid II:

$$
\begin{array}{cc}
\underset{\displaystyle \mathbf{I}}{\text{CH}_3-\overset{\displaystyle \overset{O}{\underset{\|}{}}}{C}-\overset{\displaystyle \overset{H}{\underset{|}{O}}}{\text{CH}_2}} & \underset{\displaystyle \mathbf{II}}{\text{CH}_3-\overset{\displaystyle \overset{O}{\underset{\|}{}}}{C}-C\overset{OH}{\underset{O}{}}}
\end{array}
\tag{7.29}
$$

According to Meyer and Bauder [1982], V_3 is equal to 0.193 and 1.02 kcal/mol in I and II, respectively. Seemingly, a similar effect takes place in the previous example (7.28).

In conclusion of this section we present a table of barrier heights and tunneling frequencies (Table 7.2). This table is based on the data collected by Clough et al. (1981a) and, in addition, includes the results from the later papers discussed above as well as the values of V_3 for some isolated molecules.

7.3. INTERACTION BETWEEN ROTATION AND LATTICE VIBRATIONS

Any one-dimensional description of tunneling rotations, even if it incorporates the effects of the environment through vibrationally adiabatic barriers, is inadequate when it comes to considering such effects as broadening of tunneling spectral lines. In this section we discuss the genuinely mutlidimensional effects that cannot be treated using any effective one-dimensional potential. To treat them quantitatively, one needs to know the constants of coupling of rotation to intra- and intermolecular vibrations. Unfortunately, neither experiment nor theory presently provide reliable numbers that would serve as a basis for a quantitative description relying, say, on a standard system–bath Hamiltonian like Eq. (2.40). In principle, the AAP method seems to be capable of determining the parameters needed, but such calculations going beyond vibrationally adiabatic potentials have not been done so far.

TABLE 7.2
Potential Barriers and Tunneling Frequencies for Methyl Groups in Organic Molecules
and Solids

Species	$\Delta/2\pi\hbar$ (Hz)	V_0 (kcal/mol)	Species	$\Delta/2\pi\hbar$ (Hz)	V_0 (kcal/mol)
4-Methyl pyridine	1.25×10^{11}	0.083	2,5-Dimethyl pyridine	3.46×10^9	0.632
Lithium acetate	6.05×10^{10}	0.184	Acetaldehyde	—	0.646[a]
Oxyacetone	—	0.193[a]	4-Methyl benzoic acid	3.15×10^9	0.652
Acetylacetone	1.01×10^{10}	0.357	2,6-Dimethyl pyridine	2.98×10^9	0.660
Lead acetate	1.09×10^{10}	0.436	3-Methyl pyridine	2.42×10^9	0.698
Acetic acid	—	0.481[a]	Nitromethane	$6.29 \times 10^{10}*$	<0.3[a]
				8.47×10^9	0.768
Methane II	1.97×10^{10}	0.503	Pentamethyl benzene	1.52×10^9	0.788
m-Xylol	6.19×10^9	0.502	Sodium acetate	1.40×10^9	0.804
Toluene	6.05×10^9	0.520	1-Chloro-2-methyl propene	—	0.809[a]
3-Fluorotoluene	3.33×10^9	0.575	2-Hexanone	1.40×10^9	0.810
Ammonia acetate	3.94×10^9	0.610	Zinc acetate	1.20×10^9	0.836
3-Methyl benzoic acid	3.63×10^9	0.624	Pyruvic acid	—	0.970[a]
Methanol	—	1.045[a]	3-Pentanone	3.6×10^6	2.34
Magnesium acetate	4.36×10^8	1.052	Acetophenone	3.43×10^6	2.36
Dimethyl acetylene	4.11×10^8	—	3-Hexanone	2.4×10^6	2.43
3,5-Dimethyl pyridine	3.9×10^8	1.078	Ethane	—	2.93[a]
4-Toluenacetic acid	2.38×10^8	1.150	Dimethyl ether	—	2.98[a]
P-Xylol	2.35×10^8	1.17	4-Heptanone	4.8×10^5	3.00
Methylmalonic acid	1.60×10^8	1.240	Propane	—	3.33[a]
Acetone	9.6×10^7	0.830[a]	n-Hexane	2.98×10^5	3.4
		1.41			
Dimethylsulfide	—	1.449	n-Octane	3.00×10^5	3.4
Tetramethyl-germanium	7.5×10^7	1.470	n-Decane	3.01×10^5	3.4
Copper acetate	7.3×10^7	1.482	n-Dodecane	3.01×10^5	3.4
Methylanime	—	1.98[a]	n-Heptane	1.46×10^5	3.4
				2.98×10^5	
Hexamethyl-benzene	1.2×10^7	2.030	n-Nonane	1.34×10^5	3.4
				2.99×10^5	
o-Xylol	8.9×10^6	2.13	n-Undecane	3.36×10^5	3.4

[a] Molecules in the gas phase.

Nevertheless, we shall attempt to identify those vibrations that affect the tunneling barrier most strongly.

We showed in Chapter 2 that the type of optimum tunneling path depends on the ratios of the characteristic barrier frequency ω_0 to the frequencies of the vibrations to which the tunneling coordinate is coupled. When the tunneling barrier is modulated by high-frequency modes, $\Omega \gg \omega_0$, the tunneling path is close to the MEP and tunneling effectively occurs in the vibrationally adiabatic potential. This situation is appropriate for tunneling rotation in isolated molecules, where the torsional vibrations have the lowest frequency of all intramolecular modes. Another extreme case that is easily tractable is when the interaction between rotation and low-frequency lattice vibrations is so weak that it practically does not change the barrier in the crystal compared to the barrier for an isolated molecule. Examples are crystals of n-alkanes. The weakness of coupling in these compounds is confirmed, in particular, by the vibrational spectrum of crystalline n-hexane, measured by Takeuchi et al. [1980] using INS. This spectrum is shown in Figure 7.12. In contrast with wide bands of acoustic phonons, intermolecular librations, and intramolecular torsional and bending vibrations, the peak of the methyl group torsional vibration is narrow and has small frequency dispersion, so that it is well separated from all other modes.

The lack of data on other molecular solids prevents us from drawing general conclusions about typical tunneling regimes in them. Satisfactory agreement between the potential calculated along the MEP and the

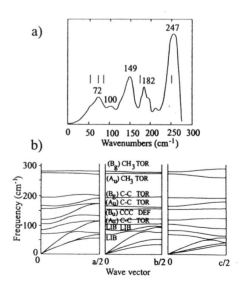

Figure 7.12. (a) Lattice low-temperature vibration spectrum of solid n-hexane. Vertical lines indicate Raman active frequencies. (b) Calculated dispersion curves of low-frequency vibrations. (From Takeuchi et al. [1980].)

experimentally measured energy levels for crystals of nitromethane and α-toluene testifies to the adequacy of the adiabatic approximation.

Although the rotation barrier is chiefly created by the high-frequency modes, it is necessary to consider coupling to low-frequency vibrations to account for more subtle effects such as temperature shift and broadening of the tunneling transition lines. Using the linear coupling approximation, the interaction between the tunneling coordinate and the vibrations q_j having masses m_j and frequencies ω_j can be written as

$$H_{int} = \sum_j C_j q_j \cos 3\phi + C'_j q_j \sin 3\phi \qquad (7.30)$$

where C_j and C'_j are coupling coefficients. The two terms are associated with different phonon polarization. The cosine term corresponds to displacements along the rotation axis or the direction $\phi = 0$. The sine contribution describes the phonons polarized along the line $\phi = \pi/2$. The modes with arbitrary polarization contribute to both terms in (7.30). Interaction (7.30) does not violate the symmetry of the ϕ potential and, in this respect, the coupling is symmetric, as defined in Sections 2.3 and 2.5. Nonetheless, the role of the cosine and sine coupled modes is different. The former, called "breathing modes," simply modulate the barrier (7.25), while the latter, called "shaking modes," displace the potential. Contour plots for both cases are presented in Figure 7.13. The MEP in the total configuration space $(\phi, \{q_j\})$ obeys the equation

$$q_j = -(C_j \cos 3\phi + C'_j \sin 3\phi)m_j^{-1}\omega_j^{-2} \qquad (7.31)$$

and the adiabatic potential along the MEP is

$$V_{ad}(\phi) = V(\phi) - \frac{1}{2}\sum_j (C_j \cos 3\phi + C'_j \sin 3\phi)^2 m_j^{-1}\omega_j^{-2} \qquad (7.32)$$

From (7.32) it follows that coupling contributes to the V_6 potential even if the latter term is absent in the bare potential $V(\phi)$. Both shaking and breathing vibration promote tunneling, but in a different way. Shaking makes the effective barrier narrower, while breathing lowers it. Similar to Section 4.2, we define a parameter that characterizes the relative modulation of the barrier by breathing modes ($C'_j = 0$):

$$b = V_0^{-1}\sum_j \frac{C_j^2}{2m_j\omega_j^2} \qquad (7.33)$$

and an analogous expression for shaking modes. For the nitromethane

a)

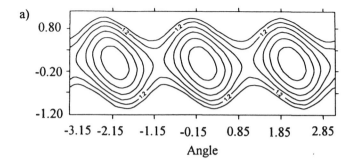

b)

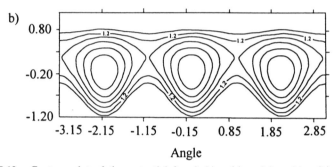

Figure 7.13. Contour plot of the potential for shaking (a) and breathing (b) vibrations (coordinate q along the y axis) coupled to hindered rotation about the threefold axis.

crystal, a rough estimate gives this parameter in the range 0.1–0.2. When the phonon coordinates are fixed at the equilibrium positions of the initial state, $(q_j^0 = C_j/m_j\omega_j^2$, when $C_j' = 0)$, the barrier height is greater than that for adiabatic path by $2bV_0$. This value is equal to the difference between bath energies in the transition and initial states, and may be thought of as a bath reorganization energy for coupling (7.30).

Consider the temperature evolution of INS spectra. When the temperature is raised, the inelastic sidebands are broadened and approach one another. The central quasielastic peak is also broadened. At sufficiently high temperature $(\beta\hbar\omega_0 \gtrsim 1)$, the sidebands collapse to a wide central line, which is typical of thermally activated hopping. As an example of this behavior, the INS spectra of $(CH_3)_2$ $SnCl_2$ measured by Wurger and Heidemann [1990] are shown in Figure 7.14. The shift of the sidebands and broadening of all the bands exhibit Arrhenius dependences with different activation energies (Figure 7.15). The quasielastic peak is narrower than the inelastic ones.

Allen [1974] proposed a phenomenological description of sideband

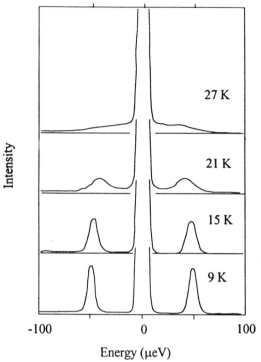

Figure 7.14. INS spectra for rotational tunneling in $(CH_3)_2SnCl_2$ crystal at different temperatures. (From Wurger and Heidemann [1990].)

temperature shift based on a general model of a randomly modulated two-frequency NMR spectrum (see, for example, Abraham [1961]). This model explains the decrease in the observed tunneling splitting by the contributions of highest multiplets with alternate signs of Δ_n (see Section 7.1). The statistically averaged splitting

$$\Delta(T) = Z_0^{-1} \sum_n \Delta_n \exp(-\beta E_n) \qquad (7.34)$$

decreases when the temperature is increased because the state $n = 1$ becomes more populated. This is an essentially one-dimensional model (cf. (7.22)), which therefore cannot explain broadening of the lines due to destruction of the coherence. Prager et al. [1987] observed a slight increase in tunneling splitting preceding its decrease in solid CH_3I, in contradiction with this model.

A consistent solution to the problem of describing the temperature

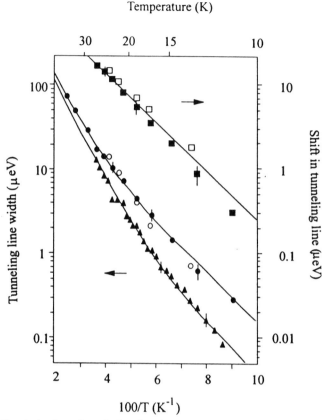

Figure 7.15. Arrhenius plot of the linewidths of inelastic (circles) and quasielastic (triangles) peaks and shifts of inelastic peaks (squares). (From Wurger and Heidemann [1990].)

evolution of the INS spectrum was given later by Hewson [1984], Whittall and Gehring [1987], and Wurger [1989]. In Section 2.3, we derived the general expression for the linewidth, noting that it appears only due to intermultiplet transitions (see Eq. (2.55)). To adapt that expression to the problem of tunneling rotation we note that the interaction matrix elements for shaking and breathing modes are different. Namely, the matrix elements

$$M'_{nn',\sigma} = \langle n\sigma | \sin 3\phi | n'\sigma \rangle \tag{7.35}$$

where n, n' denotes the multiplet numbers and σ is symmetry index (A or

E), are very small for even $n + n'$, while the cosine matrix elements

$$M_{nn',\sigma} = \langle n\sigma | \cos 3\phi | n'\sigma \rangle \tag{7.36}$$

are small for odd $n + n'$ [Wurger, 1989]. At low temperatures when only $n' = 1$ is accessible, shaking vibrations with sine matrix elements (7.35) are dominant. The ensuing expressions for the linewidth of AE and $E_a E_b$ transitions are

$$\Gamma_{AE} = (|M'_{01,A}|^2 + |M'_{01,E}|^2)J(E_{01}) \exp(-\beta E_{01})$$

$$\Gamma_{E_a E_b} = 2|M'_{01,E}|^2 J(E_{01}) \exp(-\beta E_{01}) \tag{7.37}$$

where $J(E_{01})$ is the spectral density of the shaking modes. Incorporating breathing modes is straightforward, and they entail larger activation energies associated with the participation of higher multiplets with $n' \geq 2$. The shift of spectral lines appears in the second order of perturbation theory and, with the assumption that the barrier is high enough, equals

$$\Delta(T) - \Delta(0) \simeq \sum_j \frac{(C'_j)^2}{\omega_j} \bar{n}(\omega_j) |M'_{01}|^2$$

$$\times \left[\frac{\Delta_1}{(E_{01} - \hbar\omega_j)^2 - \Delta_1^2} + \frac{\Delta_1}{(E_{01} + \hbar\omega_j)^2 - \Delta_1^2} \right] \tag{7.38}$$

where Δ_1 is the tunneling splitting in the first excited multiplet ($n = 1$), which, as argued before, is negative, and $n(\omega_j)$ is the average phonon occupation number, defined by (2.56). As seen from (7.38), the shift is caused mainly by the phonons whose frequencies are close to the torsion frequency ($\hbar\omega_j \sim E_{01}$). It is necessary to take the highest multiplets into consideration when the matrix elements of breathing modes $n \geq 2$ are much greater than $|M'_{01}|$ or the spectral density $J(E_{on})$ for $n \geq 2$ exceeds $J(E_{01})$. Seemingly, these features of coupling and spectral density are realized in solid CH_3I, leading to an unusual temperature dependence of spectral line shift [Wurger, 1989].

Huller and Baetz [1988] carried out a numerical study of the role of shaking vibrations. The effect of these vibrations is to replace the static potential (7.1) by $V[\phi - \alpha(t)]$. The phase $\alpha(t)$ is a stochastic classical variable obeying the Langevin equation:

$$I_\alpha \frac{d^2\alpha}{dt^2} + \eta \frac{d\alpha}{dt} + \frac{dU}{d\alpha} + \left\langle \psi(t) \left| \frac{\partial V}{\partial \alpha} \right| \psi(t) \right\rangle = f(t) \tag{7.39}$$

where I_α and $V(\alpha)$ are the moment of inertia and angular potential for the α coordinate and $f(t)$ is a stochastic torque satisfying the FDT (2.24). The rotor wave function ψ is subject to the Schrödinger equation:

$$i\hbar \frac{\partial \psi(\phi, t)}{\partial t} = \left\{ -\frac{\hbar^2}{2I} \frac{\partial^2}{\partial \phi^2} + V[\phi - \alpha(t)] \right\} \psi(\phi, t) \qquad (7.40)$$

The evolution of a state of a certain symmetry is described as

$$\langle \psi_\sigma(t) | \psi_\sigma(0) \rangle = A_\sigma(t) \exp(i\Lambda_\sigma t) \qquad (7.41)$$

Time dependences of the phases Λ_A and Λ_E are shown in Figure 7.16. Despite the erratic nature of these phases, it is seen that the states of different symmetry evolve in parallel. The shapes E_a and E_b show almost no difference in the realization. Moreover, at $T \sim T_c/8$, the difference of phases in the E and A states is to a good accuracy the same as at $T = 0$. In other words, coupling to a stochastic reservoir violates coherence only slightly, and the tunneling splitting may be observed at temperatures up to T_c.

According to Eqs. (7.37) and (7.38), the dependences $\Delta(0) - \Delta(T)$ and $\Gamma(T)$ are characterized by apparent activation energies that are close to E_{01}. After the sidebands collapse at $T > T_c$, the width of the resulting single quasielastic line is determined by the rate of thermally activated hopping (Eq. (7.22)). When temperature is high enough and the highest multiplets are involved in transition, the apparent activation energy approaches the barrier height. Indeed, Heidemann et al. [1989] observed for acetoamide crystals that the activation energy increases from E_{01} to V_0 after the sidebands merge (Figure 7.17). Non-Arrhenius behavior of τ_h at $T < T_c$ was observed by Punkkinen [1980] in solid methane (curve 2 in Figure 1.1). A similar dependence (Figure 1.1, curve 3), was found by

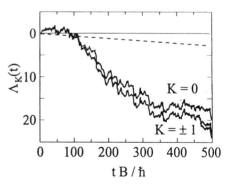

Figure 7.16. Time dependence of phases Λ_A ($K = 0$) and Λ_E ($K = \pm 1$) for a realization of stochastic force at $T = T_c/8$. Also shown are the straight lines of the zero temperature behavior of Λ_A (solid line) and Λ_E (dashed line). Time is measured in units of $\hbar/B$, $B = \hbar^2/2I$. (From Huller and Baetz [1988].)

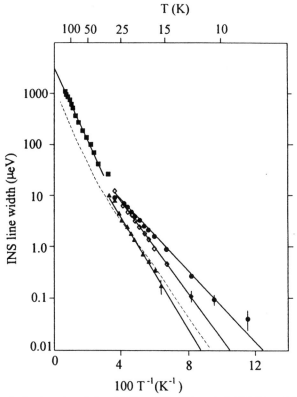

Figure 7.17. Arrhenius plot of the AE line shift (●) and of AE(◇) and $E_a E_b$ (▲) line broadening before and after collapse (■) in acetamide. (From Heidemann et al. [1989].)

Geoffroy et al. [1979] for the radical

$$(CH_3)_2C \begin{array}{c} O \\ \parallel \\ O-C \\ \\ O-C \\ \parallel \\ O \end{array} \diagdown \begin{array}{c} \cdot \\ C-CH_3 \end{array}$$ (7.42)

by using electron–electron double resonance (ELDOR) spectroscopy. The low-temperature limits of τ_h^{-1} turn out to be several orders smaller than $\Delta/\hbar$, as should be expected from a comparison of formulae (2.18) and (2.20) (see also (7.23)).

7.4. COUPLED TUNNELING ROTATION

In most organic solids, rotations of different methyl groups are independent of one another despite the comparatively short distances between the rotors. When the neighboring methyl groups are inequivalent and their frequencies are different, the only appreciable effect of their interaction is to renormalize the potential (7.25). Because coupling to lattice vibrations produces a similar effect, it is hard to distinguish between the rotor–rotor and rotor–phonon couplings. Also, for different rotors the rotor–rotor coupling is usually masked by resonant phonons.

The characteristic features of coupled rotation show up for rotors that have identical frequencies and orientations. In this case, the rotation Hamiltonian assumes the form

$$\hat{H} = -\frac{\hbar^2}{2I}\left(\frac{\partial^2}{\partial\phi_1^2} + \frac{\partial^2}{\partial\phi_2^2}\right) + V(\phi_1, \phi_2) \qquad (7.43)$$

$$V(\phi_1, \phi_2) = -V_3(\cos 3\phi_1 - \cos 3\phi_2) + g_{12}^{(-)}\cos 3(\phi_1 - \phi_2) \qquad (7.44)$$

We have made the additional assumptions that $V_3 \gg |V_6|$ and that the rotor–rotor coupling is described by only the first Fourier harmonic. The difference between the phases of the rotor–rotor interaction and the potential of a single rotor takes two values, depending on the sign of $g_{12}^{(-)}$. In the stable "staggered" configuration of rotors the phase shift between the rotors is $\pi/3$ (at $g_{12}^{(-)} = 0$), so that the signs of the rotor potentials are taken to be opposite in (7.44). Clough et al. [1984a] used the Hamiltonian (7.43) and (7.44) to describe coupled rotation of two methyl groups in lithium acetate.

To proceed, one may use an ansatz similar to Eq. (7.17) for a single rotor. Namely, assuming

$$\psi_{rs}(\phi_1, \phi_2) = \sum_{m_1, m_2} A_{m_1, m_2}^{rs} \exp[i(3m_1 + K_1)\phi_1]\exp[i(3m_2 + K_2)\phi_2]$$

$$(7.45)$$

where r and s numerate single rotor states A, E_a, and E_b in which K_1 and K_2 are equal to 0, +1, −1, respectively, one obtains an infinite system of recurrent equations for coefficients $A^{rs}_{m_1,m_2}$ and eigenvalues E_{rs}:

$$BA^{rs}_{m_1 m_2}[(3m_1 + k_1)^2 + (3m_2 + k_2)^2 - E_{rs}]$$

$$+ \frac{V_3}{2}(A^{rs}_{m_1-1,m_2} + A^{rs}_{m_1 m_2-1} + A^{rs}_{m_1+1,m_2}$$

$$+ A^{rs}_{m_1 m_2+1}) + \frac{1}{2}g^{(-)}_{12}(A^{rs}_{m_1+1,m_2-1} + A^{rs}_{m_1-1,m_2+1}) = 0 \qquad (7.46)$$

where $B = \hbar^2/2I$. One observes from these equations that the interaction of the form (7.44) mixes the free rotor states $\Delta m_1 = 1$, $\Delta m_2 = -1$, whereas the interaction of form $\cos 3(\phi_1 + \phi_2)$ mixes the states $\Delta m_1 = -1$, $\Delta m_2 = 1$. When the interaction includes the next Fourier harmonics, $\cos 6(\phi_1 \pm \phi_2)$, the coefficients with $\Delta m_1 = \pm 2$, $\Delta m_2 = \pm 2$ appear in (7.46). Haussler and Huller [1985] showed that the problem can be solved numerically by using the fact that the coefficients $A^{rs}_{m_1,m_2}$ decrease rapidly for m_1 and m_2 lying outside the circle with a certain radius N. The value of N grows when the barrier height increases, but for typical V_3 (see Section 7.2) N does not exceed 30–40. Solving the truncated system (7.46) with $m_1^2 + m_2^2 \le N^2$ amounts to diagonalization of a block $(2N + 1) \times (2N + 1)$ matrix, which gives the eigenvalues E_{rs} and coefficients $A^{rs}_{m_1,m_2}$. The lowest energy levels of two coupled C_3 rotors are shown in Figure 7.18. The INS spectrum corresponding to this quartet consists of four lines: the high frequency triplet ω_{01}, ω_{12}, and ω_{13} and low frequency line ω_{23}, whose relative intensities are $8:4:4:1$, respectively. The dependence of these frequencies on V_3 and $g^{(-)}_{12}$, found by Haussler and Huller [1985], are presented in Figure 7.19. When $|g^{(-)}_{12}|$ increases, ω_{01} also

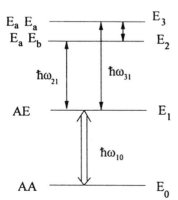

Figure 7.18. The lowest tunneling multiplet of two coupled methyl groups. Symmetry labels and intramultiplet transition frequencies are indicated.

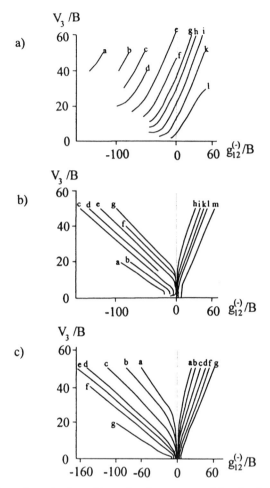

Figure 7.19. Lines of equal values of tunneling frequency ω_{10} (a) and ratios ω_{21}/ω_{10} (b) and ω_{32}/ω_{10} (c) in the plane V_3, $g_{12}^{(-)}$. In (a), ω_{10}/B values are (a) 10^{-7}, (b) 10^{-6}, (c) 5×10^{-6}, (d) 2×10^{-5}, (e) 10^{-4}, (f) 5×10^{-4}, (g) 2×10^{-3}, (h) 5×10^{-3}, (i) 2×10^{-2}, (k) 10^{-1}, (l) 5×10^{-1}. In (b), ω_{21}/ω_{10} values are (a) 0.5, (b) 0.75, (c) 0.909, (d) 0.952, (e) 0.980, (f) 0.990, (g) 0.995, (h) 1.005, (i) 1.020, (k) 1.053, (l) 1.111, and (m) 1.429. In (c), ω_{32}/ω_{10} values are (a) 0.005, (b) 0.002, (c) 0.01, (d) 0.05, (e) 0.1, (f) 0.2, (g) 0.6, and (h) 1.0. (From Haussler and Huller [1985].)

increases; this is a direct consequence of symmetric coupling between rotors.

Assuming a high barrier, two CH$_3$ groups have 9 localized states corresponding to 3×3 minima on the PES. In the basis of the site functions of these states the Hamiltonian is a supermatrix:

$$\hat{H} = \begin{vmatrix} \hat{A} & \hat{B} & \hat{B} \\ \hat{B} & \hat{A} & \hat{B} \\ \hat{B} & \hat{B} & \hat{A} \end{vmatrix}$$

with the elements

$$\hat{A} = \begin{bmatrix} D & T_0 & T_0 \\ T_0 & D & T_0 \\ T_0 & T_0 & D \end{bmatrix} \qquad \hat{B} = \begin{bmatrix} T_0 & T_1^{(+)} & T_1^{(-)} \\ T_1^{(+)} & T_0 & T_1^{(-)} \\ T_1^{(+)} & T_1^{(-)} & T_0 \end{bmatrix} \qquad (7.47)$$

where D is the diagonal matrix element that accounts for intrawell interaction and is irrelevant for splitting. The matrix element

$$T_0 = \left\langle \frac{2\pi}{3}, \frac{\pi}{3} \middle| \hat{H} \middle| 0, \frac{\pi}{3} \right\rangle = \left\langle 0, \frac{\pi}{3} \middle| \hat{H} \middle| 0, -\frac{\pi}{3} \right\rangle \qquad (7.48)$$

corresponds to tunneling of one of the rotors, and the elements

$$T_1^{(+)} = \left\langle \frac{2\pi}{3}, \pi \middle| \hat{H} \middle| 0, \frac{\pi}{3} \right\rangle \qquad T_1^{(-)} = \left\langle \frac{2\pi}{3}, -\frac{\pi}{3} \middle| \hat{H} \middle| 0, \frac{\pi}{3} \right\rangle \qquad (7.49)$$

correspond to coupled rotations in the same and opposite directions, respectively. When $T_1^{(+)} = T_1^{(-)} = 0$, the energy spectrum consists of three equally spaced levels, AA, AE, and EE, with energies $-4T_0, -T_0, 2T_0$. The highest EE level is split as a result of the coupling between rotors, so that this splitting is proportional to $g_{12}^{(-)}$ and equals $3(T_1^{(+)} + T_1^{(-)})$.

Contour plots of PES (7.44) for various values of $g_{12}^{(-)}/V_3$ are shown in Figure 7.20. If $g_{12}^{(-)}/V_3 < 1$, the saddle points are associated with T_0 transitions, while the path of the coupled rotation goes through a maximum. $T_1^{(\pm)}$ transitions go through the saddle points when $g_{12}^{(-)}/V_3 \geq 1$.

Using a basis set method, Peternelj et al. [1987] calculated the spectrum of coupled rotation for a potential of a more general form than (7.44):

$$V(\phi) = \frac{V_3}{2} \sum_{n=1}^{N} [1 - \cos 3(\phi_n - \delta_n)] + \frac{V_3}{4} \sum_{n,n'}^{N} [\tilde{g}_{nn'}^{(+)} \cos 3(\phi_n + \phi_{n'})$$

$$+ \tilde{g}_{nn'}^{(-)} \cos 3(\phi_n - \phi_{n'})] \qquad (7.50)$$

The coupling coefficients $g_{nn'}^{(\pm)}$ are determined by dispersion and coulomb

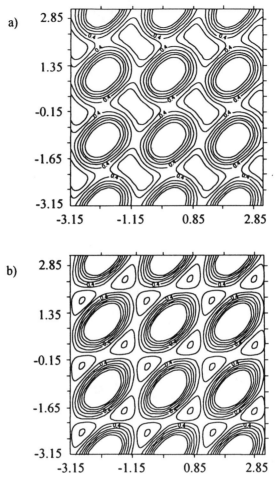

Figure 7.20. Two-dimensional PES of coupled rotation of two methyl groups with common axis: (a) $g_{12}^{(-)}/V_3 = 0.6$, (b) $g_{12}^{(-)}/V_3 = 1.0$.

interactions of XY_3 groups (see Section 7.2) and depend on the relative orientation of rotor axes. For rotors situated in the nodes of an orthorhombic lattice with elementary cell parameters $\ell_1 = \ell_2 = \ell_\perp$, $\ell_3 = \ell_\parallel$, the nonzero components of coupling are $g_\parallel^{(-)}$ and $g_\perp^{(+)}$. The energy minimum corresponds to $\phi_n^0 = \delta_n$, and the phase differences between the neighboring rotors are $\pm\pi/3$ and 0 along the axes ℓ_1, ℓ_2, and ℓ_3, respectively. That is, the rotors are arranged along ℓ_3 in an eclipsed configuration.

The rotor–rotor interaction leads to the formation of a band of torsional levels with dispersion law

$$\omega^2(\mathbf{k}) = \omega_0^2[1 + 2g_\perp^{(+)}(1 + \cos k_1\ell_1) + 2g_\perp^{(+)}(1 + \cos k_2\ell_2)$$
$$+ 2g_\parallel^{(-)}(1 - \cos k_3\ell_3)] \tag{7.51}$$

where the wave vector components k_1, k_2, k_3 range from 0 to $\pm\pi/\ell_j$. Tunneling splitting of the E states in the band appears in the second order of perturbation theory when the rotor–rotor interaction does not destroy the aforementioned translational symmetry of equilibrium rotor orientation. For isolated metal-tetramethyls, $M(CH_3)_4$, situated in the cubic lattice, $g^{(-)}/g^{(+)} = 2.8 \times 10^{-3}$. Neglecting $g^{(-)}$, the minima of (7.50) at $\delta_n = 0$ are

$$\phi_\sigma^0 = \begin{cases} \dfrac{2\pi}{3}\sigma & g^{(+)} < \dfrac{1}{6} \\[3mm] \dfrac{2\pi}{3}\sigma \pm \dfrac{1}{3}\cos^{-1}(6g^{(+)})^{-1} & g^{(+)} > \dfrac{1}{6} \end{cases} \tag{7.52}$$

i.e., each minimum splits in two at $g^{(+)} > \frac{1}{6}$.

In the lithium acetate crystal, CH_3 groups of anions located in neighboring nodes of the orthorhombic lattice form pairs with internal distances shortened to 2.65 Å. The distance between pairs is 6.6 Å. It is this structure of the crystal that results in the coupled rotation. The INS spectrum, measured by Clough et al. [1984a] (Figure 7.21), contains a high-frequency triplet with intensity ratio $2:1:1$ and a low-frequency line with a smaller intensity. This is in agreement with the theoretical spectrum of two coupled rotors described above. The calculated values of V_3/B and $g_{12}^{(-)}/B$ are 2.96 and 6.58, respectively. Positive internal rotor–rotor interaction in the pairs is dominant; this interaction conspires with the potential for a single rotor to localize the two methyl groups in a staggered configuration. The matrix elements T_0, $T_1^{(+)}$, and $T_1^{(-)}$ are of the same order of magnitude, so that two types of tunneling motion, rotation of a single rotor and concerted rotation of two rotors, are both evident.

In the crystal of tin dimethyldichloride, studied by Prager et al. [1986], the observation of coupled rotation is also due to close packing of the methyl groups. In this case, the values V_3/B and $g_{12}^{(-)}$ are 27.0 and 16.8, respectively. NMR spectra of 2,5-dimethylpyrazine and durene manifest coupled rotations [Takeda and Chihara, 1983]. Sridharan et al. [1985] studied NMR spectra of $Ge(CH_3)_4$ and showed that the rotor–rotor interactions in this species result in anomalous temperature dependence

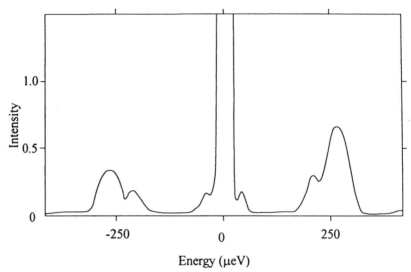

Figure 7.21. INS spectrum of lithium acetate dihydrate at 2.6 K. (From Clough et al. [1984a].)

of linewidths. Below 26 K, the lines are broadened when temperature is decreased. Above 26 K, thermally activated hopping "averages" this interaction and recovers the ordinary behavior of spectra. The frequency ω_{23} estimated from these data is 30 kHz. In the $Sn(CH_3)_4$ crystal, the methyl groups are inequivalent and the rotor–rotor interaction is masked (see, for example, Zhang et al. [1991]). This suppression of coupled rotation can be avoided by studying INS spectra of the same species in Ar and Kr matrices [Prager and Langel, 1986, 1989], where matrix isolation causes the CH_3 groups to be equivalent, and the weakening of inter-molecular interactions leads to a significant increase in tunneling frequencies (72 μeV in matrix in comparison with 1.72 and 13.3 μeV in solids). However, the fine structure expected for coupled rotations is not observed.

Rotational levels of tetrahedral NH_4^+ and CH_4 in cubic and tetrahedral crystalline fields are classified according to irreducible representations of the direct product $\bar{O} \times O$ and $\bar{T} \times T$ octahedral and tetrahedral groups, respectively [King and Hornig, 1969; Smith, 1973, 1975]. The potential is expanded in a series in Wigner rotational functions $D_{MM'}^J(\Omega)$, where Ω represents the Euler angles (ϕ, θ, ψ) [Wigner, 1959]. For a tetrahedral field, functions with $J = 3$ are dominant. Neglecting all other harmonics, the potential takes the form

$$V(\Omega) = \frac{V_0}{2} (D^3_{2,2} + D^3_{2,-2} - D^3_{-2,2} - D^3_{2,-2})$$

$$= \frac{V_0}{2} (\sin 2\phi \sin 2\psi \cos \theta (1 - 3\cos^2\theta) + 2 \cos 2\phi \cos 2\psi \cos 2\theta)$$

$$\tag{7.53}$$

Making the transformation to new coordinates

$$x = \theta \cos \phi \qquad y = \theta \sin \phi \qquad z = \phi + \psi \tag{7.54}$$

one sees that in the limit of high barriers (7.53) becomes a symmetric three-dimensional potential:

$$V(x, y, z) \simeq \frac{V_0}{2} \left[2 - 4(x^2 + y^2 + z^2) + \frac{4}{3}(x^4 + y^4 + z^4) \right.$$

$$\left. + 8(x^2y^2 + x^2z^2 + y^2z^2) \right] \tag{7.55}$$

In the lowest order of perturbation theory, the energy levels of the three-dimensional anharmonic oscillator are

$$E(n_1, n_2, n_3) = -V_0 + \hbar\omega_0 \left(\frac{3}{2} + n_1 + n_2 + n_3 \right) - \frac{B}{4} \left[8(n_1n_2 + n_1n_3 + n_2n_3) \right.$$

$$\left. + \sum_{k=1}^{3} (2n_k^2 + 10n_k + 3) \right]$$

where ω_0 is related to V_0 by Eq. (7.3). Tunneling splitting in the ground state $(n_1 = n_2 = n_3 = 0)$ creates a triplet of levels A, T, and E. The spacings between them are 2Δ and Δ, respectively. Smith [1975] calculated the correlation between Δ and the barrier height. Rotational potentials of NH_4^+ in crystalline fields of different symmetry were calculated by Smith [1985], who took into account the contributions from Wigner functions with $J > 3$.

Tunneling rotation of three-dimensional coupled rotors in solid methane is dealt with in many experimental and theoretical studies. The phase transition at 20.4 K turns this species from orientationally disordered phase I to partially oriented phase II. According to neutron

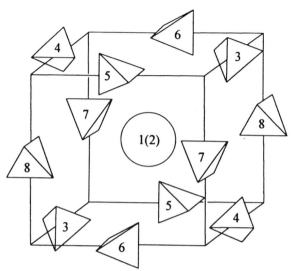

Figure 7.22. Crystal structure of solid methane phase II. The orientationally disordered molecules 1 and 2 and ordered molecules 3–8 occupy O_h and D_{2d} positions, respectively. (From Kobashi et al. [1984].)

diffraction data [Press, 1972], the cubic elementary cell of phase II comprises eight nodes of different sublattices (Figure 7.22). Sublattices 1 and 2 contain the octahedral positions in which the molecules CH_4 do not have preferred orientation and rotate almost freely. Orientational order takes place in the other sublattices, which have symmetry D_{2d}. The height of the rotation barrier is about 0.5 kcal/mol, and the libration amplitude is 20°. Eigenstates of both types are classified according to the irreducible representations of direct products $\bar{O}_h \times O_h$ and $\bar{T}_d \times D_{2d}$ of the symmetry groups assigned to the crystalline field and to the molecules in nodes. Therefore, solid methane II contains O_h and D_{2d} molecules in three states, A, T, and E, with total spin of the 4-proton system 2, 1, and 0, respectively. The potential of rotation can be represented in the form

$$V = \sum_j V_c(\Omega_j) + \frac{1}{2} \sum_{j,j'} W_{jj'}(\Omega_j, \Omega_{j'}) \qquad (7.57)$$

James and Keenan [1959] showed that the octupole–octupole interactions are dominant in the multipole expansion of V_c in (7.57):

$$V_c(\Omega) = B[\beta_4 V_4(\Omega) + \beta_6 V_6(\Omega)] \qquad (7.58)$$

where V_4 and V_6 are expressed through the rotational Wigner functions $J = 4$ and $J = 6$ [Yamamoto et al., 1977].

Expansion (7.58) becomes obvious if we consider the lattice symmetry. Two nearest neighbors belonging to the same sublattice are situated on the diagonals of the opposite vertices and their contributions to the crystalline field should be equal, so that the odd harmonics vanish. Coefficients β_4 and β_6 can be calculated using the AAP method. Yamamoto et al. [1977] found $\beta_4 = 5.43$, $\beta_6 = 7.18$; these numbers were later revised by Smith [1990] so as to give the best fit to the libration spectrum ($\beta_4 = 6.50$, $\beta_6 = -5.33$). The intermolecular potential is expanded in a series in rotational tetrahedral functions u_ν, which can be expressed in terms of Wigner functions at $J = 3$:

$$W_{jj'}(\Omega_j\Omega_{j'}) = \sum_{\nu\nu'} C^{jj'}_{\nu\nu'} u_\nu(\Omega_j) u_{\nu'}(\Omega_{j'}) \qquad (7.59)$$

James and Keenan [1959] used a mean-field approximation for potential (7.57). Yamamoto et al. [1977] proposed an expanded JK method (EJK), which takes into account higher-order terms in expansion (7.59). In the self-consistent mean-field approximation we have (see also Kobashi et al. [1984])

$$\hat{H}_{MFA} = BJ_\alpha^2 + V_c(\Omega_\alpha) + \overline{W_\alpha(\Omega_a)}$$

$$\bar{W}_\alpha = \sum_{\alpha'} \langle \Phi_{\alpha'} | W_{\alpha\alpha'}(\Omega_\alpha, \Omega_{\alpha'}) | \Phi_{\alpha'} \rangle \qquad (7.60)$$

$$\hat{H}_{MFA} |\Phi_\alpha\rangle = E_\alpha |\Phi_\alpha\rangle$$

where the index $\alpha = 1, \ldots, 8$ enumerates the sublattices. Functions $\Phi_\alpha(\Omega_\alpha)$ are linear combinations of rotational Wigner functions corresponding to J from 0 to 8 and from 0 to 10 for D_{2d} and O_h molecules, respectively. Equations (7.60) mean that the rotation occurs in the static potential formed by crystalline field (7.58) and by the average field $\bar{W}_\alpha$ created by the nearest neighbors. From the lattice symmetry it follows that $\bar{W}_\alpha$ is zero for O_h molecules; six nonzero fields for D_{2d} molecules have the same magnitude and differ by signs:

$$\overline{W(O_h)} = 0 \qquad \bar{W}_3 = -\bar{W}_4 = \bar{W}_5 = -\bar{W}_6 = \bar{W}_7 = -\bar{W}_8 \qquad (7.61)$$

so that the mean-field potentials are

$$V(O_h) = V_c \qquad V(D_{2d}) = V_c + \delta_7 u_7 \qquad (7.62)$$

where V_c and u_7 are defined by (7.58) and (7.53), respectively, with

$$V_0 = \frac{\sqrt{7}}{2} B$$

The coefficient δ_7 and the coefficients at rotational Wigner functions entering into $|\phi_\alpha\rangle$ are found by solving self-consistent equations (7.60). According to [Smith, 1990], δ_7 is equal to 19.65 for CH_4 and 48 for CD_4. If one neglects the crystalline field potential V_c [Smith, 1985; Huller and Raich, 1979], then the O_h molecules rotate freely, as follows from (7.62).

The libration spectrum of methane II was calculated by Kobashi et al. [1984] in the EJK approximation. The levels of the ground and first excited states have the symmetry $\bar{A}_1 A_1$ and $\bar{A}_2 T_1$ for O_h molecules and $\bar{A}_2 B_1$ and $\bar{A}_2 E$ for D_{2d} molecules, respectively. The excitations are librational Frenkel excitons, whose states constitute bands, as follows from the translational invariance of the system. The exciton spectrum consists of six branches: $2T_{1g}$, $2T_{1u}$, T_{2g}, and T_{2u}. The results of these calculations are in good agreement with the experimental data of Calvani and Lupi [1987, 1989].

The energies of the two lowest excited states of the O_h molecules measured by Kappula and Glaser [1972) are 1.06 and 1.8 meV (8.55 and 14.52 cm^{-1}), whereas the values expected for a free rotor are 1.3 and 2.6 meV. Press and Kollmar [1975] and Heidemann et al. [1981, 1984] measured tunneling splittings in the ground states of the D_{2d} molecules. The INS spectrum taken from the latter paper is presented in Figure 7.23. Since tunneling splitting is comparable with $k_B T$ at 5 K, only the lowest A state is populated at lower temperatures and only one line of the AT transition is observed in the INS spectrum. The peaks of TE, ET, and TA transitions show up at higher temperatures. Transitions AE and EA are spin-forbidden.

At finite temperatures, solid methane II is an "alloy" of molecules in A, E, and T states; the statistically averaged mean-field depends on temperature and is given by

$$\langle W_\alpha \rangle = \mathrm{Tr}(\rho_\alpha \bar{W}_\alpha) \qquad (7.63)$$

where the density matrix ρ_α is expressed in terms of the mean-field Hamiltonian (7.60). The solution of self-consistent field equations at finite temperatures given by Yamamoto et al. [1977] predicts an increase in tunneling frequencies Δ_{AT} and Δ_{TE} when temperature is decreased. More spherical A molecules have a smaller effective octupole moment compared to molecules in T and E states, and the mean field $\langle W_\alpha \rangle$

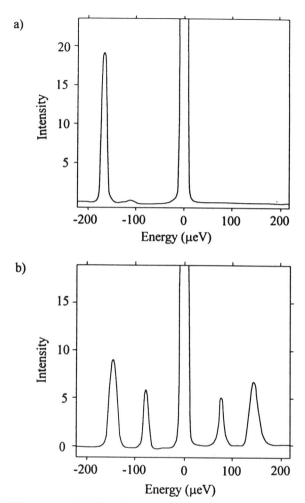

Figure 7.23. INS spectrum of solid methane phase II. (a) Spin temperature $T_s < 0.2$ K. (b) $T_s = 5$ K. (From Heidemann et al. [1984].)

decreases when the population of the A state increases at low temperatures. The experimental temperature dependence of Δ_{AT} and Δ_{TE} is in excellent agreement with the calculations (Figure 7.24).

When temperature is increased, tunneling peaks not only shift, but also exhibit an increasing inhomogeneous broadening. An explanation of this effect was given by Heidemann et al. [1981, 1984], based on a simple model, which is a version of the Bragg-Williams molecular field theory (see, for example, Kubo [1965]). According to this model, the inhomoge-

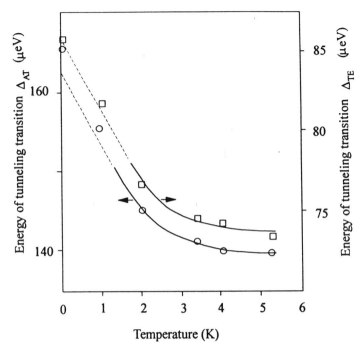

Figure 7.24. Temperature dependences of AT and TE transition frequencies in solid methane phase II. The solid curves are calculated using the statistically averaged mean field approximation. Experimental and calculated values are assumed to be equal at 2 K. (From Yamamoto et al. [1977].)

neous broadening is caused by very slow spin relaxation, whose characteristic time exceeds 10^4 s at $T \leq 1$ K. When all three states are populated, solid methane II is a statistical mixture of elementary cells with different concentrations of molecules frozen in A, T, and E states. The distribution of molecular field over these configurations leads to the distribution of barrier heights. If the barrier heights in the lattices containing only A and only T (E) molecules are V_{0A} and $V_{0A}(1 + \epsilon)$, respectively, then, provided that $\epsilon \ll 1$, the effective barrier height in the cell with m molecules in state A and $(8 - m)$ molecules in state T or E is a linear function of m:

$$V_0(m) = V_{0A}\left[1 + \left(1 - \frac{m}{8}\right)\epsilon\right] \qquad (7.64)$$

The Bragg-Williams molecular field approximation assumes a binomial distribution of configurations

$$P(m) = \binom{8}{m} C_A^m (1 - C_A)^{8-m} \tag{7.65}$$

where C_A is the equilibrium population of state A:

$$C_A = [1 + \tfrac{9}{5} \exp(-\beta \Delta_{AT}) + \tfrac{2}{5} \exp(-\beta \Delta_{AE})]^{-1} \tag{7.66}$$

According to distribution (7.65), the position of tunneling peak $\Delta(C_A)$ and its inhomogeneous width $\delta\Delta(C_A)$ are

$$\Delta(C_A) = \Delta_0[1 - \epsilon(1 - C_A)]$$

$$\delta\Delta(C_A) = \frac{\Delta_0}{\sqrt{8}} \epsilon_1 \sqrt{C_A(1 - g_A)} \tag{7.67}$$

where ϵ_1 is proportional to ϵ. Formulas (7.66) and (7.67) explain the temperature evolution of INS spectra described above.

Grondey et al. [1986] used this method to analyze the INS spectra of mixtures $(CH_4)_{1-x}Kr_x$, in which the available configurations include not only various spin-rotational states of methane molecules, but also the states where these molecules are partially substituted by krypton atoms.

In the context of chemical reactions that are subject to dispersive kinetics as a result of structural disorder, the above model suggests that a widening of the intermediate region between the Arrhenius law and low-temperature plateau should occur. The distribution of barrier heights should also lead to nonexponential kinetic curves (see Section 6.5).

7.5. COOPERATIVE ROTATION

In some molecular crystals, pairs of CH_3 groups are ordered in regular chains. One such crystal is 4-methylpyridine, whose structure is shown in Figure 7.25. Fillaux and Carlile [1989, 1990] interpreted the INS spectra of this crystal using the model of cooperative tunneling rotation. The chains are arranged along the **a** and **b** axes, and the rotation axes are parallel to the **c** axis. A short distance between adjacent methyl groups of the chain leads to a strong rotor–rotor coupling, compared to in-tramolecular interactions. In other words, if it were not for interaction between the rotors, they would rotate almost freely. This interaction leads to long wavelength collective excitations along the chain, for which the angular coordinate of a rotor changes by a little through each lattice period. This circumstance invites replacing the rotational Hamiltonian of

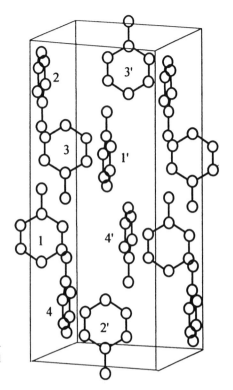

Figure 7.25. Crystal structure of 4-methylpyridine at 4 K. (From Fillaux and Carlile [1990].)

the chain by its continuous analogue, as a lattice of atoms is modeled by continuous elastic medium (or string in one dimension) to describe the low-frequency phonon excitations (see, for example, Feynman, 1972):

$$\hat{H} = \frac{1}{2} \int dx \left[\left(\frac{\partial \phi}{\partial \tau} \right)^2 + c_0^2 \left(\frac{\partial \phi}{\partial x} \right)^2 + \omega_0^2 (1 - \cos 3\phi) \right] \qquad (7.68)$$

where $c_0 = L\omega_c$ is the velocity of a wave propagating along the chain with period L and with the rotor–rotor interaction $g_{12}^{(-)} = \frac{1}{9} I \omega_c^2$. The Euler-Lagrange equation corresponding to the Hamiltonian (7.68) is

$$\frac{\partial^2 \phi}{\partial t^2} = c_0^2 \frac{\partial^2 \phi}{\partial x^2} - 3\omega_0^2 \sin 3\phi \qquad (7.69)$$

The nonlinear wave equation thus obtained is the famous sine-Gordon equation, which is well known from soliton theory (see, for example, Dodd et al. [1982] and Rajaraman [1982]). The long wave approximation used to replace the discrete rotor angle ϕ_j by continuous variable $\phi(x, t)$

is valid when

$$c_0 \gg \omega_0 \sqrt{3} \tag{7.70}$$

This condition is equivalent to $g_{12}^{(-)} \gg \frac{3}{2} V_0$ and implies that the rotor–rotor interaction should be greater than the hindered rotation barrier.

The simplest solution to (7.69), is a soliton, i.e., a solitary wave with unchanged profile, which runs at velocity u:

$$\phi_S(x, t) = \pm \frac{4}{3} \tan^{-1} \left\{ \exp \left[\pm \left(1 - \frac{u^2}{c_0^2} \right)^{1/2} \frac{x - ut}{L} \frac{\omega_0}{\omega_c} \right] \right\} \tag{7.71}$$

Soliton (7.71) and the corresponding antisoliton, which is a similar wave propagating in the opposite direction, are characterized by the energy

$$E_S = E_S^0 \left(1 - \frac{u^2}{c_0^2} \right)^{-1/2} \tag{7.72}$$

where the soliton (antisoliton) energy at rest is equal to

$$E_S^0 = 4 (V_0 g_{12}^{(-)})^{1/2} \tag{7.73}$$

This energy corresponds to the rest mass $M_S^0 = E_S^0 / C_0^2$. In the context of the model of cooperative rotation, the soliton describes a correlated change in phases of the rotors of the chain; appreciable phase changes are accumulated through the distance $L \, \omega_c / \omega_0$, which is greater than the lattice period because of condition (7.70).

To create a soliton, a finite energy E_S is needed, which is greater than the barrier height V_0. Therefore, when $\beta V_0 > 1$, the appearance of a soliton is a rare event which can be neglected when analyzing the motion of the chain.

Aside from solitons, there is a continuous spectrum of torsional modes, or "rotons." These excitations are the eigenstates of the linearized Hamiltonian. To obtain their spectrum, one replaces the last term in (7.68) with a harmonic potential. This approximation implies that the vibrational amplitude of a rotor must be small enough compared with the large amplitude motion of a rotor participating in a soliton. The frequencies of rotons obey the dispersion equation:

$$\omega^2(k) = \omega_0^2 + 4\omega_c^2 \sin^2 \frac{kL}{2} \tag{7.74}$$

At low temperatures, where the density of solitons is small, the two types of excitations are well separated. In between solitons, rotons behave as if there were no solitons at all; the effect of rotons on the soliton profile results in its harmonic oscillations [Currie et al., 1980].

Another type of excitation, the breather (doublet excitation), is described by

$$\phi_B(x, t) = \frac{4}{3} \tan^{-1} \left\{ \frac{\left(\frac{\omega_0^2}{\omega_B^2} - 1 \right)^{1/2} \sin\left[\omega_B \left(t - \frac{u}{c_0^2} x \right) \left(1 - \frac{u^2}{c_0^2} \right)^{-1/2} \right]}{\cosh\left[\left(1 - \frac{\omega_B^2}{\omega_0^2} \right)^{1/2} \left(\frac{x - ut}{L} \right) \left(\frac{\omega_0}{\omega_c} \right) \left(1 - \frac{u^2}{c_0^2} \right)^{-1/2} \right]} \right\}$$

(7.75)

The energy of the breather may be parametrized by the frequency of oscillations of its envelope, ω_B:

$$E_B(u, \omega_B) = 2E_S^0 \left(1 - \frac{u^2}{c_0^2} \right)^{-1/2} \left(1 - \frac{\omega_B^2}{\omega_0^2} \right)^{1/2}$$

(7.76)

When $\omega_B \to 0$, the amplitude of breather tends to $2\pi/3$; its width approaches twice the width of soliton. In the opposite limit $\omega_B \to \omega_0$, the breather width becomes infinite while its amplitude tends to zero. The energy of the breather ranges from zero to twice the soliton energy.

In fact, a breather may be thought of as a result of the attractive interaction between soliton and antisoliton. Inside the breather envelope, the soliton and antisoliton oscillate with respect to each other with period $2\pi/\omega_B$. Because creating a breather requires an arbitrarily small energy, excitations of this type should persist at low temperature and should dominate the dynamics of the rotor chain.

Our discussion of solitons and breathers has been purely classical, since we implied the existence of excitations corresponding to the classical solutions (7.71) and (7.75). To see this correspondence it is natural to invoke the quasiclassical approximation. Dashen et al. [1975] quantized the sine-Gordon problem and showed that in the WKB approximation for a discrete chain with periodic boundary condition the soliton mass is renormalized to

$$M_k = M_S^0 \left(1 - \frac{g^2}{8\pi} \right)$$

(7.77)

where g is the order of the rotation symmetry group ($g = 3$ for C_3 group).

For the breather, the renormalized mass is

$$M_B(\ell) = 2M_S^0\left(1 - \frac{g^2}{8\pi}\right)\sin\left[\frac{\ell g^2}{16(1 - g^2/8\pi)}\right]$$

$$\ell = 1, \ldots, \frac{8\pi}{g^2} - 1 \qquad (7.78)$$

That is, the simple classical picture we described applies to the quantum mechanical case provided the masses of solitons and breathers are changed. This remarkably simple result is due to the special characteristics of the sine-Gordon equation. The quantization reduces to factorization of the action to the classical action and a constant factor that is independent of the soliton velocity and the breather frequency.

When $\ell \geq 8\pi/g^2 - 1$, the breather breaks up into a soliton and antisoliton. The fundamental state of the breather is $\ell = 1$, and $2E_S - E_B(1)$ is the activation energy needed to dissociate the breather and create a soliton–antisoliton pair. When $g^2/8\pi > 1$, the breather does not exist.

For a threefold potential, only the fundamental state of the breather exists. The classical frequencies of the breather are

$$\omega_B(\ell) = \omega_0 \cos\left[\frac{\ell g^2}{16(1 - g^2/8\pi)}\right] \qquad \ell = 1, \ldots, \frac{8\pi}{g^2} - 1 \qquad (7.79)$$

Finally, the WKB spectrum of the breather is

$$E_{B,n} = (E_B^2(\ell) + n^2\omega_c^2)^{1/2} \qquad n = 0, 1, \ldots \qquad (7.80)$$

The appearance of the breather replaces the transition in the torsional spectrum of a single well by the transitions $n = 1 \leftarrow 0, 2 \leftarrow 0, \ldots$.

Tunneling rotation in the chain gives rise to tunneling states, which can be labeled by the wave vector ranging within the first Brillouin zone:

$$-\frac{g}{2} \leq k \leq \frac{g}{2} \qquad (7.81)$$

The eigenfunctions are the Bloch waves:

$$\Psi_{\sigma,k}(\phi) = \exp(ik\phi)F_{\sigma,k}(\phi) \qquad (7.82)$$

where the site functions $F_{\sigma,k}$ have the same period as the rotation potential. For the C_3 group, there are only two types of solutions. In the center of the Brillouin zone, $k = 0$, the methyl groups rotate in a concerted fashion so that the rotor–rotor interactions are zero. At the boundaries of the zone, $k = \pm\frac{3}{2}$, the adjacent methyl groups have opposite phases of rotation. Consequently, the two types of eigenstates are the solutions of the Mathieu equations for $k = 0$ and $k = \pm\frac{3}{2}$:

$$\left[-\frac{\hbar^2}{2I}\frac{\partial^2}{\partial\phi^2} + \frac{V^0}{2}(1 - \cos 3\phi)\right]\Psi(\phi) = E\Psi(\phi)$$

$$\left[-\frac{\hbar^2}{2I}\frac{\partial^2}{\partial\phi^2} + \frac{V_0}{2}(1 - \cos 3\phi) + \frac{g_{12}^{(-)}}{2}(1 - \cos 6\phi)\right]\Psi(\phi) = E\psi(\phi) \quad (7.83)$$

In contrast with the chain of coupled oscillators, the translational invariance of a chain of coupled rotors leads not to a continuous spectrum, but to two branches of tunneling states determined by Eqs. (7.83). These states are coherent, whereas the space-localized breather states (7.80) are incoherent. In this respect, the transitions between breather states are similar to thermally activated rotation of a single group, though the number of rotors lying within the breather envelope and participating in the collective motion is greater than unity ($\sim 2\omega_c/\omega_0$). The above discussion of collective rotation, which is based on the paper of Fillaux and Carlile [1990], demonstrates that the spectrum of a chain of coupled rotors is much richer than the spectrum one can expect from the traditional band model.

Coherent and incoherent transitions have different intensities in INS spectra. For 4-methylpyridine crystals, the 4 MP-h_7 and 4 MP-d_7 incoherent transitions are 10 and 5 times more intense than coherent transitions. INS spectra of various mixtures of 4 MP-h_7 and 4 MP-d_7 measured by Fillaux and Carlile [1990] are presented in Figure 7.26. The intense peak at 520 μeV (4.19 cm^{-1}) in pure 4 MP-h_7 is assigned to the $1 \leftarrow 0$ transition of the breather. The less intense peaks, 535 and 468 μeV, are assigned to tunneling in-phase and out-of-phase rotation. The parameters of the sine-Gordon potential corresponding to these frequencies are found by solving the Mathieu equations (7.83), yielding $V_0 = 29.5$ cm^{-1} and $g_{12}^{(-)} = 44$ cm^{-1}. The characteristics of the soliton and breather in a chain with these parameters are given in Table 7.3. The breather width (~ 20 Å) in the fundamental state is about 5 times greater than the distance between neighboring CH$_3$ groups. The calculated frequencies of breather transitions in higher excited states ($n = 3, 4, 5$) are in agreement with the Raman and INS spectra. Mixed crystals of 4 MP-h_7 and 4 MP-d_7 are very inhomogeneous media for propagation of a breather, since the values of ω_0 and ω_c are different for the two isotopomers. Thus, the regions of space that breathers can sample are confined to isotopomer clusters

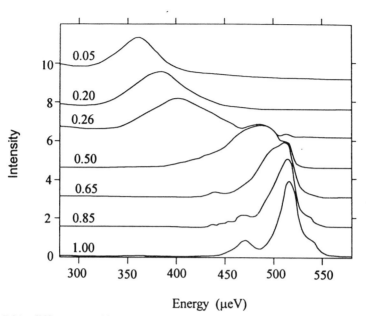

Figure 7.26. INS spectra of isotopic mixtures of 4-methylpyridine and its perdeuterated derivative at 2.5 K. (From Fillaux and Carlile [1990].)

TABLE 7.3
Soliton and Breather Modes in 4-Methylpyridine Crystals:
$V_0 = 29.5\ \text{cm}^{-1}$, $g_{12}^{(-)} = 44\ \text{cm}^{-1}$

	4 MP-h_7	4 MP-d_7
$\hbar\omega_0$ (cm^{-1})	27.27	18.79
$\hbar\omega_c$ (cm^{-1})	33.30	22.95
ω_c/ω_0	1.22	1.22
$E_s^{(0)}$ (cm^{-1})	144.1	144.1
$M_k c_0^2$ (cm^{-1})	92.5	92.5
M_k (amu)	6.89	13.78
$E_\beta(1)$ (cm^{-1})	142.16	142.16
$M_\beta(1)$ (amu)	10.59	21.18
$\hbar\omega_\beta$ (cm^{-1})	17.45	12.03
Amplitude of breather envelope (rad)	1.17	1.17

whose boundaries act as reflecting walls. The energy of the fundamental state and transition frequencies of the breather decrease when the cluster size decreases; this results in the low-frequency shift of the main peaks in the INS spectra of isotopomer mixtures (see Figure 7.26). The statistical distribution of $4\,MP\text{-}h_7$ and $4\,MP\text{-}d_7$ in the bulk leads to inhomogeneous broadening of these peaks, according to the mechanism described previously for solid methane at finite temperatures.

VIBRATION–ROTATION TUNNELING SPECTROSCOPY OF MOLECULES AND DIMERS

CONTENTS

This chapter is devoted to tunneling effects observed in vibration–rotation spectra of isolated molecules and dimers. The relative simplicity of these systems permits one to treat them in terms of multidimensional PES's and even to construct these PES's by using the spectroscopic data. Modern experimental techniques permit the study of these simple systems at superlow temperatures where tunneling prevails over thermal activation. The presence of large-amplitude anharmonic motions in these systems, associated with weak (e.g., van der Waals) forces, requires the full power of quantitative multidimensional tunneling theory.

8.1. SUPERSONIC COOLING

To successfully use high-resolution molecular spectroscopy to study tunneling, two conditions have to be met: suppression of hot bands and removal of inhomogeneous broadening. In the traditional technique of equilibrium sample preparation these conditions are mutually exclusive: To decrease the hot band intensity one needs to lower the temperature, which entails the condensation of a sample and, consequently, appearance of inhomogeneous spectral effects which are due to intermolecular interactions in the solid. To some extent, a compromise is achieved in the matrix isolation method, where the intermolecular interactions between the guest and host molecules are minimized by using the noble gas matrix. However, even in this case the asymmetry of the potential is usually comparable to or greater than the tunneling splitting (see, for example, Sections 6.2 and 6.3). Only some special methods, such as hole burning, permit further reduction of inhomogeneous broadening in solids.

A universal method that satisfies both requirements is supersonic cooling. For this reason, this method is a cornerstone of modern low-temperature studies. The results described in this chapter are mainly obtained with this method. For special reviews devoted to applications of supersonic cooling in molecular spectroscopy we refer the reader to the papers by Levy [1980] and Cohen and Saykally [1992]. Technical details of the method can be found in the papers we cite in this chapter.

Supersonic isoentropic expansion of a gas stream leads to its cooling, as follows from the well-known thermodynamic relations

$$\frac{T}{T_0} = \left(\frac{P}{P_0}\right)^{1-1/\gamma} = \left(1 + \frac{\gamma - 1}{2} M^2\right)^{-1} \qquad (8.1)$$

where P_0 and T_0 are the initial pressure and temperature in reservoir, $\gamma = C_p/C_v$ is the ratio of heat capacities, and M is the Mach number, which depends on the nozzle diameter D and the downstream distance from the nozzle x:

$$M = A\left(\frac{x}{D}\right)^{\gamma - 1} \qquad (8.2)$$

where A is a constant equal to 3.26 for a monatomic gas. Formulae (8.1) and (8.2) are valid when collisions provide the cooling. In the collision-less regime, the values M and T become constant and $M \sim 133(P_0 D)^{0.4}$. According to (8.1), an expansion from 10 atm to 10^{-3}–10^{-4} torr results in cooling of translational degrees of freedom from 300 to $\sim$0.1 K; the Mach number is $\sim 10^2$. When molecules of interest are seeded into an expanding carrier gas, the cold translational bath acts as a refrigerant for the other degrees of freedom in the post-nozzle region. A peculiarity of supersonic expansion that distinguishes it from stationary refrigerants is that seeded molecules are in contact with the bath only during a finite time. As the expansion proceeds, the density of the carrier gas decreases, so eventually it is unable to provide the number of collisions needed to achieve equilibrium. This means that in the low-density collisionless regime the system is frozen in a highly nonequilibrium state. Those degrees of freedom that equilibrate most rapidly with the translational bath are the coldest. Because condensation is a much slower process than rotational or even vibrational relaxation, effective internal cooling is achieved before condensation. Therefore, the use of supersonic expansion permits the preparation of internally cold isolated gas-phase molecules. The relaxation rates increase when the mass of the molecules of the carrier increases. Measurements of hot band intensities show that rotational temperature of seeded molecules are close to the corresponding translational temperatures; in many cases they do not exceed 5–8 K. This rotational cooling is sufficient to suppress the hot bands connected with population of higher rotational levels. The vibrational temperature is usually less than 30–50 K, and thermal excitation of low-frequency vibrations is effectively overdamped (see, for example, Amirav et al. [1980]).

Deep cooling of translational degrees of freedom leads to the formation of dimers between seeded molecules as well as between the seeded molecules and the atoms of the carrier gas. The bond energy in these dimers usually does not exceed 1–3 kcal/mol, so that they dissociate at higher temperatures.

8.2. INVERSION SPLITTING

Inversion splitting in the ammonia molecule is one of the best-known examples of spectroscopical manifestation of tunneling. In this molecule, a planar transition-state structure with energy of 5.94 kcal/mol separates two equilibrium pyramid-shaped mirror symmetric configurations. The inversion splitting is

$$\nu_{inv} = \nu_5 \exp\left(-\frac{1}{\hbar} S\right) \qquad (8.3)$$

where the frequency of the umbrella vibration ν_5 is 950 cm^{-1} for NH_3 and 745 cm^{-1} for ND_3. For NH_3, ν_{inv} equals 0.8 cm^{-1} in the ground state and 36.5 cm^{-1} in the first vibrationally excited state (see, for example, Townes and Shawlow [1955]). The inversion may be described as tunneling of reduced mass $m = 3m_H(m_N + 3m_H \sin^2\alpha)/(m_N + 3m_H) = 2.54$ amu (where α is the equilibrium pyramid angle 21.8°) through the distance 0.77 Å, which is equal to the relative displacement of the N atom and H_3 plane between the two potential minima. The inversion splitting in ND_3, where the tunneling mass is 4.45 amu, is equal to 0.053 cm^{-1} for $n = 0$ and 3.9 cm^{-1} for $n = 1$. The spectrum of rotational inversion levels of NH_3 is shown in Figure 8.1. The allowed rotational transitions satisfy selection rules for a symmetric top $\Delta J = 0, \pm 1, \Delta K = 0, \langle +|\leftrightarrow|-\rangle$ and $\Delta J \neq 0$ for $K = 0$. The influence of rotation on the inversion spectrum is similar to the effect of either symmetric or antisymmetric vibrations (see Section 2.5), depending on the direction of the rotation axis. Both effects are caused by change of potential due to centrifugal forces. Rotation

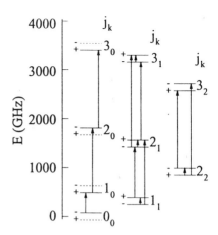

Figure 8.1. Rotation inversion energy levels and allowed transitions of the free NH$_3$ molecule. The selection rules are $\Delta J = 0, \pm 1$, $\Delta K = 0$ $(\Delta J \neq 0$ when $K = 0)$, $\langle + | \leftrightarrow | - \rangle$. The $K = 0$ levels that are forbidden by the Pauli principle are shown with dotted lines. The lower A and E states are $J = 0$, $K = 0$ and $J = 1$, $K = 1$, respectively.

about the symmetry axis flattens the pyramid and reduces the barrier for tunneling, while rotation around the perpendicular axis increases the barrier.

Inversion of ND$_3$ is tunneling of a heavy mass $(m \gg m_H)$ through a barrier whose height is typical of radical chemical reactions. As argued in Chapter 2, the probability of $\exp(-2S/\hbar)$ rather than $\exp(-S/\hbar)$ an incoherent transition is proportional to $\exp(-2S/\hbar)$ rather than $\exp(-S/\hbar)$ as in (8.3), so that the rate constant of the tunneling reaction associated with the transfer of the same reduced mass through the same barrier would be $\sim \Delta_{\mathrm{inv}}^2 / \nu_5 \simeq 10^5\,\mathrm{s}^{-1}$. For this reason it is natural that in cryochemical reactions, whose measured rate constants are 5–10 orders of magnitude smaller than $\Delta_{\mathrm{inv}}^2 / \nu_5$, it is possible to observe tunneling of reduced masses as large as 15–20 amu.

For NH$_3$-like molecules, which exhibit only one large amplitude motion (i.e., tunneling), the PES can be constructed using the spectroscopic data. Such a procedure is well known for rigid molecules, but in the present case additional assumptions about the coupling of tunneling to other degrees of freedom are needed. Bunker and Landberg [1977] introduced a semirigid Hamiltonian based on the assumptions that (1) the frequencies ω_r and anharmonicity constants $X_{r'r}$ of small-amplitude vibrations depend on the tunneling coordinate Q and (2) the potential $V(Q, \{n_r\})$ depends on the quantum numbers of nontunneling modes n_r. The potential function

$$V(Q, \{n_r\}) = V(Q) + \sum_r \hbar \omega_r(Q)(n_r + \tfrac{1}{2}) \sum_{r,r'} \hbar X_{rr'}(Q)(n_r + \tfrac{1}{2})(n_{r'} + \tfrac{1}{2})$$

$$(8.4)$$

is nothing but the vibrationally adiabatic potential. Expanding $\omega_r(Q)$ and $X_{rr'}(Q)$ in a power series in Q allows one to obtain information about the PES in the vicinity of the minimum energy path. Bunker [1983] presented a comprehensive review in this field. Using this method Spirko et al. [1976] calculated the PES of NH_3 molecule.

Steckler and Truhlar [1990] studied the inversion of NH_3 as a good example to examine two numerical methods to construct a reaction path: the intrinsic reaction coordinate (IRC) and MEP. It should be noted that the term MEP is interpreted differently by different authors. The interpretation adopted in this book can be stated loosely by considering the MEP to follow a single *reaction coordinate* while allowing all other transverse degrees of freedom to relax to their equilibrium positions (see Chapter 4 for formal definitions). Such a definition is, however, not invariant with respect to transformations to other coordinate systems. In fact, we have not been differentiating between a number of similar paths, IRC and MEP among them. The reason for this is intuitively obvious: To the extent that the vibrationally adiabatic approximation is valid, all of them represent the motion along the bottom of a narrow channel, so they coincide when they are applicable to calculation of a tunneling rate. Not all of them are equally feasible numerically for a realistic PES, though, so the question of defining a good reaction path is nevertheless worthwhile. When the vibrationally adiabatic approximation breaks down, none of these artificially constructed paths coincide with the actual optimum tunneling path, although one of them may turn out to be better than the others.

The methods of constructing different reaction paths are described in numerous papers and reviews (see, for example, Truhlar and Garrett [1984, 1987], Garrett et al. [1988], Ischtwan and Collins [1988], and references therein). In the IRC method proposed by Fukui [1970], the steepest descent path from the saddle point of a multidimensional PES $V(\dot{\mathbf{X}})$ to the reactant and product valleys is found by numerically solving the equation

$$\frac{d\mathbf{X}}{dt} = -\frac{\nabla V(\mathbf{X})}{|\nabla V(\mathbf{X})|} \tag{8.5}$$

where $\mathbf{X}$ is the N-dimensional position vector in mass-weighted coordinates, ∇ is the N-dimensional gradient operator, and t is a parameter. The IRC is defined as the length of reaction path arc:

$$ds^2 = \sum_{k=1}^{N} (dx_k)^2 \tag{8.6}$$

The PES in the vicinity of IRC is approximated by an $(N-1)$-dimensional parabolic valley, whose parameters are determined by using the gradient method. Specific numerical schemes taking into account p previous steps to determine the $(p+1)$th step render the Euler method stable and allow one to optimize the integration step in Eq. (8.5) [Schmidt et al., 1985]. When the IRC is found, the changes of transverse normal vibration frequencies along this reaction path are represented as

$$\omega_{\perp,r}(\lambda) = \omega_{\perp,r}^{\#} + \alpha_{1r}(\lambda - \lambda^{\#}) + \alpha_{2r}(\lambda - \lambda^{\#})^2 + \cdots \qquad (8.7)$$

In the case of inversion, the natural coordinate λ is the angle between the N–H bond and the C_3 axis. In addition, there are three transverse vibrations, two of which are degenerate. The one-dimensional potential, bond lengths, and frequencies along the IRC calculated by Steckler and Truhlar [1990] are listed in Table 8.1. The transverse vibrations have high frequencies compared to ν_5. This fact supports the initial assumption about vibrational adiabaticity of the problem.

The asymmetry of the potential caused by intermolecular interaction suppresses inversion in complexes of NH_3 with various molecules (see the review by Nelson et al. [1987]). An exception is the Ar–NH_3 complex, which has been studied by Fraser et al. [1985], Gwo [1990], and Schmuttenmaer [1991]. In this complex, the line connecting the argon atom with the center of mass is perpendicular to the C_3 axis. The distance between them is 3.83 Å, and the resulting change in the inversion potential is small. The ammonia inversion frequency is also suppressed in solid matrices of noble gasses [Abouaf-Marguin et al., 1977; Girardet et al., 1984] for similar reasons. According to the data of Girardet and Lakhlifi [1985, 1989], inversion splitting in a NH_3 guest molecule in a

TABLE 8.1

Potential, N–H Bond Length, and Inverse Vibration Frequencies as Functions of Intrinsic Reaction Coordinate for Inversion of the Ammonia Molecule

λ (deg)	$V(\lambda)$ (kcal/mol)	R_{N-H} (Å)	ν_1 (A_1) (cm^{-1})	ν_2 (E) (cm^{-1})	ν_4 (E) (cm^{-1})
0	6.21	0.994	3752	3980	1642
1.25	6.16	0.994	3751	3978	1643
2.50	6.05	0.994	3749	3977	1644
3.75	5.85	0.994	3745	3971	1647
5.00	5.58	0.995	3738	3964	1650
10.00	3.92	0.998	3704	3922	1671
15.00	1.85	1.002	3653	3856	1697
Equilibrium	0.00	1.012	3564	3721	1728

nitrogen crystal at 5.5 K is $0.017\,\mathrm{cm}^{-1}$ for the ground state and $1.40\,\mathrm{cm}^{-1}$ for the first vibrationally excited state.

Inversion splittings can be observed for the formaldehyde molecule in its electronically excited $\tilde{A}^1A_2$ and $\tilde{a}^3A_2$ states because the molecule adopts a nonplanar equilibrium structure. In the former, the angle between the C–O bond and CH_2 plane is 34°, and the barrier height for inversion is $350\,\mathrm{cm}^{-1}$ ($1.00\,\mathrm{kcal/mol}$), while in the latter this angle increases to as much as 41° and the barrier height, correspondingly, increases to $775\,\mathrm{cm}^{-1}$. The potentials obtained by Jensen and Bunker [1982] are represented in Figure 8.2 (see also the review by Bunker [1983]). Inversion splitting in the triplet state $\tilde{a}^3A_2$ is 36 and $\sim10\,\mathrm{cm}^{-1}$ for H_2CO and D_2CO, respectively.

In the HCNO molecule the HCN bending vibration has a large amplitude, and the barrier height depends on the quantum numbers n_1 and n_2 of the CH and CN stretching modes. When $n_1 = n_2 = 0$, the barrier is $11.5\,\mathrm{cm}^{-1}$, while for $n_1 = 1$, $n_2 = 0$ and $n_1 = 0$, $n_2 = 1$, it is 41 and $35\,\mathrm{cm}^{-1}$, respectively. The potential found by Bunker et al. [1979] shows that the equilibrium (classical) configuration is linear, but the zero-point energies of aforementioned vibrations, whose frequencies decrease under bending, lead to a bent configuration.

Inversion doubling has been observed in microwave spectrum of methylamine CH_3NH_2. This splitting depends on the quantum numbers of rotation and torsion vibrations [Shimoda et al., 1954; Lide, 1957; Tsuboi et al., 1964]. Inversion of NH_2 alone leads to the eclipsed configuration corresponding to the maximum barrier for torsion. Thus, the transition between equilibrium configurations involves simultaneous NH_2 inversion and internal rotation of CH_3; that is, inversion appears to be strongly coupled with internal rotation. The inversion splits each rotation–vibration (n, k) level into a doublet, whose components, in turn, are split into three levels with $m = 0$, ±1 by internal rotation of the

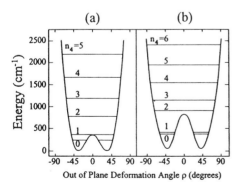

Figure 8.2. Vibrationally adiabatic potential for formaldehyde conversion in excited electronic (a) $\tilde{A}'A_2$ and (b) $\tilde{a}^3A_2$ states. The levels of the ν_4 vibration are indicated. (From Jensen and Bunker [1982].)

methyl group. Therefore, instead of a single line of $1 \leftarrow 0$ wagging vibration transition, six lines are observed in accordance with the selection rule for nuclear spin functions of protons in the CH_3 and NH_2 groups ($\Delta M = 0$ and $s \leftrightarrow s$, $a \leftrightarrow a$, respectively). Splitting in the ground vibrational state, as a result of tunneling rotation, is less than $0.2\,cm^{-1}$. The inversion splitting in the excited $n = 1$ state is $6.8\,cm^{-1}$, which is considerably smaller than that in NH_3 due to the strong coupling between inversion and internal rotation. The potential of internal rotation in CH_3NH_2 depends on the relative positions of protons of both groups and can be described by

$$V_r = \tfrac{1}{4}V(\lambda)[2 - \cos 3(\phi_1 - \phi_a) - \cos 3(\phi_1 - \phi_b)] \qquad (8.8)$$

where ϕ_1, ϕ_a, and ϕ_b are the angles of CH_3, H_a, and H_b with respect to the z axis and λ is the inversion coordinate (Figure 8.3a). By introducing the angles

$$\rho = \tfrac{1}{2}(\phi_a - \phi_b) \qquad \theta = \phi_1 - \tfrac{1}{2}(\phi_a + \phi_b) \qquad (8.9)$$

relation (8.8) becomes

$$V_r(\theta, \lambda) = \tfrac{1}{2}V(\lambda)\{1 + \cos[3\rho(\lambda)] \cos 3\theta\} \qquad (8.10)$$

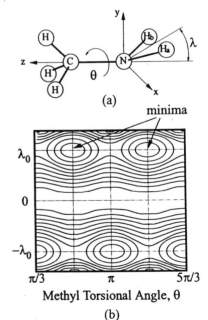

Figure 8.3. (a) Coordinate system for methyl-amine molecule. For simplicity the CN bond has been drawn along the symmetry axis of the CH_3 group, although it actually lies about $3°$ away from this axis. (b) The schematic contour map of the two-dimensional PES for internal motions in methylamine. (From Tsuboi et al. [1964].)

where θ is the angle between C–H bond and yz plane. The angle ρ is related to λ by

$$\cos \rho = \sin \lambda (\sin^2\lambda + \tan^2\lambda_0)^{-1/2} \tag{8.11}$$

In the equilibrium configuration $\lambda = \lambda_0 = \pm 68.2°$ and $\cos \rho = \pm 1$. Formula (8.10) demonstrates that inversion ($\lambda = \lambda_0 \to -\lambda_0$) shifts the equilibrium value of θ from 0 to 60°. The rotation barrier appearing in (8.10) is an even function of ρ and has a minimum at the transition state. Tsuboi et al. [1964] assumed that the rotation barrier vanishes when $\lambda = 0$ and represented $V(\lambda)$ as

$$V(\lambda) = V_0\left(\frac{\lambda}{\lambda_0}\right)^2 \tag{8.12}$$

The two-dimensional potential of inversion and internal rotation may be modeled by

$$V(\theta, \lambda) = \frac{V_0}{2}\left(\frac{\lambda}{\lambda_0}\right)^2\{1 + \cos[3\rho(\lambda)]\cos\theta\} + V_1\left[\left(\frac{\lambda}{\lambda_0}\right)^2 - 1\right]^2 \tag{8.13}$$

where the second term represents a double-well potential for inversion, which differs only slightly in CH_3NH_2 and NH_3. The contour plot of potential (8.13) is shown in Figure 8.3b. Since conversion can occur when CH_3 rotates through $\pm 60°$, the MEP is doubly degenerate. We have already encountered such a situation in the problem of two-proton transfer, described by the model potential (6.37). In the present case the optimum tunneling trajectory is two-dimensional over all values of the parameters and can be found by using the instanton analysis, as explained earlier.

An example of inversion in a large molecule is the inversion of 9,10-dihydroanthracene in its excited singlet state. This is observed in the fine structure of its spectrum in a supercooled jet. In the excited state the molecule bends out of plane with respect to the short axis, and its symmetry decreases from D_{2d} to C_{2v}, which allows the 00 transition. The fluorescence excitation spectrum measured by Chakraborthy and Choudhury [1990] contains a progression of low-frequency bands turning into diffuse spectrum when energy exceeds the energy of 00 transition by ~ 150 cm^{-1}. The existence of this converging progression indicates that the surface of the excited state is a double well and there is an inversion between these two equilibrium configurations (see Figure 8.4). These

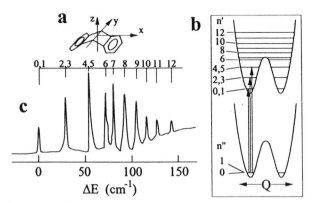

Figure 8.4. Inversion of 9,10-dihydroanthracene in the excited singlet state. (a) The equilibrium bent shape of excited molecule, (b) the double-well potential for inversion, and (c) the fluorescence excitation spectrum of the jet-cooled molecule. (From Chakraborthy and Choudhury [1990].)

authors approximated the conversion potential as

$$V(Q) = \frac{A_1}{2} Q^2 + A_2 \exp(-a^2 Q^2) \qquad (8.14)$$

The parameters in Eq. (8.14) were found by comparing its spectrum with the experimental one: $A_1 = 3.49 \times 10^{25} \text{ s}^{-2}$, $A_2 = 2.48 \times 10^{-13}$ erg, $a = 1.05 \times 10^{19} \text{ g}^{-1/2} \text{ cm}^{-1/2}$. These values correspond to an inversion barrier height of 94 cm^{-1} and an equilibrium mass-weighted coordinate $Q_0 = \pm a^{-1}[\ln(2a^2 A_2/A_1)]^{1/2}$, which is equal to $\pm 6.38 \times 10^{-20}$ cm g$^{1/2}$, i.e. ± 9.36 a.u. If it is assumed that the bending angle of the molecule is 16°, then $2Q_0$ relates to the displacement of a reduced mass equal to the mass of the benzene ring over 1.1 Å. The frequencies of the intrawell and upside-down barrier vibrations are

$$\omega_0 = \left(2A_1 \ln \frac{2a_2 A_2}{A_1}\right)^{1/2} \qquad \omega^* = (2a^2 A_2 - A_1)^{1/2} \qquad (8.15)$$

and equal 28 and 24 cm^{-1}, respectively. Tunneling splitting in the ground vibration state is estimated by

$$\Delta_0 = \frac{\hbar \omega_0}{\pi} \exp\left[-\frac{\pi}{\hbar \omega_\#}\left(V_0 - \frac{\hbar \omega_0}{2}\right)\right] \qquad (8.16)$$

which is the usual semiclassical formula (2.20) for a parabolic barrier.

This splitting is $1.3 \times 10^{-3} \, \text{cm}^{-1}$. Inversion splitting of the third vibrational level was measured to be $\sim 6 \, \text{cm}^{-1}$.

The diphenyl molecule in its ground state is nonplanar; the dihedral angle between benzene rings planes is $42 \pm 2°$. Im and Bernstein [1988] determined the potential of internal hindered rotation of the benzene rings with respect to each other by analyzing the IR spectrum in the same manner as described for 9,10-dihydroanthracene. The barrier height was calculated to be $122 \, \text{cm}^{-1}$, $\omega_0 = 50 \, \text{cm}^{-1}$. Both examples demonstrate the possibility of coherent tunneling of a reduced mass that far exceeds the mass of a hydrogen atom. The value of $\omega_{\#}$ mentioned above is associated with a crossover temperature equal to 12 K. Note that the inversion splittings associated with the coherent regime of inversion transitions is observed only because of the low vibrational and rotational temperatures that are achieved by supersonic cooling.

Tunneling rocking motion in vinyl radical

$$
\begin{array}{c}
\text{H}_s \\

\end{array}
\text{C}{=}\text{C} \rightleftharpoons \text{C}{=}\text{C} \qquad (8.17)
$$

is usually considered an inversion of nonplanar structure. The barrier height corresponding to the planar transition state with C_{2h} symmetry is 4.4–6.2 kcal/mol, according to the calculations of Paddon-Row and Pople [1985]. However, the high-resolution IR spectra obtained by Kanamori et al. [1990] have shown that the radical is planar and thus the inversion is equivalent to the rotation around a C_2 axis perpendicular to the radical plane. The measured splitting $\Delta = 0.05 \, \text{cm}^{-1}$ corresponds to a barrier no higher than 4 kcal/mol. From the relationship between the rotational constants for symmetric and antisymmetric states of the tunneling doublet it follows that the barrier height is more likely to be ~ 3.3 kcal/mol.

Deycard et al. [1987, 1988, 1990] studied the conversion of nonplanar three-member ring radicals using electron paramagnetic resonance. These data demonstrate a rich variety of tunneling regimes. The inversion of the cyclopropyl radical is associated with α-hydrogen transfer between equilibrium positions at which the C_α–H_α bond forms a $\pm 39.3°$ angle with the ring plane. In the transition state, the radical becomes planar without considerable deformation of the molecular skeleton. The frequency of coherent tunneling exceeds $10^{11} \, \text{s}^{-1}$ at 42 K. The barrier height for inversion is equal to 3 kcal/mol.

The inversion of the methylcyclopropyl radical, characterized by the same barrier height, becomes thermally activated above 90 K, and its rate constant increases from 3×10^5 to 10^9 Hz in the range 90–170 K. Quantum chemical calculations by Zebretto et al. [1989] demonstrate that the inversion is accompanied by a relative displacement of the ring and the methyl group, situated in its equilibrium configuration at the angle $\pm 39°$ to the ring plane.The geometry of initial and transition states is shown in Figure 8.5. Inversion is possible under $60°$ rotation of the CH_3 group. Strong coupling of hindered rotation with the out-of-plane bending vibration leads to a two-dimensional tunneling path. The frequencies of the bending vibration ($293\,cm^{-1}$) and upside-down barrier ($1298\,cm^{-1}$) exceed the torsion frequency, which is $183\,cm^{-1}$ in the ground state and $94\,cm^{-1}$ in transition state. It was suggested that the transition rate is limited by thermally activated surmounting of the inversion barrier when the rotation is fast. When the temperature decreases, the hindering of rotation starts to play a significant role and reduces the overall transition rate. From structures shown in Figure 8.5, the CH_3 group in the transition state is turned by $30°$, so that the MEP represents concerted inversion and rotation.

In the oxyranyl radical, the barrier for inversion

$$\underset{H_a}{\overset{H_s}{>}}C_\beta-C_\alpha\underset{H_\alpha}{\overset{O}{<}} \quad \rightleftarrows \quad \underset{H_s}{\overset{H_a}{>}}C_\beta-C_\alpha\underset{O}{\overset{H_\alpha}{<}} \qquad (8.18)$$

is larger, $9.0\,kcal/mol$. The inversion frequency determined from hyperfine structure of the EPR spectrum is $7\,MHz$. A temperature-independent transition rate is observed below $140\,K$, while at higher temperatures thermally activated hopping is predominant. The temperature dependence of the rate constant was demonstrated in Figure 1.1 (curve 4) as an example of the transition from thermal activation to a low-temperature plateau. In the equilibrium configuration of the radical (8.18), the α-hydrogen atom is situated at the angle $\pm 45.1°$ with respect to the ring plane. Inversion proceeds along the coordinate of the bending

Figure 8.5. The methylcyclopropyl radical in various conformations. (a) The nonplanar-bisected conformation is stable. (b) The non-planar-straddled conformation corresponds to methyl group rotation by $60°$. (c) The planar eclipsed configuration is the transition state. (d) The planar-bisected structure corresponds to conversion without rotation. (From Zebretto et al. [1989].)

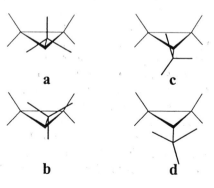

vibration $C_\beta C_\alpha H_\alpha$; this vibration has frequency $883\,\text{cm}^{-1}$. Tunneling involves not only the motion of the hydrogen in the plane perpendicular to the ring plane, but also deformation of the ring itself and stretching of the $C_\alpha H_\alpha$ bond. Because these motions are characterized by frequencies that are high compared to the frequency of the bending vibration, the transition turns out to be nearly adiabatic. There are no low-frequency vibrations that could modulate the barrier. The vibrationally adiabatic potential calculated by analyzing normal vibrations in the ground and transition states is in good agreement with the experimentally observed temperature dependence of the rate constant. The crossover temperature calculated using the one-dimensional model agrees with experiment.

Inversion of the dioxanyl radical

$$\text{(8.19)}$$

is associated with a transition between the equilibrium configurations with symmetry C_1 through the transition state with symmetry C_2, in which the H_α and C_α atoms are situated in a straight line with the center of mass of the ring. In the ground state the CH_α bond is directed at an angle $\pm 42°$ with respect to the $OC_\alpha O$ plane. The barrier height is $7.3\,\text{kcal/mol}$. The H_α and C_α atoms are displaced in opposite directions in the process of inversion. Since the inversion coordinate interacts strongly with the out-of-plane bending vibration of the ring, the tunneling path is two dimensional. The bending vibrational frequencies are 163 and $206\,\text{cm}^{-1}$ in the ground state and 168 and $244\,\text{cm}^{-1}$ in the transition state. They are considerably smaller than the frequency of the vibration along the inversion coordinate ($973\,\text{cm}^{-1}$), so applicability of the sudden approximation may be suggested in this case. The symmetrically coupled low-frequency bending vibrations promote the tunneling. Thermally activated inversion takes place above $120\,\text{K}$. The low-temperature limit of the inversion frequencies are 54 and $1.1\,\text{MHz}$ for H_α and D_α radicals, respectively. A strong coupling of the inversion to the bending vibration is seen from the fact that the mass-weighted displacements along the coordinates of the inversion and out-of-plane deformation in the transition state ($0.620\,\text{a.u.}$ for inversion, 0.173 and $0.155\,\text{a.u.}$ for two bending vibrations) are comparable.

8.3. PSEUDOROTATION

The internal steric strains inherent to planar rings of cyclic organic saturated molecules disturb the planarity of the ring and create a set of isoergetic configurations whose energies are lower than the energy of the planar structure. The transitions between these stable configurations, puckering, occur because of the change in dihedral torsional angles when the changes in the lengths of chemical bonds and in the angles between them are negligible, so that the barriers for puckering are comparatively small. For this reason, the frequencies of puckering are available for spectroscopic measurements despite large amplitude displacements of some heavy groups in these transitions. Many organic molecules, including many biologically important species, have sets of stereoisomers formed due to the steric strains, because puckering is a rather common phenomenon. Puckering is described in detail in reviews by Legen [1980], Strauss [1983], and Laane [1987]. Kilpatrick et al. [1947] proposed a special mechanism of puckering, called pseudorotation. The dynamics of out-of-plane displacements of the carbon atoms in an N-member ring is characterized by $N - 3$ normal vibrations. In particular, the cyclopentane molecule has two such modes, symmetric and antisymmetric, with the mass-weighted displacement of the kth atom along the C_5 axis equal to

$$Z_k^a = Z_0 \cos\left[\frac{4\pi}{5}(k-1)\right] \qquad Z_k^s = Z_0 \sin\left[\frac{4\pi}{5}(k-1)\right] \qquad k = 1, \ldots, 5$$

(8.20)

There are two types of equilibrium structures: envelope (E) and twist (T), with symmetries C_2 and C_s, respectively, shown in Figure 8.6. The conformation transition, puckering, is determined by a linear combination of Z_k^a and Z_k^s, and is described by a continuous variation of amplitude and phase

$$Z_k = \frac{\sqrt{2}}{5} q \cos\left[\phi + \frac{2\pi}{5}(k-1)\right]$$

(8.21)

where the factor $\sqrt{2}/5$ comes from the condition $\sum_1^5 Z_k^2 = q^2$. From (8.21)

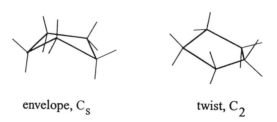

envelope, C_s twist, C_2

Figure 8.6. Two conformers of cyclopentane. (From Strauss [1983].)

it follows that there are ten E and T structures in the range from 0 to 2π corresponding to $\phi = 0°$, $36°, \ldots$ and $\phi = 18°$, $54°, \ldots$, respectively. Taking into account that the planar structure has at least axial symmetry, the potential can be expressed in terms of the generalized puckering coordinates, q and ϕ, as

$$V(q, \phi) = V_0(q) + V_r(q, \phi) \tag{8.22}$$

$$V_0(q) = V_{02}q^2 + V_{04}q^4$$

$$V_r(q, \phi) = (V_{22}q^2 + V_{24}q^4)\cos 2\phi + V_{44}q^4 \cos 4\phi \tag{8.23}$$

Using rectangular coordinates, $q_1 = q \cos \phi$, $q_2 = q \sin \phi$, this function becomes

$$V(q_1, q_2) = a_{11}q_1^2 + a_{12}q_1^4 + a_{21}q_2^2 + a_{22}q_2^4 + c_{12}q_1^2q_2^2 \tag{8.24}$$

From (8.23) and (8.24) one can see two special cases when the potential becomes separable. In the first case $c_{12} = 0$, we have two independent anharmonic modes, each having two equilibrium positions. In the second case, the angular part of the potential (8.23) V_r is zero, and the motion breaks up into radial vibration in the double well $V_0(q)$ and a free rotation, i.e. propagation of the waves of transverse displacements along the ring. The latter case is called free pseudorotation. Since the displacements of atomic groups in the wave are purely transverse, they do not contribute to the total angular momentum.

To introduce the generalized coordinates q and ϕ (or q_1 and q_2), one needs first to define the reference plane. Cremer and Pople [1975b] and Cremer [1990] showed that this plane can be defined unambiguously by the conditions of the constant position of the center of mass and the absence of total angular momentum. Instead of displacements Z_k, the torsional angles of each fragment θ_k can be used as variables, which are expressed as functions of q_1 and q_2. The PES for free pseudorotation is a ring valley with symmetry $C_{\infty v}$. The eigenfunctions are products of free rotor and anharmonic oscillator eigenfunctions

$$\Psi_{mn}(q, \phi) = \varphi_n(q) \exp(im\phi) \tag{8.25}$$

The eigenvalues of pseudorotation for various vibrational states $n = 0, 1, \ldots, E_{mn} = B_n m^2$, are characterized by rotational constants:

$$B_n = \frac{\hbar^2}{2\mu \langle q^2 \rangle_n} \tag{8.26}$$

where $\mu \langle q^2 \rangle_n$ is the reduced moment of inertia of the pseudorotor, and $\langle q^2 \rangle_n$ is the mean square amplitude of pseudorotation in the nth vibrational state. For cycloalkane puckering ($\mu = 14$ amu, $q_0 = 0.47$ Å) $B_0 = 3.25$ cm^{-1}. Due to small values of the rotational constant, the levels of pseudorotation make a large contribution to the vibrational entropy. Based on this fact, Kilpatrick et al. [1947] explained the excess entropy of cyclopentane molecule.

When the double-well potentials for both coordinates in (8.24) are slightly different and $c_{12} \simeq 2(a_{12}a_{21})^{1/2}$, the pseudorotation is hindered, but the barrier along ϕ is smaller than the radial barrier. The reaction path is inserted in a narrow ring valley with $q = q_0$. The maxima and minima of hindered pseudorotation potential correspond to transition states and stable ring configurations, for example, to the E and T types. Other types of PES are realized for different interrelations between the coefficients in (8.24) [Harris et al., 1969]. These authors considered the case when the contribution of higher harmonics to $V_r(q, \phi)$ is small, as for usual rotation, $V_{44}q_0^4 \ll V_{22}q_0^2 + V_{24}q_0^4$. Since $V(q, \phi)$ should increase at $q \to \pm\infty$, V_{04} and V_{24} are positive. The potential is a double well for all ϕ if $V_{02} < 0$, $V_{02} - 2V_{22} < 0$. This case corresponds to pseudorotation. Harris et al. [1969], Engelholm et al. [1969], and Kim and Gwinn [1969] reconstructed PES's of this type for oxalane, 1,3-dioxalane, and 1,1-difluorocyclopentane using microwave spectroscopy data. These are shown in Figures 8.7b and c. When $V_{22}/V_{02} < 0$, the potential remains a double well only for a limited range of ϕ corresponding to either

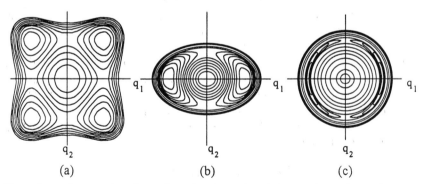

(a)   (b)   (c)

Figure 8.7. Contour plot of two-dimensional PES (8.24) for pseudorotation and conversion. (a) Hypothetical molecule with two independent double-well vibrations, $V = -550.0q_1^2 - 400.0q_2^2 + 56.6q_1^4 + 30.2q_2^4$. The contours are at 200-cm^{-1} intervals. (b) $V = -575.5q_1^2 - 430.0q_2^2 - (5.6q_1^2 - 7.1q_2^2)^2$. The contours are at 150-cm^{-1} intervals. (c) $V = -478.3q_1^2 - 470.0q_2^2 - (6.83q_1^2 + 6.83q_2^2)^2$, where $V(q_1, q_2)$ is given in cm^{-1}. The first five contours along the minima are spaced at 10 cm^{-1}, while the remaining contours are at 150-cm^{-1} intervals. (From Harris et al. [1969].)

transition or stable configuration. In the last case pseudorotation disappears, being replaced by inversion via the planar transition state $q = 0$. Assuming the limit of a high barrier for pseudorotation, we arrive at the problem of two-dimensional tunneling considered in Chapter 4. It is really the case of a tunneling coordinate coupled to only one vibration.

When the coefficient V_{44} is not small, $4V_{44}q_0^4 > V_{22}q_0^2 + V_{24}q_0^4$, the PES has four minima and four saddle points (Figure 8.7a). For positive V_{44}, the minima are equivalent and correspond to the configurations with symmetry C_1, while T and E structures are the transition states. The minima are situated at the pseudorotation angles ϕ_0, $-\phi_0$, $\pi - \phi_0$, $\pi + \phi_0$, where $\phi_0 = \frac{1}{2}\cos^{-1}(-V_2/V_4)$, $V_2 = V_{22}q_0^2 + V_{24}q_0^4$, $V_4 = V_{44}q_0^4$. For negative V_{44} the PES has two global and two local minima corresponding to T and E structures, the corresponding energies differ by V_2. The local minima turn into transition states (saddle points) when V_{44} is varied. At low temperatures ($\beta V_2 \gg 1$) the pseudorotation spectrum contains only the bands assigned to the structure with lower energy; the hot bands of the other structure have low intensity. A PES with $V_4 > 0$ was found to correspond to hindered pseudorotation in 1,3-dioxalane [Baron and Harris, 1974]. The case $V_4 < 0$ is realized for chlorocyclopentane, whose microwave spectrum relates only to the E conformer because the energy of the T structure is 0.8 kcal/mol higher [Loyd et al., 1978]. Choe and Harmony [1980] observed both structures in cyanocyclopentane, where the energy difference does not exceed 0.2 kcal/mol. Cremer and Pople [1975a,b] showed that all types of PES listed above are realized in oxygen-contained 5-member cycles. The results of quantum chemical calculations by those authors are in satisfactory agreement with spectroscopic data.

The transition between E and T structures in cyclopentane is almost free; the barrier height for pseudorotation does not exceed 20 cm^{-1}, whereas the energy of the planar structure is 1950 cm^{-1} higher than that for the bent E and T structures. For this reason, the angular and radial motions separate well [Ikeda et al., 1972; Carreira et al., 1972]. The pseudorotation amplitude q_0 was not only calculated from spectroscopic data, but also found from the NMR spectrum of a cyclopentane guest molecule in a liquid crystal matrix, $q_0 = 0.463 \text{ Å}$ [Poupko et al., 1982]. Quantum chemical calculations show that pseudorotation is not accompanied by any significant deformation of C–H and C–C bonds (see the review by Laane [1987]); i.e., it is actually a synchronous motion of rigidly coupled CH_2 fragments described by a single angular coordinate. MacPhail and Variyer [1989] and Besnard et al. [1991] established that the pseudorotation remains very weakly hindered in liquid and in plastic phases of cyclopentane.

In oxalane, which has C_{2v} symmetry, the minima of the PES correspond to E forms; the energy of the planar structure is $1220\ cm^{-1}$ greater than that of E form, and the barrier height for pseudorotation is $50\ cm^{-1}$. The tunneling splitting in the ground and first excited states of the radial vibration (with frequency $260\ cm^{-1}$) is 0.67 and $1.5\ cm^{-1}$, respectively. The two-dimensional potential in terms of the mass-weighted rectangular coordinates is

$$V(q_1 q_2) = -360.06q_1^2 - 5500q_2^2 + 30.25q_1^4 + 56.61q_2^4 + 82.16q_1^2 q_2^2$$

$$(8.27)$$

where energy is expressed in cm^{-1}. The amplitude of pseudorotation q_0 is 0.37 Å. In 1,3-dioxalane the barrier height is $45\ cm^{-1}$. The separability of the radial and angular motions breaks down in cyclopentanone and 1,1-difluorocyclopentane, where the barriers for pseudorotation and radial interconversion are comparable. The PES's for these molecules found by Kim and Gwinn [1969] and Ikeda and Lord [1972] are

$$V(q_1 q_2) = 1301.5q_1^2 - 14530.8q_2^2 + 365.9q_1^4 + 70351.6q_2^4 + 14765q_1^2 q_2^2$$

$$(8.28)$$

$$V(q_1 q_2) = -575.5q_1^2 - 430.0q_2^2 + (15.6q_1^2 + 7.1q_2^2)^2 \qquad (8.29)$$

These PES's are shown in Figures 8.8 and 8.7b. In cyclopentanone the transition state with energy $750\ cm^{-1}$ corresponds to $q = 0$. The transition is interconversion through the planar configuration rather than pseudo-

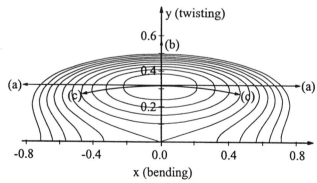

Figure 8.8. Contour plot of PES (8.28) for cyclopentanone ring inversion. The contours are at $75\text{-}cm^{-1}$ intervals. (From Ikeda and Lord [1972].)

rotation. Exact analytical solution shows that only a one-dimensional trajectory along the twisting coordinate, $q_1 = 0$, satisfies the equations of classical motion in upside-down potential (8.28). This trajectory (that is, instanton) coincides with the MEP. The transverse mode is the bending vibration, whose frequency depends on the reaction coordinate (96 and 51 cm^{-1} at the minimum and transition state, respectively). The frequencies of small vibrations in the well and of upside-down barrier for the potential along the trajectory (237 and 170 cm^{-1}, respectively) are several times greater than the bending vibration frequency, so that the sudden approximation may be used. The softening of the transverse mode along the path results in a decrease of the vibrationally adiabatic barrier height compared with the bare one at $q_1 = 0$. However, as we noted, one should rather use the sudden approximation in this case. The tunneling splitting increases from 0.01 to 0.55 cm^{-1} when the quantum number of the transverse mode is increased from 3 to 8, demonstrating the promoting role of the vibration.

The opposite case, when the coefficient a_{11} in (8.24) is negative, produces a "squeezed" potential

$$V(q_1, q_2) = V_0(q_2^2 - 1)^2 + \tfrac{1}{2}[\omega_\perp^2 + \gamma(q_2^2 - 1)]q_1^2 \qquad (8.30)$$

In this case, again only a one-dimensional instanton exists. The transverse frequency increases when moving along the MEP from the well to the transition state. So we have an effect of dynamically induced barrier formation in which the height of the vibrationally adiabatic barrier exceeds $V_\#$. The analysis of this "squeezed" potential by Auerbach and Kivelson [1985] shows that the vibrationally adiabatic approximation is valid when

$$\left\langle \frac{1}{\omega_\perp^2} \frac{d\omega_\perp}{dt} \right\rangle \ll 1 \qquad (8.31)$$

where averaging is performed over the period of the trajectory in the vibrationally adiabatic potential. We mentioned this condition in Section 4.1 where we derived the vibrationally adiabatic approximation from the instanton formulae.

Note that in the present case, neither the Fukui theorem [Fukui, 1970] mentioned in Section 8.4 nor the similar statement concerning the initial direction of the instanton path (see Section 4.1) are valid. Both state that the reaction coordinate (IRC in the case of Fukui theorem or instanton) near the minimum of the surface is directed along the coordinate of the vibration with the lowest frequency. That would mean for our case that the tunneling should proceed along the bending rather than twisting coordinate. However, this is not true because the reaction path is purely one dimensional.

Substitution impedes pseudorotation in cyclopentane. The relative stability of various conformers of 5-member cycles $(CH_2)_4X$ depends on the symmetry of the X group. When this symmetry is C_{2v} ($X = C{=}CH_2$, CO, CF_2, SiH_2), the T structures are more stable. For the groups with C_s symmetry ($X = CH{-}C{\equiv}CH$, CHCN, CHCl, $NC{\equiv}N$) the E structures have lower energy than T [Pfafferott et al., 1985]. In accordance with this rule, the E and T conformers of chlorocyclopentane are the equilibrium and transition states, respectively. The barrier height for pseudorotation (1.17 kcal/mol) is much smaller than the energy of the planar structure (5.2 kcal/mol). The frequency of the bending vibration relating to puckering is 115 cm^{-1} [Hildebrandt and Shen, 1982].

In pyrrolidine and its chlorine and N-methyl derivatives the barrier for pseudorotation is 1.66, 2.98, and 2.8 kcal/mol, respectively. The E structures are stable. Puckering dynamics in pyrrolidine was considered by Pfafferott et al. [1985].

In cyclohexane there are three puckering modes described by the generalized coordinates q_2, q_3, and ϕ. Cremer and Pople [1975a,b] introduced the amplitude Q ($q_2 = Q \sin \theta$, $q_3 = Q \cos \theta$) to portray the set of conformers in spherical coordinates (Figure 8.9). The axial conformer C ($\theta = 0°$, 180°) has a chair-like shape. Transverse displace-

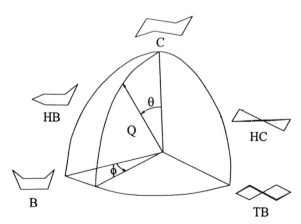

Figure 8.9. One octant of a sphere on which the conformations of cyclohexane can be mapped. Special conformations: C, chair; B, boat; TB, twist boat; HB, half-boat; and HC, half-chair. (From Cremer and Pople [1975a,b].)

ments of CH_2 fragments in this form are related with q_3 by relation

$$Z_1 = -Z_2 = Z_3 = -Z_4 = Z_5 = -Z_6 = \frac{1}{\sqrt{6}} q_3 \qquad (8.32)$$

The value of $q_3 = (6)^{-1/2}R$ (R is the CC bond length) is 0.63 Å. Under pseudorotation the equatorial boat-shaped structures B ($\theta = 90°$, $\phi = 0$, 60°, 120°, . . .) turn into a twist-boat structure TB ($\theta = 90°$, $\phi = 30°$, 90°, . . .). The transitions between the chair and twist boat structures involve the intermediate formation of half boat (HB) and half chair (HC) structures. Quantum chemical calculations carried out by Dixon and Komornicki [1990] show that the axial structure C with symmetry D_{3d} is stable. The energies of structures B and TB are 7.9 and 6.8 kcal/mol higher than C. The barrier for transition from C to TB is 12.2–12.4 kcal/mol. Because of the high barriers for pseudorotation, only thermally activated conformational transitions occur in cyclohexane.

8.4. INTERCONVERSION OF HYDROGEN-BONDED DIMERS OF DIATOMIC MOLECULES

Among hydrogen-bonded dimers the most thoroughly investigated example is $(HF)_2$, in which tunneling interchange of dimer subunits results in a well-resolved splitting of rotational levels of nearly symmetric rotor. This splitting was discovered in molecular beam electric resonance microwave spectra by Dyke et al. [1972] and later studied in detail in the ground and vibrationally excited states using IR tunable difference frequency laser spectroscopy, far IR spectroscopy, and double IR-microwave resonance [Pine and Lafferty, 1983; Pine et al., 1984; Pine and Howard, 1986; Quack and Suhm, 1990; Dayton et al., 1989]. A special place in these studies is occupied by the supersonic cooling technique, which enables one to obtain tunneling rotational spectra at rotational temperatures as low as 1 K. It was established that the molecules in the dimer are bonded by a single linear hydrogen bond. Two planar configurations, $H_aF_a-H_bF_b$ and $F_aH_a-F_bH_b$, which have C_s symmetry with inequivalent molecules, belong to the doubly degenerate ground state. These configurations convert into one another via a transition state in which both molecules become equivalent and which has C_{2h} symmetry [Curtiss and Pople, 1976]. The PES of the dimer has been constructed [Quack and Suhm, 1991], based on the experimental data combined with quantum chemical calculations [Hofranek et al., 1988; Bunker et al., 1990; Rijks and Wormer, 1989]. Configurations of the dimer and kinematics of normal modes obtained by Quack and Suhm [1991] are shown in Figure 8.10.

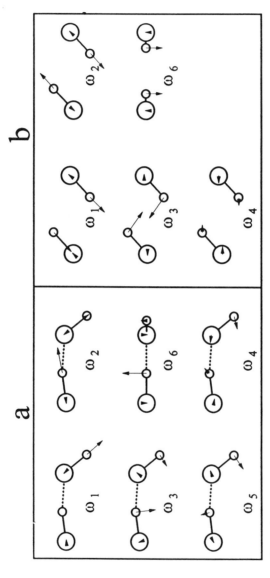

Figure 8.10. Normal coordinate displacement amplitudes and phases (arrows) at (a) the C_s minimum and (b) the C_{2h} saddle point of the $(HF)_2$ potential energy surface. Intramolecular HF bonds are drawn in thick lines. The hydrogen bond is indicated as a dashed line. (From Quack and Suhm [1991].)

282

TABLE 8.2
Bond Lengths, Bond Angles, and Normal Vibration Frequencies in the Ground and
Transition States of HF Dimer

	Ground State, C_s	Transiton State, C_{2h}	Transiton State, $C_{\infty v}$	Frequency	C_s (cm^{-1})	C_{2h} (cm^{-1})
r_1 (Å)	0.913	0.912	0.912	ω_1	4104	4082
r_2 (Å)	0.910	0.912	0.909	ω_2	4044	4082
R (Å)	2.765	2.669	2.054	ω_3	487	570
					(588)[a]	(719)
θ_1 (deg)	9.0	54.9	0	ω_4	153	141
					(174)	(164)
θ_2 (deg)	116	54.9	180	ω_5	211	−(197i)
					(221)	
ϕ_1 (deg)	180	180	180	ω_6	401	395
					(447)	(419)

[a] The frequencies calculated by Hancock et al. [1986] are given in parentheses.

The bond lengths, angles, and vibrational frequencies in the ground and transition states are listed in Table 8.2. Apart from the transition state with symmetry C_{2h}, concerted rotation of molecules is realized in the linear conformation $C_{\infty v}$, which can be obtained by mirror reflection in a plane perpendicular to the dimer plane and corresponds to rotating the hydrogen-bonded molecule through an angle of 9°. As follows from Table 8.2, the internal rotation practically does not change the lengths of HF bonds and preserves planar structure of the dimer ($\psi = 180°$). Hence, in practice the PES for tunneling-rotation motion can be considered to be three dimensional, $U = V(\theta_1, \theta_2, R)$. Such a model PES was proposed earlier by Barton and Howard [1982]. For this PES, the interconversion barrier height is 302 cm^{-1}. Later another PES was used with the barrier height 385 cm^{-1} and a smaller tunneling distance [Hancock et al., 1988; Hancock and Truhlar, 1989]. A two-dimensional cut of the PES in the (θ_1, θ_2) plane [Quack and Suhm, 1991] is shown in Figure 8.11b. The path of interconversion involving the transition $C_s \rightarrow C_{2h} \rightarrow C_s$ is nearly linear; $\theta_1 + \theta_2$ is constant, corresponding to the concerted rotation of the two molecules. Note that the *trans* configuration of the C_{2h} transition state associated with such a rotation can be derived directly by analyzing the intensity distribution in rotational spectra of the dimer.

The MEP for dissociation of the dimer is defined by the condition that at every intermolecular distance R, θ_1 and θ_2 take the values so as to minimize the energy; the potential along this MEP is modeled by a Morse potential with $D_e = 4.3–4.5$ kcal/mol and $\alpha = 1.45–1.55$ Å^{-1}. The one-dimensional barrier for interconversion and the corresponding correlation between θ_2 and θ_1 are presented in Figure 8.11a. The concerted rotation

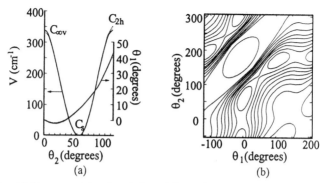

Figure 8.11.   (a) The potential and variation of the angle θ_1 along the MEP as functions of θ_2. Ab initio and semiempirical data are shown by solid and dashed lines, respectively. (b) 2D cut of the multidimensional PES for $(HF)_2$. The contours are at 100-cm^{-1} intervals. $R = 2.81$ Å. (From Quack and Suhm, 1991].)

is preceded by an initial stage near equilibrium, where the change in θ_1 is small. Paths of transition $C_s \rightarrow C_{2h} \rightarrow C_s$ were considered by Hancock et al. [1986] using two versions of the vibrationally adiabatic approximation, the most advanced of which is the small curvature approximation developed by Skodje et al. [1981a,b] and Truhlar et al. [1982]. In this method, one determines the Cartesian mass-weighted coordinates describing the motion of reduced mass μ (in the present case $\mu = \frac{1}{2}m_{HF}$). Then, the reaction path s is determined by steepest descent from the saddle point. At every point of the path local normal vibrations orthogonal to s are found [Isaakson and Truhlar, 1982]. The transverse vibrations are quantized at every fixed s using Bohr-Sommerfeld quantization rules (3.49). The tunneling splitting is given semiclassically by Eq. (3.103); the energy of zero-point longitudinal vibration in the vibrationally adiabatic potential, E_0, is found from the semiclassical quantization condition

$$\frac{1}{\hbar} \int_{s_<}^{s_>} \{2\mu[E_0 - V_{rad}(s)]\}^{1/2} = \frac{\pi}{2} \tag{8.33}$$

The effects of curvature of the MEP are incorporated in this method through mass renormalization. Figure 8.12 shows how θ_1, θ_2, and R change when moving along the reaction path. According to the Fukui theorem, the initial part of the reaction path near the minimum of the potential, $S_0 = \pm 1.09$ Å, coincides with the direction of R, because this direction corresponds to the lowest frequency mode (174 cm^{-1}), whereas the disrotatory combination of in-plane librations ($\theta_1 + \theta_2 = $ const) has a larger frequency (221 cm^{-1}). Outside this initial segment, at $|s| < s_1 =$

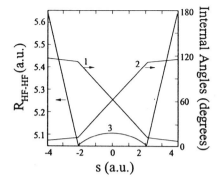

Figure 8.12. The intermolecular angles θ_1 (1) and θ_2 (2) and the intermolecular distance (3) of $(HF)_2$ along MEP as functions of the reaction coordinate. Distances s and R are given in atomic units. (From Hancock et al. [1986].)

1.087 Å, the direction of concerted libration becomes the reaction coordinate with the upside-down barrier frequency $\omega_{\#} = 197\,\text{cm}^{-1}$, while the stretching vibration along R is transverse with respect to the reaction path, with $\omega_{\perp} = 164\,\text{cm}^{-1}$. This leads to a sharp drop in the transverse frequency when going from the initial to transition state. According to predictions of Section 4.1, the switching of the transverse frequency reduces the prefactor so that the vibrationally adiabatic approximation overestimates the tunneling probability. The effect is still insignificant in this case because the change in $\omega_{\perp}$ is sufficiently small. The tunneling splitting calculated by Hancock et al. [1988] and Hancock and Truhlar [1989] is $\Delta = 0.6\text{–}0.8\,\text{cm}^{-1}$, in excellent agreement with the experimental value $0.66\,\text{cm}^{-1}$ (19.776 GHz). In the totally deuterated dimer, Δ decreases to $0.053\,\text{cm}^{-1}$ (1.580 GHZ). The interconversion barrier in $(DF)_2$ is $\sim 270\,\text{cm}^{-1}$.

Pine and Lafferty [1983] discovered that tunneling splitting decreases by a factor of 3 when intramolecular vibrations of hydrogen bonded (ω_1) or free (ω_2) HF subunits are excited to the $n = 1$ level (Δ is measured to be 0.215 and $0.234\,\text{cm}^{-1}$ in $(HF)_2$, and $0.016\,\text{cm}^{-1}$ for both excitations in $(DF)_2$). Since the difference in frequencies of these modes (about $60\,\text{cm}^{-1}$, see Table 8.2) is much greater than Δ, exciting either one of them should lead to an energy bias that destroys any resonant tunneling, and it is surprising that splitting is still observed. Mills [1984] and Fraser [1989] suggested that a vibrational exchange occurs between the two HF subunits, which symmetrizes the potential. In essence, the initial asymmetric terms are analogous to diabatic terms in the electronically nonadiabatic tunneling problem, while the vibrational exchange is an analogue of diabatic coupling V_d in Hamiltonian (3.110). The diabatic and adiabatic terms are shown in Figure 8.13a. As we saw in Section 3.7 and Appendix C, diabatic coupling makes the adiabatic potential symmetric, thus recovering the resonance. However, nonadiabaticity decreases

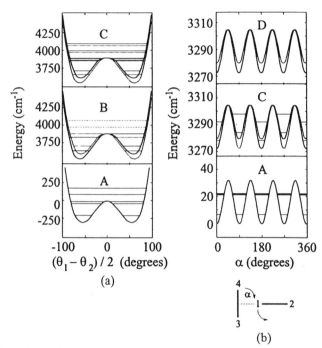

Figure 8.13. (a) HF dimer potential as a function of $(\theta_1 - \theta_2)/2$ in the ground state (A), diabatic V_a and V_b (B), and adiabatic V_1 and V_2 (C) terms of excited state. Diabatic terms correspond to the excitation of the stretching vibration of hydrogen-bonded and free HF molecules. The energy levels for these potentials are shown. (b) Acetylene dimer potential as a function of concerted hindered rotation coordinate in the ground state (A), diabatic V_a and V_b (C), and adiabatic $V_\parallel$ and $V_\perp$ (D) terms of excited states. The coupled rotation coordinate is shown schematically at the lower right. The diabatic terms correspond to the excitation of asymmetric stretches of hydrogen-bonded and free acetylene molecules. (From Fraser [1989].)

the tunneling splitting, which is reflected in the prefactor in (3.116). Since the tunneling splitting in the adiabatic potential Δ_{ad} is weakly dependent on coupling V_d, the decrease in Δ should be ascribed to the Landau-Zener factor.

A straightforward calculation of the tunneling splitting has been performed by Fraser [1989] within the framework of the one-dimensional model of concerted interconversion. The diabatic terms were taken in the form

$$V_{1,2}(Q) = V(Q) + \frac{\hbar}{2}(\omega_1 - \omega_2) \pm \frac{\hbar}{2}(\omega_1 + \omega_2)\frac{Q}{Q_0} \qquad (8.34)$$

where $V(Q)$ is the symmetric ground-state potential, $Q = \theta_1 - \theta_2$, and $\pm Q_0$ are equilibrium positions. It was assumed that the major contribution to the diabatic coupling results from the dipole–dipole interaction of transition momenta. It can be represented by a Fourier series:

$$V_d = \frac{M_{01}^2}{R^3} \sum_n C_n \cos(nQ) \qquad (8.35)$$

When V_d was varied within the interval 1–8 cm^{-1}, the tunneling splitting was found to be nearly linear with V_d, in agreement with the semiclassical model of Section 3.7 (see, e.g., Eq. (3.116)). The prefactor Δ/Δ_{ad} ranged from 0.1 to 0.3 indicating nonadiabatic tunneling. Since this model is one dimensional it fails to explain the small difference between splittings in the states with ω_1 and ω_2 vibrations excited.

Ordinarily, nonadiabatic tunneling is considered to occur between electronic terms. We do not know any direct experimental evidence of nonadiabatic regime for tunneling, and it is difficult to estimate V_d and to separate the Landau-Zener prefactor from the leading exponential term, which is strongly dependent on the parameters of the problem. The present case involves different vibrational (not electronic) states. Moreover, the barrier is relatively low, and coupling, having an electrostatic origin, can be estimated directly. A similar decrease of Δ in vibrationally excited states is observed in many other dimers such as $(H_2O)_2$ [Huang and Miller, 1988], $(HCl)_2$ [Blake et al., 1988], and $(C_2H_2)_2$ [Ohshima et al., 1988].

The planar dimer of acetylene has four equilibrium configurations:

$$
\begin{array}{cccc}
4 & 2 \quad 3 & & 1 \\
| \; 1{-}2 & 3{-}4 \; | & | \; 2{-}1 & 4{-}3 \; | \\
3 & 1 \quad 4 & & 2 \\
\end{array}
\qquad (8.36)
$$

which subsequently turn into each other as a result of concerted rotation of both the subunits. The one-dimensional potential has C_4 symmetry:

$$V(Q) = \frac{V_4}{2}(1 - \cos 4Q) \qquad (8.37)$$

where Q is the coordinate of concerted rotation similar to that in $(HF)_2$. The rotation-tunneling spectra correspond to $V_4 = 33.2$ cm^{-1}; the tunneling splitting is 0.07 cm^{-1} (2.1 GHZ). When a CH stretching vibration of one of the inequivalent subunits is excited, the symmetry decreases from C_4 to C_2. The potential minima at 0 and 180° are 8.07 cm^{-1} lower than at

90 and 270° because of the difference in stretching frequencies of hydrogen bonded and free molecules. As in the previous example of $(HF)_2$, diabatic coupling caused by vibrational exchange symmetrizes this potential and makes coherent tunneling possible. The diabatic and adiabatic terms considered by Fraser [1989] are shown in Figure 8.13b.

8.5. COUPLED TUNNELING MOTIONS IN HYDROGEN-BONDED DIMERS

A molecule is considered rigid when it never departs very far from a unique symmetrical configuration. Obviously, this is not the case of polyatomic molecules or dimers containing large amplitude motions along some of the degrees of freedom. To classify the states of nonrigid molecules that can be in various conformations, Longuet-Higgins (1963) constructed special symmetry groups. The necessity of special considera- tion for such molecules is seen from the examples of ammonia conversion and internal rotation in ethane. The ground state of the pyramidal ammonium molecule has symmetry C_{3v}, but the symmetry of the planar transition configuration is D_{3h}, and it is this group that one can use to classify the quantum states when tunneling is taken into account. The stable staggered configuration of ethane belongs to the D_{3d} group, whereas in eclipsed configuration the symmetry is D_{3h}. That is, certain symmetry elements of a nonrigid molecule do not belong to the symmetry point group of the ground state. This statement holds for any molecules that exhibit wide amplitude motions, when deformation of the molecular skeleton changes the symmetry group compared to that of the ground state. Since the Hamiltonian is invariant under any permutations of positions and spins of any set of identical nuclei, and under any combination of permutation with inversion, the permutation-inversion groups can be used for classifying tunneling-rotation states. To consider the limited number of feasible conformations, these groups include only a restricted number of elements, which correspond to the symmetry of the path of the molecular transformation. Therefore, the choice of the inversion-permutation group is determined not only by the symmetry of the ground state, but also by the symmetry of the reaction path. Subsequent assignment of the vibration–rotation tunneling spectra based on group-theoretical analysis allows one to verify the choice of this path. For example, comparing intensities of several bands in the $(HF)_2$ spectrum with the selection rules derived by Hougen and Ohashi [1985], one concludes that the interconversion occurs via the *trans* path.

The structure of the water dimer found by Dyke et al. [1977] from microwave electric resonance spectra of a supersonic molecular beam is shown in Figure 8.14A. The dimer consists of a hydrogen-bonded donor molecule and free acceptor molecule, which together constitute a *trans*-linear complex with C_s symmetry. The plane of the donor molecule is at the same time the symmetry plane for the acceptor. In the ground state of the dimer, angles θ_a and θ_b are 58 and 51° and the O–O distance is 2.98 Å. The intramolecular parameters are essentially the same as those for nonbonded water molecules. Dyke [1977] classified the rotation-tunneling states using permutation-inversion group analysis. Apart from the identity operation E, the group includes the intramolecular H-atom permutation (12), (34) in each molecule and in both molecules (12) (34),

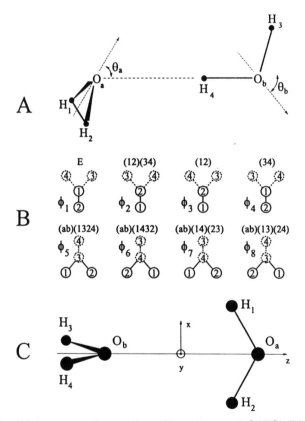

Figure 8.14. (A) A structure of water dimer. (From Dyke et al. [1977].) (B) Permutations transforming one stable configuration of water dimer into another. The numbers 1, 2, 3, 4 in circles indicate the hydrogen atom positions. The oxygen atoms are not shown. Solid and dashed circles correspond to H atoms situated in front of and behind the figure plane, respectively. (From Coudert et al. [1987].) (C) Hypothetical water dimer configuration used by Hougen [1985] to introduce three large-amplitude coordinates.

as well as the intermolecular H-atom permutations combined with exchange of O atoms (ab)(13)(24), (ab)(14)(23), (ab)(1234), and (ab)(1423). The configurations created by these eight permutations are shown in Figure 8.14B. These permutations and their products with the inversion operation form the group G_{16}, which is isomorphic with the point symmetry group D_{4h}, as can be easily checked in a straightforward way, comparing symmetry operations and permutations. In view of the high barrier for direct intermolecular hydrogen exchange, permutations (13), (24), (ab) (12), . . . corresponding to this kind of process, are not included in the group.

In the limit of an infinitely high barrier, all 16 nonsuperimposable isoergetic configurations of water dimer have the same vibration–rotation spectrum of a prolate asymmetric top. Tunneling between various configurations splits each vibration–rotation level according to the irreducible representations of the G_{16} group. Because the total rotation–internal tunneling rotation wave functions are antisymmetric, there are 6 rotation–tunneling sublevels for each J with $K = 0$, and 12 sublevels for each J with $K > 0$. For $K = 0$, the correlation between levels of a rigid top and the dimer with internal rotation is shown in Figure 8.15. The dipole-allowed transitions indicated by Coudert et al. [1987] are given in the same figure. The states of rigid dimer belong to A′ and A″ representations of the C_2 group and are split into sextets A_1^+, B_1^+, E^+, A_2^-, B_2^-, E^- and B_1^-, A_1^-, E_1^-, B_2^+, A_2^+, E^+, corresponding to the irreducible

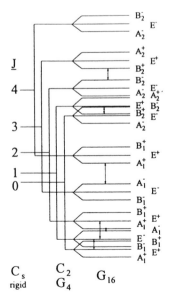

Figure 8.15. Correlation diagram between levels of a rigid rotor $K = 0$ (water dimer with C_s symmetry in the nontunneling limit), a rotor with internal rotation of the acceptor molecule around the C_2 axis (permutation-inversion group G_4), and group G_{16}. The arrangement of levels is given in accordance with the hypothesis by Coudert et al. [1987]. The arrows show the allowed dipole transitions observed in the $(H_2O)_2$ spectrum. The pure rotational transitions $E^+(J = 0) \leftrightarrow E^-(J = 1)$ and $E^-(J = 1) \leftrightarrow E^+(J = 2)$ have frequencies 12 321 and 24 641 MHz, respectively. The frequencies of rotation–tunneling transitions in the lower triplets $A_1^-(J = 1) \leftrightarrow A_1^+(J = 2)$ and $A_1^-(J = 3) \leftrightarrow A_1^+(J = 4)$ are equal to 4863 and 29 416 MHz. The transitions $B_2^+(J = 0) \leftrightarrow B_2^+(J = 1)$ and $B_2^+(J = 2) \leftrightarrow B_2^+(J = 3)$ with frequencies 7355 and 17 123 MHz occur in the higher multiplets.

representations of the G_{16} group. The transitions $A_i^{\pm} \leftrightarrow A_i^{\mp}$, $B_i^{\pm} \leftrightarrow B_i^{\mp}$ ($i = 1, 2$) are rotation–tunneling ones, whereas the transitions between doubly degenerate levels $E^{\pm} \leftrightarrow E^{\mp}$ are pure rotational and have essentially the same frequencies as those in the hypothetical dimer without tunneling. The high barrier condition is well satisfied for the water dimer, where the frequency of the intermolecular stretching vibration (150 cm^{-1}) is about two orders of magnitude greater than the tunneling splitting. Under this condition the vibration tunneling wave functions can be represented as linear combinations of the eight configurations shown in Figure 8.14b. The coefficients of these linear combinations, calculated by Dyke [1977], are given in table 8.3. Using this table, Odutola et al. [1988] found the energies of the eigenstates:

$$E(A_1^+) = -\tfrac{1}{2}(V_a + V_d + V_{ad} + \Delta_1 + \Delta_2)$$

$$E(E^+) = -\tfrac{1}{2}(V_a - V_d - V_{ad})$$

$$E(B_1^+) = -\tfrac{1}{2}(V_a + V_d + V_{ad} - \Delta_1 - \Delta_2)$$

$$E(A_2^-) = \tfrac{1}{2}(V_a + V_d - V_{ad} + \Delta_1 - \Delta_2) \qquad (8.38)$$

$$E(E^-) = \tfrac{1}{2}(V_a - V_d + V_{ad})$$

$$E(B_2^-) = \tfrac{1}{2}(V_a + V_d - V_{ad} - \Delta_1 + \Delta_2)$$

where $V_a = \langle \phi_1^+ | H | \phi_4^+ \rangle$ and $V_d = \langle \phi_1^+ | H | \phi_3^+ \rangle$ are the matrix elements of hydrogen exchange in acceptor and donor molecules, respectively, $V_{ad} = \langle \phi_1^+ | H | \phi_2^+ \rangle$ relates to synchronous intramolecular exchange in both molecules of the dimer, and $\Delta_1 = \langle \phi_1^+ | H | \phi_5^+ \rangle = \langle \phi_1^+ | H | \phi_6^+ \rangle$ and $\Delta_2 =$

TABLE 8.3
Coefficients of Various Configurations in Expansion of Water Dimer Eigenfunctions:
$J = 0, K = 0$

State	Φ_1	Φ_2	Φ_3	Φ_4	Φ_5	Φ_6	Φ_7	Φ_8
A_1^+	1	1	1	1	1	1	1	1
B_1^+	1	1	1	1	-1	-1	-1	-1
E^+	$2^{1/2}$	$-2^{1/2}$	$2^{1/2}$	$-2^{1/2}$	0	0	0	0
	0	0	0	0	$2^{1/2}$	$-2^{1/2}$	$-2^{1/2}$	$2^{1/2}$
A_2^-	1	1	-1	-1	1	1	-1	-1
B_2^-	1	1	-1	-1	-1	-1	1	1
E^-	$2^{1/2}$	$-2^{1/2}$	$-2^{1/2}$	$2^{1/2}$	0	0	0	0
	0	0	0	0	$-2^{1/2}$	$2^{1/2}$	$-2^{1/2}$	$2^{1/2}$

Note. All coefficients are multiplied by $8^{1/2}$.

$\langle \phi_1^+ | H | \phi_7^+ \rangle = \langle \phi_1^+ | H | \phi_8^+ \rangle$ characterize two inequivalent intermolecular exchanges (Figure 8.14b). The superscript $+$ in the wave functions refers to the parity with respect to reflection in the plane of symmetry and is indicated because the observed states correlate with the vibrational ground state of the frozen conformation of the water dimer. Since the acceptor molecule rotates almost freely, the tunneling splitting V_a is the largest and, according to the estimate of Coudert et al. [1987] (see also Zwart et al. [1990]), exceeds 200 GHz. Each sextet for $K = 0$ breaks up into the lower and upper triplets, as shown in Figure 8.15. In the absence of tunneling motions but in the presence of rotation of acceptor molecule, the dimer levels belong to G_4 group representations. Proton exchange in the donor molecule is associated with the rupture of the hydrogen bond, whose energy is about 5 kcal/mol. For this reason, Coudert et al. [1987] suggested that V_d and V_{ad} are smaller than Δ_1 and Δ_2, so that the tunneling path that results in splitting to the lower triplet corresponds to the cyclic transition $1 \to 5 \to 2 \to 6 \to 1$ shown in Figure 8.16.

The dynamics of the water dimer proposed by Hougen [1985] and Coudert and Hougen [1988, 1990] is represented by hierarchy of three tunneling motions: internal rotation of the acceptor molecule (transition $1 \leftrightarrow 4$), intermolecular exchange ($1 \leftrightarrow 5$), and internal rotation of the donor molecule ($1 \leftrightarrow 2$). The frequencies of these motions are ~200, 19.53, and <1 GHz, respectively, and only the first of them has nothing to do with rupture of intermolecular hydrogen bond. Using the semiempirical PES proposed by Coker and Watts [1987], Coudert and Hougen

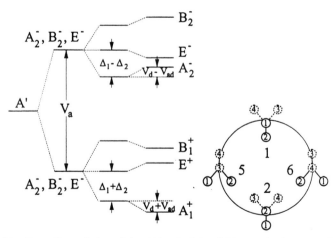

Figure 8.16. Tunneling splitting of the A' level of the rigid rotor in the water dimer (*left*) results from permutations between the stable configurations shown at the lower right.

[1989] introduced three large-amplitude coordinates, ζ_1, ζ_2, ζ_3, describing the above set of tunneling motions (see also Hougen [1985]). To obtain one of the stable configurations shown in Figure 8.14b, subsequent transformations of a hypothetical configuration that consists of two water molecules with a common C_2 axis directed along z (Figure 8.16) are performed: (1) rotation of molecules about the z axis in opposite directions through angle ζ_3, (2) rotation of the dimer as a whole about the z axis through angle ζ_1, and (3) geared rotation of molecules through angle ζ_2 with respect to the axis parallel to the y axis and passing through the center of mass of each molecule. The angle of geared rotation of the donor and acceptor molecules is the same for each of them:

$$\zeta_{2a} - \zeta_{2d} = 0 \tag{8.39}$$

This choice of coordinates does not take into account the small difference between θ_a and θ_d (Figure 8.14A) and ensures that the intermolecular distance is constant. Linear combinations

$$\zeta_{1a} = \zeta_1 + \zeta_3 \qquad \zeta_{1d} = \zeta_1 - \zeta_3 \tag{8.40}$$

describe the rotations of donor and acceptor molecules about their C_2 axes, whereas ζ_{2a} and ζ_{2d} define the rotations with respect to the y axis. The equilibrium values of these large-amplitude coordinates for various dimer configurations are given in Table 8.4. A contour plot of a model PES $U(\zeta_1, \zeta_2)$ and MEPs for transitions $1 \leftrightarrow 2$, $1 \leftrightarrow 5$, and $1 \leftrightarrow 6$ are presented in Figure 8.17A. Variation of the abscissa ζ_1, when all other coordinates are fixed at their equilibrium values, corresponds to a geared simultaneous rotation. The path $1 \leftrightarrow 2$ corresponds to the variation of ordinate when condition (8.39) is fulfilled. When proceeding along this path, the acceptor molecule is inverted and the donor molecule exchanges the positions of its protons. The paths $1 \leftrightarrow 5$ and $1 \leftrightarrow 6$ are similar to the *trans* path of intermolecular exchange in $(HF)_2$. Internal rotations of the donor and acceptor molecules are shown on the two-dimensional PES

TABLE 8.4
Equilibrium Values of Large-Amplitude Coordinates for Configurations Shown in Figure 8.15b

	1	2	3	4	5	6	7	8
ξ_1^0	0	π	$-\pi/2$	$\pi/2$	$\pi/2$	$3\pi/2$	0	π
ξ_2^0	θ_0^a	θ_0	θ_0	θ_0	$\pi - \theta_0$	$\pi - \theta_0$	$\pi - \theta_0$	$\pi - \theta_0$
ξ_3^0	0	0	$-\pi/2$	$-\pi/2$	0	0	$\pi/2$	$\pi/2$

[a] θ_0 is the angle between the C_2 axis of the molecule and the z axis, $\theta_a = \theta_d = \theta_0$.

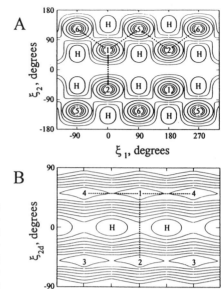

Figure 8.17. Two-dimensional cuts of the PES for water dimer. (A) $U(\zeta_1, \zeta_2)$. Contours are at 119.4-cm^{-1} intervals. The dashed-line path $6 \leftrightarrow 1 \leftrightarrow 5 \leftrightarrow 2 \leftrightarrow 4$ corresponds to that shown in Figure 8.16. (B) $U(\zeta_{1a}, \zeta_{2d})$. Contours are at 127-cm^{-1} intervals. Maxima occur at the positions labeled H. (From Coudert and Hougen [1988].)

$U(\zeta_{1a}, \zeta_{2d})$ (Figure 8.17B). If the rotation barrier for the donor molecule were infinite, the tunneling path $1 \leftrightarrow 4$ would be the rotation of the acceptor molecule at constant value of $\zeta_{2a} = \theta_a$. If the barrier for rotation of the donor molecule were low, another path would be followed, in which the donor molecule rotates through 180° simultaneously with the inversion of the acceptor molecule; this path is similar to the coupled inversion and rotation in methylamine (see Section 8.2). It has a segment where $\zeta_{2a} = 0$. Since the barrier height is finite, path $1 \leftrightarrow 4$ is intermediate between these two limits.

Our analysis of tunneling in the water dimer is only qualitative because of the complicated topology of the PES, which has many stable configurations and transition states. Quantum chemical calculations of transition-state structures were undertaken by Smith et al. [1990]. Several structures are shown in Figure 8.18. Path $1 \leftrightarrow 4$ goes through the transition states corresponding to the cyclic C_i structure with energy 0.59 kcal/mol. This energy is larger than the value 0.37 kcal/mol estimated by Coudert and Hougen [1988]. Nonreactive modes play a significant role in reducing the barrier along the MEP, as seen from the deformation of the dimer in its transition state. The energy of the cyclic structure, in which the molecular C_2 axes form angles $\pm 27°$ with respect to xz plane, is lower than the energy of the planar structure. The C_2 axis of the acceptor molecule

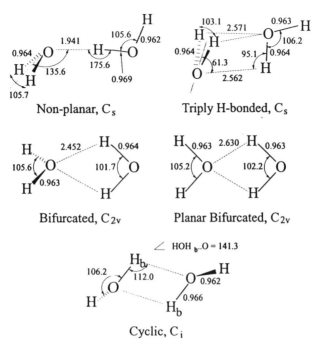

Figure 8.18. Optimized parameters for stationary points on water dimer and hydrogen fluoride dimer PES. Bond lengths are in angstroms and bond angles are in degrees. (From Smith et al. [1990].)

rotates through 50° in the xz plane and ζ_{2d} is ~8° at the saddle point. This picture agrees with the above path $1 \leftrightarrow 4$, which was described by Coudert and Hougen [1988].

Intermolecular exchange proceeds through the transition state with another cyclic structure (not shown) which has an energy of 0.87 kcal/mol. Instead of one linear hydrogen bond in the ground state, two nonlinear bonds with a O–O distance of 2.265 Å are formed in this transition state. The barrier height for rotation of the donor molecule corresponds to the bifurcated C_{2v} structure (Figure 8.18) and is 1.88 kcal/mol. A cut of PES along this path is shown in Figure 8.19. Apart from this saddle point (shown at 240°), this path involves several nonplanar structures with one single and one double nonlinear hydrogen bond. The local minimum of the structure shown at 45° has energy very close to the energy of that at 240°. This indicates a bifurcation of the path: The transition can proceed either directly through the transition state at 240° or through the minimum at 45° by surmounting an additional barrier

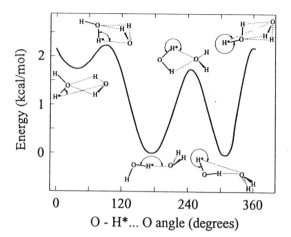

Figure 8.19. The potential along two different paths for donor molecule rotation in water dimer. (From Smith et al. [1990].)

associated with the planar bifurcated C_{2v} structure. It is relevant to point out that the latter is a second-order saddle point and can be thought of as a linkage of two C_i first-order saddle points with the cyclic C_i structure shown in Fig. 8.18. Generally speaking, high-order saddle points do not affect the tunneling path in the adiabatic approximation because the MEP bypasses these points.

The spectrum of normal vibrations of the stable configuration of water dimer was calculated by Frisch et al. [1986]. It includes torsional (661, 198, 164, and 135 cm^{-1}) and bending (371 and 102 cm^{-1}) modes, in addition to the stretching intermolecular mode mentioned above.

The two-dimensional PES shown in Figure 8.17 (as well as in Figures 8.3b and 8.7c) is typical of internal rotation coupled to inversion of the other part of the system. This situation is also realized in methylamine inversion, where the rotation barrier is modulated not by a harmonic oscillation but by motion in a double-well potential. The PES for these coupled motions can be modeled as follows:

$$V(\theta, q) = \tfrac{1}{2}(V_0 + Cq^2)(1 - \cos n\theta) + V_1(q^2 - 1)^2 \qquad (8.41)$$

where θ and q are the rotation and inversion coordinates, respectively. When $C/2V_1 < 1$, there is a single MEP for rotation, defined by

$$q_{\mathrm{ad}}(\theta) = \left[1 - \frac{C}{4V_1}(1 - \cos n\theta)\right]^{1/2} \qquad (8.42)$$

and the potential along this MEP is

$$V_r[\theta, q_{ad}(\theta)] = \frac{V_0 + C}{2}(1 - \cos n\theta) - \frac{C^2}{16V_1}(1 - \cos n\theta)^2 \quad (8.43)$$

When $C/2V_1 > 1$, this MEP becomes piecewise. The segment described by Eq. (8.42) exists only for $|\theta| < \theta_1 = 1/n \cos^{-1}(1 - 4V_1/C)$; in the range $\pi/n - \theta_1 > \theta > \theta_1$, $q = 0$ is the only path. The height of the rotational barrier is equal to

$$V_r^{\#} = \begin{cases} V_0 + C\left(1 - \dfrac{C}{4V_1}\right) & \dfrac{C}{2V_1} < 1 \\[2ex] V_0 + V_1 & \dfrac{C}{2V_1} \geq 1 \end{cases} \quad (8.44)$$

The MEP for inversion corresponds to $\theta = 0$ and is characterized by the barrier height V_1. When $C/2V_1 > 1$, apart from this MEP, there is a path that includes two segments described by Eq. (8.42) and a second-order saddle point. The barrier along this path is greater than V_1 and equal to $V_1(1 + 2V_0/C)$. The transverse frequency along the straight-line MEP for inversion has a minimum at the saddle point $q = 0$, $\theta = 0$; consequently, the vibrationally adiabatic barrier is lower than the static one.

Unexpectedly, the ammonia dimer does not contain a linear hydrogen bond NH---N directed along the C_3 axis of acceptor molecule. The structure of $(NH_3)_2$, shown in Figure 8.20, was derived by Nelson et al. [1985a,b 1987a,b] from the analysis of pure rotational transitions. The rotational constant corresponds to distance 3.337 Å between the centers of mass of the subunits. The angles between the dimer axis and the C_3 axes of the molecules were determined from dipole moments and quadrupole coupling constants to be 48.6 and 64.5°. With an additional assumption about the symmetry plane and center for conversion, Nelson and Klemperer [1987] classified the rotation–tunneling states of $(NH_3)_2$ by using the permutation-inversion group. Taking into account the rigidity of intramolecular bonds and the asymmetry of the potential for the umbrella vibration of inequivalent molecules, these motions were

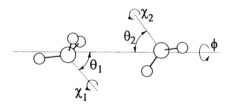

Figure 8.20. The structure of ammonia dimer. The centers of mass of the two subunits lie on the z axis. θ_1 and θ_2 are the angles between the C_3 axes of the subunits and the z axis.

excluded from consideration. The symmetry operations that correspond to the feasible dimer configurations form a G_{36} group. According to Bunker [1979], this group is isomorphic to the molecular symmetry group of the acetone molecule, when tunneling exchange is taken into account. The rigid dimer with this structure has C_s symmetry, so its rotational levels with $K_a = 0$ even and odd J and belong to the A' and A'' irreducible representations, respectively. Tunneling splits each of these levels into an octet. The correlation diagram presented in Figure 8.21 compares the states of the G_{36} group and its subgroups. Internal rotation of only one subunit of the dimer corresponds to the G_6 group and leads to splitting of the A' level into A_1 and E levels with the internal rotation quantum numbers $m_1 = 0$ and $m_2 = 1$, respectively. Combining rotations of both molecules forms G_{18} group, which has one A state and four E states. In state A neither of the two internal rotors is excited, $m_1 = m_2 = 0$. One of them is excited in the E_1 ($m_1 = 1$, $m_2 = 0$) and E_2 ($m_1 = 0$, $m_2 = 1$) states. These levels are nondegenerate because the rotors are inequivalent. In states E_3 both rotors are excited and rotate in the opposite directions ($m_1 = 1$, $m_2 = -1$, and $m_1 = -1$, $m_2 = 1$). In state E_4, $m_1 = m_2 = 1$. Intermolecular exchange transforms the G_{18} group into the G_{36} group, adding inversion to the operations of the $2C_3$ group. The A_1, E_3, and E_4

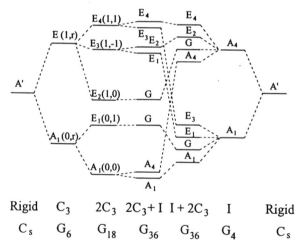

Rigid	C_3	$2C_3$	$2C_3 + I$	$I + 2C_3$	I	Rigid
C_s	G_6	G_{18}	G_{36}	G_{36}	G_4	C_s

Figure 8.21. Correlation diagram for NH_3 dimer. On the far left and right the system is in the nontunneling limit. The correlations from left to right lead to the nonrigid limit, with internal rotation splitting much greater than interchange tunneling splitting. The correlations from right to left describe the reverse situation. The designation Γ (m_1, m_2) indicates that the total tunneling wave function transforms as Γ, and m_1 and m_2 are the rotational angular momenta of subunits in the free rotor limit. (From Nelson and Klemperer [1987].)

levels are split into doublets, but the levels of nondegenerate states E_1 and E_2 correlate with the G levels of the G_{36} group. When $|m_1| \neq |m_2|$, the rotors are in different spin states and tunneling is forbidden, at least in the first order of perturbation theory. That is, the $G \leftrightarrow G$ transitions are pure rotational. This picture, described by the left part of the diagram, applies to the case $2C_3 + I$, when internal rotation tunneling splittings are greater than inversion splittings. The opposite case is shown in the right part of the diagram. When a rigid dimer transforms into the dimer with intermolecular exchange, the C_s group is replaced by the G_4 group, in which the A' level is split into an A_1, A_4 doublet, as in $(HF)_2$. The diagram column $I + 2C_3$ corresponds to the levels of the G_{36} group, when the inversion splitting exceeds the tunneling—rotation splitting. Nelson et al. [1987a,b] and Havenith et al. [1991] pointed out that the rotation—tunneling and vibration—rotation—tunneling spectra of $(NH_3)_2$ are in agreement with $2C + I$ case rather than with $I + 2C_3$. Similar consideration of the A" level, $J = 1$, $K_a = 0$, shows that this level is split into the octet A_2, A_3, E_1, E_2, E_4, E_3, G, G, since the rotational wave function belongs to the A_2 representation of the G_{36} group and $A_1 \times A_2 = A_2$, $A_4 \times A_2 = A_3$, $E_1 \times A_2 = E_1$, $E_2 \times A_2 = E_2$, $E_3 \times A_2 = E_4$, $E_4 \times A_2 = E_3$. $G \times A_2 = G$. Therefore, the expansion of the eigenfunctions of the C_s group in the G_{36} group is

$$A' = A_1 + A_4 + E_1 + E_2 + E_3 + E_4 + 2G$$
$$A'' = A_2 + A_3 + E_1 + E_2 + E_4 + E_3 + 2G$$

(8.45)

The total wave function of the dimer in the ground vibrational state can be represented as a product of wave functions for dimer rotation, internal rotation, and intermolecular exchange. The last of these depends on the polar angles θ_1 and θ_2 between the dimer axis and the molecular C_3 axes and has the form

$$\Phi_a = \Phi_1(\theta_1)\Phi_2(\theta_2) \qquad \Phi_b = \Phi_2(\theta_1)\Phi_1(\theta_2)$$

(8.46)

where Φ_1 and Φ_2 correspond to configurations of inequivalent subunits with equilibrium angles 48.6 and 64.5°, respectively. Intermolecular exchange transforms Φ_a into Φ_b and vice versa. Functions (8.46) belong to the $A_2 + A_4$ representation, so that the A_1 and A_4 states are described by their symmetric and antisymmetric combinations. Because rotation of the dimer as a whole separates from internal motions, the tunneling wave function for internal rotation and exchange depends on four indices, m_1, m_2, m_1', and m_2', and the sign of the linear combination of Φ_a and Φ_b. A space fixed z component of the transition dipole moment can be expressed in terms of molecular transition dipole components as

$$\mu_z = \alpha_{az}\mu_a + \alpha_{bz}\mu_b + \alpha_{cz}\mu_c \tag{8.47}$$

where $\alpha_{\lambda z}$ ($\lambda = a, b, c$) are the direction cosines of inertial axes in a fixed coordinate system. The values of μ_a, μ_b, and μ_c are proportional to the displacement along the large amplitude coordinate. As in the case of diatomic dimers, this coordinate is the antisymmetric combination of polar angles:

$$\theta_a = \theta_1 - \theta_2 \tag{8.48}$$

Neglecting small contributions from other motions to μ_a and μ_b, one gets

$$\mu_a = c_a\theta_a \qquad \mu_b = c_b\theta_a \qquad \mu_c = 0 \qquad \mu_z = (c_a\alpha_{az} + c_b\alpha_{bz})\theta_a \tag{8.49}$$

A direct calculation shows that α_{az}, α_{bz} and θ_a are transformed as A_2 and A_4, respectively, so the representation for μ_z is A_3. Using the character table for the G_{36} irreducible representations, Nelson and Klemperer [1987] found the selection rules for the transitions $J = 1 \leftarrow 0$, $K_a = 0$:

$$A_1 \leftrightarrow A_3 \qquad A_2 \leftrightarrow A_4 \qquad E_1 \leftrightarrow E_2 \qquad E_3 \leftrightarrow E_3 \qquad E_4 \leftrightarrow E_4 \qquad G \leftrightarrow G$$

$$\tag{8.50}$$

These rules show that the $G \leftrightarrow G$ transition, in contrast with the others, is purely rotational. In the coordinate system shown in Figure 8.20, the transition states for the *cis* and *trans* paths of interconversion have symmetry axes C_{2x} and C_{2y} and relate to the symmetry groups C_{2v} and C_{2h}, respectively. The different symmetries of the transition states results from the fact that the same permutation relates to different symmetry operations in C_{2v} and C_{2h}. For example, (ab)(14)(28)(36) is equivalent to inversion in C_{2h}, while in C_{2v} it corresponds to the reflection in the σ_{xy} plane. The symmetry of the reaction path does not affect the symmetry of states with even K_a (and $K_a = 0$). However, the selection rules for transitions $K_a = 1 \leftarrow 0$ are different for *cis* and *trans* paths. The classification of states with even and odd K_a and K_c, which relate to the prolate and oblate top limits, was given by Havenith et al. [1991] and is presented in Table 8.5. For the *cis* path, all transitions $K_a = 1 \leftarrow 0$ are pure rotational, while for the *trans* path all transitions except $G \leftrightarrow G$ are tunneling–rotation transitions.

Finally, we mention a very spectacular system, the mixed dimer $H_2O–NH_3$, studied by Fraser and Suenram [1992]. One proton of water molecule forms a hydrogen bond OH---N while the other is free.

TABLE 8.5
Symmetry of Rotational Wave Functions for *trans* and
cis Paths of Ammonia Dimer Interconversion

		Ψ_{rot}	
K_a	K_c	*trans*	*cis*
Even	Even	A_1	A_1
Even	Odd	A_2	A_2
Odd	Even	A_2	A_3
Odd	Odd	A_1	A_4

Exchange of these protons leads to the splitting of rotation–internal rotation levels. The tunneling exchange frequency is 56 MHz in the ground state and increases to 2.12 THz (71 cm^{-1}) when the NH_3 umbrella vibration ν_5 is excited. This increase is caused by the almost free rotation of the NH_3 molecule and by the weakening of the hydrogen bond following vibrational excitation. Since the N atom participates in the hydrogen bond, the asymmetry of the potential suppresses inversion of NH_3. However, excitation of the ν_5 vibration makes it possible. Hydro-

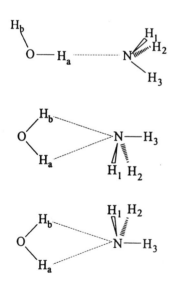

Figure 8.22. A possible tunneling pathway leading to the NH_3 inversion in the excited state. (From Fraser and Suenram [1992].)

gen exchange in the water molecule symmetrizes the potential for inversion. A proposed scheme of coupled tunneling motions of the dimer subunits is represented in Figure 8.22. The transition state is doubly degenerate as in the case of two-proton transfer, considered in Sections 6.2 and 6.4.

8.6.   LARGE-AMPLITUDE MOTIONS IN VAN DER WAALS COMPLEXES

Among van der Waals complexes, the complexes of simple molecules with rare gas atoms has received a great deal of attention because the PES of these species can be reconstructed from spectroscopic data. Studies of isolated complexes in supersonic jets provide a unique opportunity to determine experimentally the parameters of complex multidimensional PES, which may have several minima, with accuracy approaching 0.01–0.001 cm^{-1} in some cases. This feat seemed to be a formidable challenge only a few years ago! On the other hand, in view of their relative simplicity these complexes are reasonable subjects for quantum chemical calculations, so reliable comparisons of theory with experiment are possible.

Novick et al. [1973] were the first to study the electric resonance spectrum of Ar–HCl and show that this complex has a linear hydrogen-bonded structure, the amplitude of angular motion of the HCl molecule reaching 41°. A multidimensional potential for this system was proposed by Holmgren et al. [1978], based on the analysis of the ground-state rotational spectrum. The studies of vibration–rotation spectra performed by Ray et al. [1986], Robinson et al. [1987], Marshall et al. [1985], Howard and Pine [1985], Lovejoy and Nesbitt [1989] allowed reconstruction of the two-dimensional PES. Hutson [1988] proposed a PES that is similar to Holmgren's potential near the global minimum, but also contains a local minimum corresponding to the linear structure Ar–ClH (Figure 8.23). The existence of the second minimum follows from ab initio calculations performed by Andzelm et al. [1985] and Douketis et al. [1984]. The saddle point for interconversion is situated at $\theta^{\#} = 100°$; the barrier height is 67.1 cm^{-1}. Motion along the angular coordinate is accompanied by a decrease in the distance between the Ar atom and the center of mass of the HCl molecule from 3.39 Å in the ground state at $\theta = 0$ to 3.87 and 3.66 Å in the transition and final ($\theta = 180°$) states, respectively. The energy in the final state is 35.2 cm^{-1}.

Similar calculations enable one to define more precisely the parameters of atom–atom potentials. Hutson [1988, 1990] parametrized the intermolecular potential using the measured angular dependences of energy

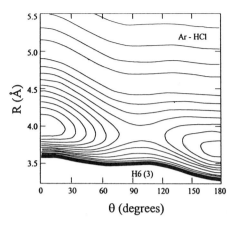

Figure 8.23. Contour plot of the PES for Ar–HCl. Contours are at 10-cm^{-1} intervals. (From Hutson [1988].)

and interatomic distance. These dependences are expanded in series of Legendre polynomials or modified spherical harmonics. Since the geometrical parameters of van der Waals complexes are close to those in molecular solids, the coefficients of these expansions can be used in the empirical AAP method described in Section 7.2. The Buckingham potential (7.26) can be modified to take into account higher order terms of the multipole interaction.

DeLeon and Muenter [1980] established that, in contrast with linear complexes of rare gases with hydrogen halides, the complex of acetylene with argon has a T-shape structure with a very low frequency of bending vibration (<10 cm^{-1}) corresponding to hindered internal rotation with large amplitude. The existence of linear and T-shape structures with similar energies is characteristic of the Ar–HCN complex [Leopold et al., 1984; Fraser and Pine, 1989]. Although the microwave spectrum corresponds closely to the hydrogen-bonded linear structure, the approximate parameters of the semi-rigid Hamiltonian appear to be incorrect. In particular, the small Ar–H distance does not agree with the small frequency of stretching vibration. This inconsistency was explained in the above papers as a result of strong coupling between rotational and translational motions with large amplitudes. From this example, we see that the concept of fixed molecular structure is to be changed for the systems whose PES's have multiple shallow minima so that the nuclear wave functions are not localized in limited regions of space near a single global minimum. When large-amplitude motions are present, the concept of a vibrationally averaged structure is more meaningful. In particular, the approximation of the semi-rigid rotor becomes invalid when the tunneling frequencies in floppy molecules are comparable with the rotation frequencies.

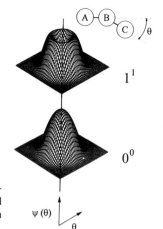

Figure 8.24. Eigenfunctions for the lowest two rovibrational ν^0_{bent} levels 0^0 and 1^1 of the bending potential (8.51), plotted in a rotating coordinate system. (From Nesbitt and Naaman [1989].)

To show how the parameters of an averaged structure can be extracted from spectroscopic data, Nesbitt and Naaman [1989] considered a model of three atoms, as shown in Figure 8.24, where the angular potential is a flat well:

$$V(\theta) = V_0 \theta^6 \qquad (8.51)$$

They numerically solved the vibration–rotation Schrödinger equation; the eigenfunctions of two lowest states are shown in Figure 8.24 in the rotating coordinate system. The centrifugal barrier in the $\ell = 1$ state results in displacement of the maximum of the vibrational wave function from $\theta = 0$, producing a bent configuration. If $V_0 = 10 \, \text{cm}^{-1}$ and the mass of each particle is 20 amu, the angular rms amplitudes are 38° in the state at $l = 0$ and 49° at $l = 1$. In spite of these large amplitude motions, the spectrum of the problem for energies up to $50 \, \text{cm}^{-1}$ can be fitted exactly using the Hamiltonian of a semirigid rotor with an interbond angle of 53.5°, This model illustrates that (1) an analysis of a vibration–rotation spectrum in a limited range of small energies does not necessarily reveal the existence of large-amplitude motions, and (2) a coincidence of the observed spectrum with the calculated one for a semi-rigid rotor does not prove that a molecule is actually rigid. Of course, this uncertainty is significantly diminished when the energies of higher levels are available from IR spectroscopic measurements.

The question regarding the statistical nature of the equilibrium configuration arises only in the case of low barriers. For chemical reactions, such barriers would correspond to very high rate constants so they would not satisfy the initial TST assumption that the barrier height should be much greater than the characteristic energies of the initial state (see inequality (1.1)). In the case of tunneling reactions this question is closely connected with the validity of the WKB approximation.

Cohen and Saykally [1991] and Lascola and Nesbitt [1991] reconstructed the multidimensional PES for the Ar–H$_2$O complex from experimental data. The intermolecular potential has the form

$$V(\theta, \phi) = \sum_{\mu, \nu} V_{\mu\nu} j_{\mu\nu}(\theta, \phi) \tag{8.52}$$

where $j_{\mu\nu}(\theta, \phi)$ are modified spherical harmonics. From the symmetry of the potential, $V(\phi) = V(\pi - \phi) = V(-\phi)$, it follows that only the harmonics with even ν enter into this expansion. The coefficients $|V_{10}|$, V_{20}, and V_{22}, which are 3.3, -5.14, and $16.53 \, \text{cm}^{-1}$, respectively, fit the measured spectrum nicely. The global minimum corresponds to a planar

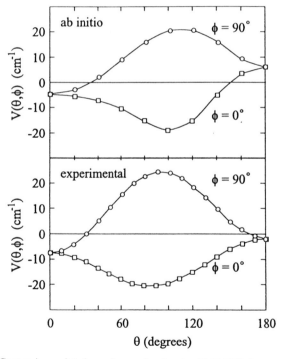

Figure 8.25. Comparison of θ dependences for the Ar–H$_2$O PES for cuts at constant ϕ obtained from ab initio calculation [Chalasinski et al., 1991] and from analysis of the near-IR vibration–rotation spectrum [Lascola and Nesbitt, 1991].

structure ($\phi = 0$) with the Ar atom situated at a distance of 3.60 Å from
the center of mass of the water molecule, so that the line between them is
directed at an angle of 90° with respect to the C_2 axis of the water
molecule. The barrier heights for planar rotation are 12.2 and 19.1 cm^{-1}
at $\theta = 0$ and $\theta = 180°$, respectively. Since the zero-point energies of *para*
and *ortho* states of the water molecule are greater than heights of both
barriers, the planar rotation is almost free. The barrier for out-of-plane
rotation situated at $\theta = \phi = 90°$ is 45.2 cm^{-1}. An important feature of this
complex is the orientational dependence of the equilibrium intermolecular distance, which follows from the fact that the radial and angular
motions cannot be separated from each other. As shown in Figure 8.25,
the PES reconstructed from the spectrum is in good agreement with
quantum chemical calculations of Chalasinski et al. [1991].

The complex H_2O–CO_2 has a T shape due to the van der Waals
C----O bond. The barrier height for internal rotation around the common

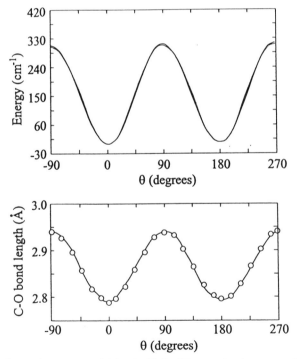

Figure 8.26. Internal rotor potential for the H_2O–CO_2 complex (upper panel). The dotted
line corresponds to ab initio scaled calculation. The lower panel shows the variation in the
intermolecular CO bond with internal rotation angle. (From Block et al. [1992].)

C_2 axis is 305 cm^{-1}, and the tunneling splitting is 0.136 cm^{-1} [Block et al., 1992]. In this complex, there is a coupling between internal rotation and the intermolecular stretching vibration. Specifically, this vibration acts as a breathing vibration, as described in Section 7.3 (see Figure 8.26). Its frequency, 113.6 cm^{-1}, is smaller than the torsion vibrational frequency, 216 cm^{-1}; therefore, the use of the sudden approximation for tunneling can be suggested.

HEAVY PARTICLE TRANSFER

CONTENTS

As shown in Section 2.5, when the tunneling mass increases much beyond the mass of an H atom, traversal of a multidimensional barrier tends to occur in the adiabatic regime (see inequality (2.86)). The tunneling trajectory approaches the MEP passing through the saddle point of the PES. This type of transition can be described as a one-dimensional tunneling through a vibrationally adiabatic barrier (1.10), and estimates of k_c and T_c can be obtained by substitution of the parameters of this barrier into Eqs. (2.6) and (2.7). It follows from these relations that k_c falls in the interval available for measurements if the increases in tunneling mass are accompanied by decreases in the barrier width and/or height such that the quantity $d(V_0 m/m_H)^{1/2}$ remains approximately constant. Thus, the tunneling transfer of heavy particles with $m \gg m_H$ can be observed if the barrier is sufficiently low and narrow.

In a typical case, the barrier widths in heavy-particle tunneling reactions correspond to transfer distances that are much smaller than that for hydrogen transfer and are not usually realized at van der Waals interreactant spacings in solids. Therefore, chemical conversions associated with heavy-particle tunneling are rare, often occurring in exoergic reactions where d is much smaller than the geometric transfer distance. A few examples of these reactions are cited in Section 9.2.

Heavy-particle tunneling is an important characteristic of degenerate impurity states in crystals and glass-like solids, and specific features of this phenomenon are similar to those discussed in previous sections. A brief description of tunneling impurity states in solids is given in Section 9.1. The third section of this chapter is devoted to low-temperature chain chemical reactions. A particularly interesting feature of these systems is that the interreactant distance for each step in the chain is decreased by strains resulting from previous steps, thereby producing a mechanism for low-temperature chain growth by heavy-particle tunneling.

9.1. TUNNELING IMPURITY STATES IN SOLIDS

Light atomic or molecular ions that are doped into alkali halide crystals as substitutional impurities frequently occupy off-center positions in the lattice site due to the mismatch in size between the impurity and the corresponding host ion it replaced. There is a small number of equivalent off-center positions, so that the impurity ground state is degenerate, and tunneling transitions between these positions become possible. In addition to tunneling splitting of the vibrational bands, these tunneling states manifest themselves as a low-temperature heat capacity that is sometimes greater than that of the pure host lattice at ~ 1 K. Reviews of the properties of such states are given by Narayanamurti and Pohl [1970] and Barker and Sievers [1975].

The vibrational frequencies of off-center impurities are usually much lower than the characteristic Debye frequency of the host lattice. They fall in a region of low phonon density, so that their coupling with the lattice vibrations is relatively weak. As a result, the bands corresponding to the local impurity vibrations are sufficiently narrow that tunneling splitting can be observed. Since transitions between different off-center positions are associated with a large charge displacement, the asymmetry of the potential wells changes upon application of an external electric or mechanical stress field. These effects enable the use of various electro-optical and electro-caloric methods as well as electric resonance methods to study the structure and dynamics of these centers. In particular, the kinetics of tunneling relaxation transition has been studied over a wide range of time scales (from 10^{-8} to 10^3 s) by observing changes in electric dichroism following switching of an external electric field.

When LiCl is doped into a KCl lattice, the Li^+ cation is incorporated as a substitutional impurity for K^+. Due to its smaller size, the impurity cation has eight equilibrium positions displaced in the [111] directions relative to the cation sublattice. These positions are laid out at the vertices of a cube inscribed in the unit cell. Each edge is 1.4 Å long, as shown in Figure 9.1A. The tunneling splitting of this feature appears as a quartet of lines, which spreads when an electric field is applied along different crystal axes, as shown in Figure 9.1B. The analogous variation of the splitting occurs when a mechanical stress field is applied along one of the crystal axes. The tunneling splitting corresponding to adjacent level spacing in zero field is 1.07 and 0.77 cm^{-1} for ^{6}Li and ^{7}Li, respectively. The librational frequencies are 17.8 and 40 cm^{-1}, and the barrier height is about 100 cm^{-1}. Even though the tunneling distance in this example is comparable to that of hydrogen transfer reactions discussed previously, the combination of a low barrier and large amplitude of the zero-point

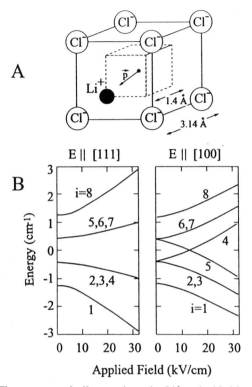

E ∥ [111]   E ∥ [100]

Figure 9.1. (A) The structure of off-center impurity Li^+ embedded in a KCl host crystal. The Li^+ cation tunnels predominantly along the edges of the dashed cube. The Cl^- ions at the centers of large cube faces are omitted for clarity; the vector **p** indicates electric dipole moment. (B) Influence of electric field on the tunneling levels in $KCl:^7Li^+$. (From Narayanamurti and Pohl [1970].)

vibration (~0.3 Å) permits the observation of coherent tunneling for a particle of mass 7 amu.

The molecular anion CN^- also occupies a [111] off-center position when present as a substitutional impurity in KCl, KBr, KI, and RbCl crystals. Librational motion of the CN^- manifests itself as a satellite on the CN stretching vibrational band. Above 60 K, the libration becomes thermally activated hopping. In $KCl:CN^-$, the librational frequency is $12\ cm^{-1}$ and the barrier height is $24\ cm^{-1}$. The tunneling splittings corresponding with this value of V_0 and a rotational constant $B = 1.25\ cm^{-1}$ in a cubic crystalline field are 1.15 and $0.85\ cm^{-1}$ for the $A_{1g}-T_{1u}$ and $T_{1u}-E_g$ transitions, respectively. For $RbCl:CN^-$ these splittings are 0.49 and $0.75\ cm^{-1}$. In $KBr:CN^-$, the tunneling splitting of the A_{1g}–T_{2g} transition was measured to be $2.3\,\text{cm}^{-1}$ by Rowe et al. [1980] using inelastic neutron scattering (INS).

Hydroxide molecular anions (OH^-) adopt [100] off-center states in alkali chlorides and bromides. These centers have six equivalent positions with different alignments of the dipoles. Electric resonance measurements give the tunneling splitting values 1.95 and $0.17\,\text{cm}^{-1}$ for NaCl:OH^- and KCl:OH^-, respectively.

In other crystals the splitting is too small for spectroscopic measurements, but the rate constants for incoherent tunneling have been found from relaxation measurements. The results obtained by Kapphan [1974] are listed in Table 9.1. The temperature dependences of the characteristic time scales for orientational relaxation of OH^- dipoles in several different crystals are depicted in Figure 9.2. Below 5 K, the relaxation times are inversely proportional to T, but scale as T^{-4} at higher temperatures.

Dick [1977] explained this behavior within the framework of a phonon-assisted tunneling mechanisms using the TLS approximation and golden rule formalism (see Sections 2.3 and 6.4). One-phonon transitions dominate the mechanism at low temperatures, resulting in a linear dependence of k with T; this follows directly from relation (6.27) when $\beta A \ll 1$. At higher temperatures, the main contribution comes from Raman processes, leading to a T^4 dependence of the rate constant. This predicted T^4 temperature dependence for RbBr:OH^- is analogous to results obtained by Silbey and Trommsdorf [1990] for two-proton transfer in benzoic acid crystals (see Section 6.4).

Table 9.1 shows that the tunneling splitting Δ decreases exponentially with increasing lattice parameter a_0. In the point-ion approximation in a cubic crystal potential, one would expect a decrease of the potential

TABLE 9.1
Rate Constants, Tunneling Matrix Elements, and Frank-Condon Factors for
Orientational Relaxation of OH Dipoles in Alkali Halide Crystals

Crystal	Nearest-Neighbor Distance (Å)	Rate Constants[a] (s^{-1})	Effective Tunneling Splitting (cm^{-1})	Φ_0
NaCl	2.79	—	—	
KCl	3.12	7×10^7	$0.17,^{[b]}\ 0.12$	7.2
KBr	3.26	3×10^4	1.4×10^{-3}	13.1
RbCl	3.22	10^5	3.3×10^{-3}	7.2
RbBr	3.39	7×10^2	1.2×10^{-4}	18.8
RbI	3.62	10^4	2.5×10^{-4}	27.0

[a] Electro-optical measurements at 1.3 K.

[b] Direct measurements of tunneling splitting.

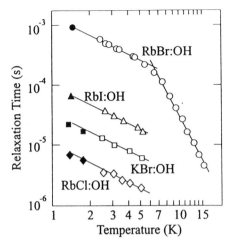

Figure 9.2. Temperature dependence of the relaxation time of OH^- in different host lattices obtained from optical (open symbols) and caloric (filled symbols) measurements. The latter have been corrected by the factor 2/3 derived from the three level model. (From Kapphan [1974].)

barrier with increasing a_0, i.e., an increase in Δ. The opposite dependence, $\Delta(a_0)$, is due to distortion of the surrounding lattice, which reduces the symmetry of the crystal field. In this case, the dipole reorientation is accompanied by reorganization of the environment in a manner similar to that discussed earlier for electron transfer in polar media and hydrogen diffusion in metals. As shown in Section 2.3, the reorganization of the medium renormalizes the tunneling matrix element, multiplying it by the Franck-Condon factor, which is sometimes referred to as a "dressing" factor. This factor diminishes Δ in comparison with the value Δ_0 for hindered rotation in a cubic crystalline field according to (2.51).

It is useful to compare the values of Φ_0 given in Table 9.1 with ones for hydrogen transfer listed in Table 6.5. Shore and Sander [1972] showed that Φ_0 can be calculated using the elastic dipole moments. The results indicate that the aforementioned exponential dependence of the splitting on a_0 is associated with an approximately linear growth of Φ_0 with interionic distance.

It should be noted that relation (2.51) is valid within the sudden approximation. However, the relaxation of heavy particle impurities typically involves motion that is slow compared with vibrations of the host lattice (i.e., the tunneling takes place in the adiabatic limit). The net effect of the adiabatic approximation is to renormalize the effective moment of inertia of the particle. This approach was used, for example, to describe vacancy diffusion in light metals. The evolution of the rate constant from Arrhenius behavior to the low-temperature plateau was described within the framework of one-dimensional tunneling of a particle with renormalized mass through an adiabatic barrier [Ranfagni et al., 1984].

We now consider two specific examples of heavy impurity relaxation. Narayanamurti et al. [1966] showed that NO_2^-, when present as an anionic substitutional impurity in KCl, rotates almost freely about its a axis, which lies in the molecular plane perpendicular to the principal axis (i.e., parallel to a line connecting the O atoms). The rotation leads to a splitting ($\Delta = 2\,\text{cm}^{-1}$) in the ν_3 fundamental band in the infrared spectrum that is comparable in magnitude with the rotational constant ($B = 4.22\,\text{cm}^{-1}$). When the host crystal is KI, the NO_2^- is displaced by 0.35 Å along the [110] direction relative to the normal I^- position. The impurity can occupy any of twelve equivalent positions, and the transitions between these positions introduce a fine structure to the ω_3 band. This spectrum, which was measured by Khatri and Verma [1983] at 1.7 K, is illustrated in Figure 9.3A, along with an energy level diagram showing the intermultiplet transitions (Figure 9.3B). The ground-state energy level splittings are determined by tunneling matrix elements for transitions between different equilibrium positions situated at the midpoints of the edges of a cube in the vacancy, viz.,

$$E(A_{1g}) = 4T_1 + 2T_2 + T_3 + 4T_4$$

$$E(E_g) = -2T_1 + 2T_2 + T_3 - 2T_4$$

$$E(T_{1u}) = 2T_1 - T_3 - 2T_4 \tag{9.1}$$

$$E(T_{2g}) = -2T_2 + T_3$$

$$E(T_{2u}) = -2T_1 - T_3 + 2T_4$$

where T_1, T_2, T_3, and T_4 relate to rotation of the dipole by 60, 90, 180, and 120° with simultaneous displacement of center of mass by x_0, $\sqrt{2}x_0$, $\sqrt{3}x_0$, and $2x_0$, respectively, where $x_0 = 0.346$ Å. The values of these matrix elements, which were assumed to be the same for the ground and excited states, are

$$T_1 = 0.039\,\text{cm}^{-1}$$

$$T_2 = 0.093\,\text{cm}^{-1}$$

$$T_3 = 0.060\,\text{cm}^{-1} \tag{9.2}$$

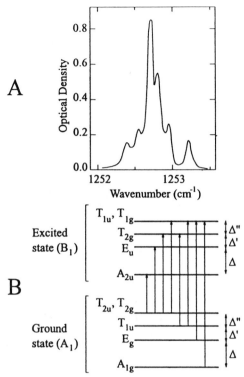

Figure 9.3. (A) IR absorption spectrum of KI crystal containing $\sim 10^{17}\,\mathrm{cm}^{-3}\,NO_2^-$ ions at 1.7 K in the ν_3 fundamental region. (B) Tunneling levels of the vibration ground and excited states for the [110] oriented NO_2^- dipoles in KI crystals. The arrows indicate allowed optical transitions, $\Delta = 6(T_1 - T_4) = 0.27\,\mathrm{cm}^{-1}$, $\Delta' = 4T_1 + 2T_2 + 2T_3 = 0.15\,\mathrm{cm}^{-1}$, $\Delta'' = 4(T_1 - T_4) = 0.13\,\mathrm{cm}^{-1}$. (From Khatri and Verma [1983].)

$$T_4 = 0.006\,\mathrm{cm}^{-1}$$

The largest values, T_2 and T_3, show that the rotation about the a axis, like in the KCl crystal, corresponds to the lowest barrier. Matrix element T_2 corresponds to a combination of concerted translation and tunneling rotation in which the oxygen atoms of NO_2^- ion are almost stationary, but the center of mass is displaced by 0.516 Å.

In connection with the aforementioned example, it is useful to note a simple method for estimating the height of a one-dimensional barrier when the values of Δ and the barrier width are known [Gomez et al., 1967]. For two displaced identical parabolic terms, the tunneling splitting is

$$\Delta = \frac{\hbar\omega_0}{\pi}\left(2\sqrt{\pi}\,\frac{a_0}{\delta_0}\right)\exp\left(-\frac{a_0^2}{\delta_0^2}\right)$$

$$\delta_0^2 = \frac{\hbar}{m\omega_0}$$

(9.3)

where a_0 is the distance between the two minima (see Section 2.5). Naturally, this expression is analogous to those described in Appendix A for more realistic potentials. When Δ and a_0 are known, (9.3) is the equation for determining ω_0, from which $V_0 = \frac{1}{2}ma_0^2\omega_0^2$ can be found.

The second example of tunneling by heavy substitutional impurities in crystals involves relaxation transitions of [110] off-center states formed by the Ag^+ ions in KCl and KBr crystals [Kapphan and Luty, 1972]. An Arrhenius plot of relaxation time is shown in Figure 9.4. The 60° transitions between the adjacent positions are thermally activated ($E_a = 0.322\,kcal/mol$) throughout the experimentally accessible range of relaxation times. The activation energy for 90° transitions between next-nearest positions is only about half as large ($0.174\,kcal/mol$) at $T > 5\,K$. However, the temperature dependence changes drastically below 5 K and corresponds to a multiphonon transition ($\tau^{-1} \sim T^{2.5}$) in the range 1.3–

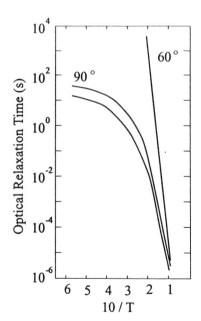

Figure 9.4. Arrhenius plot of the optical relaxation time of RbBr:Ag^+. 60 and 90° transitions are measured for electric directions [100] and [110], respectively. (From Kapphan and Luty [1983].)

3.5 K. The transition time is $\sim 10^2$ s at 1.3 K. Kapphan and Luty proposed a phonon-assisted tunneling mechanism for this process. Although the notion that a mass as heavy as Ag^+ can tunnel 0.9 Å through a ~ 0.17-kcal/mol barrier at 1.3 K is quite surprising, the experimental data and analysis clearly support this conclusion.

The wide bands of dielectric losses occur in the region from 10 Hz to 10 MHz for many organic species doped into a predeuterated poly-ethylene matrix at 4.2 K [Gilchrist, 1991]. The relaxation is caused by tunneling transitions between several equilibrium positions of OH or NH_2 groups. The bandwidth is determined by the distribution of barrier parameters in amorphous matrix.

Studies of impurity tunneling states ultimately led to the hypothesis that some of the low-temperature properties of amorphous solids (glass-es) are dominated by the presence of structural defects having at least two configurations with similar energies. A multitude of investigations of thermal, electrical, and acoustic characteristics of various glass materials in the range 0.1–10 K revealed several characteristics that distinguish the amorphous state of a substance from its crystalline form.

Phillips [1972] and Anderson et al. [1972] suggested that the low-temperature properties of glasses are determined by the existence of TLS tunneling states having a continuous distribution of splittings. At present this phenomenological description of low-temperature amorphous solids is generally accepted. Although many studies have confirmed the exist-ence of these states with tunneling splitting from 0.01 to 10 cm^{-1}, the detailed relationships between energy levels and specific structural fea-tures of glasses remain unknown in most cases. It is usually implied that the continuum of tunneling states is associated with collective translation-al and rotational motions with large amplitudes due to the distribution of lattice parameters that is inherent to disordered solids. The frequencies of these motions do not usually exceed 100 cm^{-1}, and it is this part of the lattice vibration spectrum where the density of states in glasses is sometimes greater than that in crystals.

Some features of the tunneling state spectrum can be illustrated by considering the structure of vitreous silica, which was studied by Buch-enau et al. [1986] using INS. The motif of this structure is tetrahedral SiO_4. The vibrational modes having frequencies greater than 50 cm^{-1} are assigned to stretching SiO vibrations. Lower frequencies are characteristic of bending O–Si–O and Si–O–Si vibrations as well as distortion of the tetrahedron as a whole. In addition to localized vibrations of each tetrahedron, the spectrum contains a set of modes corresponding to coupled hindered rotations of several tetrahedra in the range 10–20 cm^{-1}. Due to the structural inequivalency of these tetrahedra, the PES has a large number of shallow minima corresponding to motions of wide amplitude.

The appearance of double-well potentials as a result of structural disorder were analyzed by Klinger [1985]. Here, we illustrate his results using the particular example of a one-dimensional potential

$$U(x, R) = U_0\left(x + \frac{R}{2}\right) + U_0\left(-x + \frac{R}{2}\right),$$

where

$$U_0 = D(1 - e^{-\alpha x})^2 \tag{9.4}$$

formed by oppositely directed Morse functions having a distance R between minima. Two minima in this potential emerge when

$$R \geq \frac{2 \ln 2}{\alpha} \tag{9.5}$$

and the barrier between them is

$$V^{\#} = \left(\frac{\zeta}{1 + \zeta}\right)^2 D$$

where

$$\zeta = \tfrac{1}{2} e^{\alpha R/2} - 1 \tag{9.6}$$

The intrawell vibrational frequency is related to the fundamental frequency of the Morse function $\omega_0 = \sqrt{2\alpha D/m}$ by the relation

$$\omega^2 = \omega_0^2 \frac{\zeta(2 + \zeta)}{(1 + \zeta)^2} \tag{9.7}$$

The condition for formation of the TLS is $V^{\#} \geq \hbar\omega/2$, which corresponds to ζ values exceeding a critical value:

$$\zeta_c = 2^{-1/3}\left(\frac{\hbar\omega_0}{D}\right)^{2/3} \tag{9.8}$$

From (9.7) and (9.8) it follows that at $\zeta \simeq \zeta_c$, low-frequency TLS's appear with $\omega_c \simeq \omega_0(\hbar\omega_0/D)^{1/3}$ and large zero-point amplitude $\delta_c \simeq \delta_0(D/\hbar\omega_0)^{1/6}$. Since the tunneling splitting varies exponentially with the barrier height, as in Eq. (9.3), even a narrow distribution of R values near the critical value ζ_c results in creation of a continuous spectrum of tunneling states.

Systems having soft double-well potentials (i.e., those with low barriers) with characteristic tunneling splittings greater than $0.1 \, cm^{-1}$ are usually observed spectroscopically. However, in glasses, the intermolecular potentials usually have higher barriers and smaller splittings. It is the relatively high concentration of these closely spaced tunneling states that governs the relaxation kinetics of disordered materials at low temperatures. In conclusion, we refer the reader to the reviews edited by Phillips [1981], which are devoted to tunneling states in glasses.

9.2. CHEMICAL REACTIONS

The only example of coherent chemical exchange associated with heavy particle transfer is automerization of cyclobutadiene:

$$\boxed{} \; \rightleftarrows \; \boxed{} \qquad\qquad (9.9)$$

Interest in this reaction was kindled after Whitman and Carpenter [1982] discovered its anomalous negative activation entropy at 250–270 K. Dewar et al. [1984] considered a tunneling mechanism for this reaction. The automerization barrier arises when passing from the initial rectangular configuration with the alternating bonds of lengths 1.56 and 1.33 Å to the square transition-state configuration, which has a bond length of 1.45 Å and a barrier height of 8–12 kcal/mol according to calculations performed by Carsky et al. [1988]. The tunneling splitting was calculated by these authors within the vibrationally adiabatic approximation. The transition occurs along the MEP on a two-dimensional PES, $U(Q, q)$, that is represented by a polynomial expansion. The coordinates Q and q are the symmetric and antisymmetric combinations of the changes in the two types of bond lengths. The effect of the antisymmetric vibration is so small that the MEP is practically a straight line and corresponds to concerted change of both bonds. The reduced tunneling mass is

$$m = \frac{2(m_C + m_H)}{(1 + \sin \alpha)^2} = 6.53 \, amu \qquad\qquad (9.10)$$

where $\alpha = 81.24°$ is the angle formed by the lines drawn from each of the two carbons on the short side of the rectangle to the center of the molecule. The tunneling splitting in the ground state is $4.2 \, cm^{-1}$ in $^{12}C_4H_4$ and $2.35 \, cm^{-1}$ for $^{13}C_4D_4$. Because of the weak coupling of the tunneling coordinate to the transverse vibration, Δ changes only slightly when this mode is excited; it is 4.2, 4.6, and 5.1 cm^{-1} for the $n = 0$, 1, and 2 levels, respectively. Excitation of the symmetric vibration at 1567 cm^{-1} corresponding to the reaction coordinate causes Δ to increase to 82 cm^{-1}. After accounting for the amplitude of zero-point vibrations, the tunneling distance in the ground state is only 0.18 Å. Since the barrier is narrow, coherent tunneling dominates over thermally activated transitions, even at 350 K.

The tunneling transitions between optically active conformers of (4,4,4)-propellahexaene (three six-member carbon rings connected by the common simple C–C bond) has been studied by Zebretto and Zgierski [1989]. At room temperature, this compound exists as a racemic mixture, but according to NMR linewidth measurements, the rate constant for racemization must be smaller than 10^6 s^{-1}. The calculated barrier height is 1.86 kcal/mol, and the vibrational frequencies in the equilibrium configuration and in the upside-down barrier are 85 and 68 cm^{-1}, respectively. The parameter $dm^{1/2}$ that characterizes the tunneling length in mass-weighted coordinates is 4.5 Å amu$^{1/2}$. The rate constant calculated from this model is 6×10^{-3} s^{-1}, more than 8 orders of magnitude less than the limit determined by NMR. No other experimental measurement is currently available.

The tunneling intramolecular rearrangement of 2-norbornyl cation was considered by Yannoni et al. [1982] and Myhre et al. [1985]. The NMR spectra show that the structure is symmetric at temperatures as low as 4.2 K, so that the interconversion rate constant between the two asymmetric structures is greater than 10^9 s^{-1}:

$$(9.11)$$

According to the calculations by Fong [1974] and Brickman [1981], the barrier height is greater than 1 kcal/mol. It is unlikely that classical transitions over the barrier could occur with the indicated rate, so heavy-particle tunneling through the barrier is indicated in this case, although there is little direct evidence to support this conclusion. The calculated vibrational frequency at the minimum of the double-well potential is 200 cm^{-1}, and the displacement of the methylene group in the transition is $\sim$0.8 Å.

The low-temperature limiting rate constant for the isomerization of the biradical 1,3-cyclopentadiyl to bicyclo-(2,10)-pentane

$$\text{(structure)} \longrightarrow \text{(structure)} \tag{9.12}$$

has been found by Buchwalter and Closs [1979]. At $T_c = 15$ K, $k_c = 6 \times 10^{-4} \, \mathrm{s}^{-1}$. Above 20 K the apparent activation energy is 2.3 kcal/mol. The small value of the prefactor, according to the discussion in Section 2, is evidence for the tunneling character of the transition. The H/D kinetic isotope effect in reaction (9.12) is small, which indicates that the contribution of individual hydrogen atom displacements in the reaction is immaterial. Instead, the relevant coordinate is the out-of-plane bending vibration of the five-member ring. The experimental value of k_c is consistent with tunneling of a particle with reduced mass $\frac{1}{2}(m_C + 2m_H) \sim 7$ amu, associated with this vibration, through a parabolic barrier with a half-width of 0.6 Å. This distance is approximately equal to the relative displacement of the CH_2 groups required for formation of the four-membered ring in the product.

There have been numerous studies of low-temperature chemical reactions of matrix-isolated reactants (see, for example, the review by Perutz [1985]). Two of the most interesting from the standpoint of this volume are the reactions of NO with O_2 and O_3 studied by Smith and Guillory [1977] and Lucas and Pimentel [1979]. The *cis* dimer $(NO)_2$ has been formed in solid oxygen at 13–29 K. Reaction of this species with the O_2 matrix forms the product N_2O_4 in an electronically excited state. The transition state structure is reportedly of the form

$$\begin{array}{c} O-O \\ \vdots \quad \vdots \\ N - N \\ O^{\diagup} \qquad {}^{\diagdown}O \end{array} \tag{9.13}$$

The rate constants for ^{14}NO and ^{15}NO at 13 K are $8 \times 10^{-4} \, \mathrm{s}^{-1}$ and $5.2 \times 10^{-4} \, \mathrm{s}^{-1}$, respectively. The isotope effect, low-activation energy (0.1 versus 0.65 kcal/mol for the gas-phase reaction) and anomalously low prefactor ($\sim 10^{-2} \, \mathrm{s}^{-1}$) all point to a tunneling mechanism for this reaction.

The chemiluminescent gas-phase reaction

$$NO + O_3 \rightarrow NO_2 + O_2 \tag{9.14}$$

is exoergic by 47.1 kcal/mol, and has an activation energy 2.3 kcal/mol.

Investigation of this reaction isolated in an N_2 matrix at 10–20 K has shown that the apparent activation energy is smaller than 0.11 kcal/mol; the unimolecular rate constant for NO–O_3 reactant complexes prepared in this way is $1.4 \times 10^{-5} s^{-1}$ at 12 K. Experiments carried out using ozone enriched with ^{18}O have revealed no observable isotope effect.

The PES for the gas-phase reaction (9.14) was calculated by Viswanathan and Raff (1983) using the London-Eyring-Polanyi-Sato (LEPS) method (see, for example, Eyring et al. [1983]). Vibrationally excited products in the ground electronic state are created when the end O atom is abstracted from the O_3 molecule as a result of the approach of the NO molecule at an ONO angle of 110°. The barrier height for this pathway is 3.57 kcal/mol. After accounting for zero-point vibrational energies, the calculated activation energy (2.13 kcal/mol) is close to the measured value in the gas phase. In the transition state, the saddle point is shifted toward the reactant valley and the distances R_{OO} and R_{NO} are 1.277 and 1.957 Å, respectively. The equilibrium bond lengths in O_3 and NO are 1.272 and 1.150 Å.

The structure of the NO–O_3 van der Waals complexes was studied by Arnold et al. [1986]. Their results are summarized in Table 9.2. The binding energies of all 5 structures are similar, and the reaction coordinate for (9.14) begins at a configuration intermediate between structures A and B, during hindered rotation of NO about the symmetry axis of O_3. Comparison of the geometry of the van der Waals complex A with the transition-state arrangement shows that the arrangement of the reactants provides the same angle of attack as in the gas-phase reaction. The O–O bond stretching does not exceed 0.2 Å, and the relative displacement of the O–NO bond in the barrier is 0.45 Å. The promoting effect is due to the intermolecular vibrations as well as the aforementioned hindered rotation. Because of the low frequencies of these modes (25–$30 cm^{-1}$) the low-temperature limit has not been reached, even at 10 K.

Photodissociation and subsequent rebinding of CO and O_2 ligands to the heme center of myoglobin was studied by Austin et al. [1975] and Alberding et al. [1976a,b, 1980]. Rates of formation of the Fe–CO and Fe–O_2 bonds have been measured by transient optical bleaching experiments over a wide range of timescales (10^{-6}–$10^3 s$) and temperatures (2–100 K). The apparent activation energy in the range 30–100 K is 0.9 kcal/mol, but at lower temperatures, the reaction occurs by heavy particle tunneling. This regime is characterized by T_c and k_c, which are ~ 20 K and $\sim 10^{-1} s^{-1}$, respectively.

A qualitative structural model has been proposed by these authors. Before the photolysis, the 6-coordinated Fe^{2+} ion is located in the plane of the porphyrin ring of the heme unit. Rupture of the coordination bond

TABLE 9.2
Structure and Energy of Various O_3–NO van der Waals Complexes

Complex	Structure	Energy (kcal/mol)	cm–cm Distance (Å)
A	(O₃–NO structure)	2.418	2.30
B	(O₃–NO structure)	2.734	2.15
C	(O₃–NO structure)	1.315	2.50
D	(O₃–NO structure)	1.952	1.85
E	(O₃–NO structure)	2.751	2.80

Note. O_3 and NO are in their separated equilibrium configuration. Only the cm–cm distance is varied.

involves not only displacement of the ligand, but also a rearrangement of the coordination sphere such that the Fe^{2+} ion leaves the plane of the heme. Rebinding of the ligand is overall exoergic, but requires surmounting a barrier. As shown by Chance et al. [1983] the change in Fe–CO bond length upon photodissociation of carboxyhemoglobin is less than 0.1 Å, so that the barrier to recombination is associated mainly with reorganization of the coordination sphere (see also Debruner and Fraunfelder [1982]).

Reactions of organic radicals with chlorine

$$R + Cl_2 \rightarrow RCl + Cl \tag{9.15}$$

have been explored by Misochko et al. [1980] and Benderskii et al. [1983] using a combination of spectrophotometric detection of molecular chlorine and EPR detection of the organic radicals R. The radicals are generated by laser photolysis of frozen mixtures of chlorine with various hydrocarbons:

$$Cl_2 + RH + h\nu \rightarrow Cl + HCl + R \tag{9.16}$$

The solid-state environment prevents diffusion of radicals, so reactions are only possible with neighboring Cl_2 molecules. Reaction (9.15) can therefore occur only in clusters $(RH-Cl_2)_n$, where $n \geq 2$. Products are observed only when the mole fraction of chlorine in the mixture is greater than 0.1. The $k(T)$ dependence includes an Arrhenius region (60–90 K) in which the activation energy (2–4 kcal/mol) is 1.5–2 times larger than in the corresponding gas-phase reactions. The low-temperature plateau occurs below 40–50 K where k_c is $5 \times 10^{-3} s^{-1}$ and $2 \times 10^{-2} s^{-1}$ for the radicals of n-butylchloride and methylcyclohexane, respectively.

Calculations show that the barrier width for this reaction corresponds to the interreactant distance 2.7–2.8 Å. This is much smaller than the equilibrium van der Waals distances between them (3.4–3.5 Å). If this van der Waals distance were the actual interreactant distance, then the predicted value of k_c would be some 15 orders of magnitude smaller than the observed value. It is clear by now that the tunneling reaction (9.15) becomes possible at low temperatures only due to the promoting effect of low-frequency intermolecular and bending vibrations.

Quantum yields for photochemically induced low-temperature reactions are also sensitive to intermolecular effects (i.e., the structure of the solid). For example, the quantum yield for reaction (9.15) in glassy mixtures of methane and chlorine at 20 K varies from 0.1 to 0.001, depending on the method of sample preparation [Benderskii et al., 1983]. The lowest values are obtained for samples prepared by thermal or ultrasonic annealing prior to UV photolysis.

Haase et al. [1989] have considered a tunneling mechanism for dissociative chemisorption of nitrogen on metal surfaces in order to explain the sticking probabilities S_d at low kinetic energies, which were measured by Rettner and Stein [1987]. Classical trajectory calculations give values of S_d in agreement with experiment at moderate energies $(E \geq 20 \, kcal/mol)$ where $S_d \geq 10^{-3}$. However, for lower kinetic energies of the incident N_2 molecules, a tunneling mechanism is required, since the values of S_d predicted by a classical mechanism are 3–4 orders of magnitude smaller than the measured values, which fall in the range 10^{-7}–10^{-6}.

The origin of the tunneling efficiency can be traced to a specific feature of the PES, specifically, the presence of an extremely narrow barrier between the initial state (physisorbed N_2 molecule) and the final state (chemisorbed N atoms). The model PES proposed by Haase et al. [1989] is characterized by two coordinates: an N–N stretching coordinate (r) directed along the surface, and the distance (z) from the surface to the N_2 center of mass. In the initial state, $r_0 = 1.10$ Å, and $z_0 = 2.48$ Å. The corresponding vibrational frequencies are 2180 and 355 cm^{-1}. At the transition state $r^{\#} = 1.14$ Å, and $z^{\#} = 2.25$ Å (Figure 9.5). Comparing

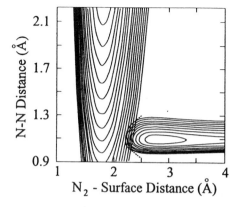

Figure 9.5. Contour plot of the two-dimensional PES for dissociative adsorption of N_2 on a rhenium metal surface. Broken line indicates the seam between valleys. (From Haase et al. [1989].)

these displacements with the zero-point amplitudes ($\delta r_0 = 0.047$ Å, $\delta z_0 = 0.055$ Å), one can see that the reaction coordinate nearly coincides with the z coordinate. The high frequency of the transverse motion (along r) means that the tunneling process occurs in the adiabatic limit. The crossover temperature, which is determined by the frequency of the upside-down barrier along the longitudinal z coordinate, is greater than 300 K. Despite the high barrier (17.5 kcal/mol) and tunneling mass, the transmission coefficient of the thin barrier is $\sim 10^{-8}$, which leads to observable values of S_d.

The quantum mechanism of dissociative chemisorption of hydrogen and hydrogen-containing molecules is widely discussed in the dynamics of catalytic reactions (see, for instance, Harris and Anderson [1985] and Chiang and Jackson [1987]).

A number of recent investigations have been concerned with the mobility of heavy atoms in rare gas matrices. Although not directly related to tunneling processes, they are concerned with important fundamental dynamics of atoms and small molecules in low-temperature solids, so we shall briefly review selected examples here. A typical experiment of this type includes the photolytic formation of atoms (see the review by Perutz [1985]) with subsequent detection of the decrease in atom concentrations due to bimolecular recombination. In most cases the rates are diffusion limited, and the temperature dependences are characteristic of thermally activated transfer.

Krueger and Weitz [1992] measured the diffusion coefficient of oxygen atoms in the ground ^3P state in solid xenon by using the fluorescence of $XeO^{\#}$ excimers to monitor the $O(^3P)$ concentration. The experimentally determined values are 5.4×10^{-18} and 2.0×10^{-17} cm^2/s at 32 and 40 K, respectively. These diffusion coefficients are only about three orders of magnitude smaller than for hydrogen atoms ($\sim 10^{-14}$ cm^2/s at 50 K), as we mentioned in Section 6.5. The diffusion is believed to take place by successive transitions between neighboring octahedral sites in the lattice. In each transition state the O atom passes through a "triangular window" between Xe atoms where the Xe–O distance is only 2.5 Å. There is a strong repulsive interaction between $O(^3P)$ and Xe in this configuration, so that the transition must be accompanied by substantial displacements of the cage atoms surrounding the impurity O atom.

The photodissociation dynamics of F_2 in solid argon were studied by Schwentner and Apkarian [1989] and Feld et al. [1990]. The fluorine molecule exists as a substitutional impurity in the fcc lattice of argon, where it can undergo rotational motion. A molecular dynamics simulation performed by Alimi et al. [1990] shows that this rotation is accompanied by a considerable displacement of the center of mass of F_2 (Figure 9.6a), so the motion is strongly coupled with translation. This is similar to the

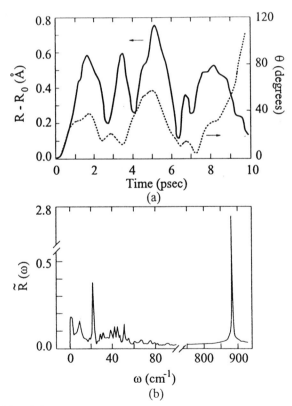

Figure 9.6. Translational and rotational motion of an F_2 molecule in solid Ar at 12 K. (a) Time evolution of coordinates; (b) Fourier transform of (a). (From Alimi et al. [1990].)

case of $KI:NO_2^-$ impurity rotation, which we discussed in Section 9.1. The Fourier transform of the R, θ motions shown in Figure 9.6a provides the vibrational spectrum in which the local low-frequency mode of 22.1 cm^{-1} is assigned to this coupled motion (Figure 9.6b). Argon matrix modes are mixed with the local mode, and we can think of rotation as occurring when the impurity molecule collides with the oscillating walls of the cage.

The photodissociation quantum yield of F_2 is determined by the relative probabilities of cage escape and recombination of the molecule (see Table 9.3. At photolysis wavelengths where the excess energy $E_k \leq 0.6 \text{ eV}$ (0.3 eV per atom) no dissociation occurs because $\frac{1}{2}E_k$ is smaller than the barrier height. The calculation of nonreactive trajectories shows that F atoms recombine in less than 2.5 ps after a number of elastic collisions with the cage walls. At higher energies, the dissociation yield increases rapidly with increasing E_k, reaching values near unity for $E_k > 2 \text{ eV}$. In this case, a symmetric exit takes place, and both the atoms leave the cage. The topology of the PES shows that there are the several directions along which the probability for exit is maximized. The cones corresponding to these reactive channels are shown in Figure 9.7. The first cone relates to escape to an octahedral site in a neighboring unit cell through the "triangular window" transition state. The second corresponds to transfer to the next nearest cell. The probability of exiting within these reactive cones depends on the amplitude of the aforementioned torsional–translational motion of the F_2 molecule, because the sudden nature of the photodissociation event causes the atoms to fly away along the original molecular axis of F_2. The amplitude of the torsional–translational motion increases with temperature such that at higher temperatures, the atoms sample increasing numbers of nonreactive paths. This interpretation is consistent with the observation that the quantum yield at 12 K is actually smaller than that at 4 K. When E_k grows, the cone angles increase to the point where the lattice becomes so "porous" that there is no inhibiting cage effect for photodissociation. A sample reactive trajectory is shown in

TABLE 9.3
Quantum Yield of F_2 Photodissociation in Argon
Crystal, Calculated by Molecular Dynamic Simulation

Excess Energy (eV)	4 K	12 K
0.5	0.06	0.00
0.75	0.40	0.12
1.0	0.75	0.55
2.0	0.86	0.90
2.8	—	0.98

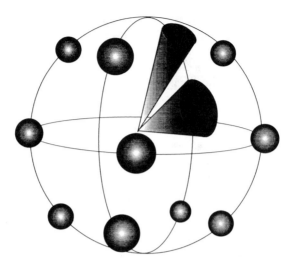

Figure 9.7. The twelve atoms of the reaction cage for F_2 in Ar corresponding to the fcc unit cell. The two cones show the initial orientations of the molecules that result in F atoms separation under photodissociation. (From Alimi et al. [1990].)

Figure 9.8. Along this path the excess energy decreases gradually until the kinetic energy is less than the barrier height. The total travel length is about 30 Å, which means that the hot atoms migrate surprisingly long distances in the lattice before being trapped.

An interesting feature of this system is that the dissipative motion of the hot atom in the randomly fluctuating lattice seems to be directional. This channeling migration is due to special features of the Ar–F interaction. The repulsive part of Ar–F potential is short range, and this provides a relatively wide channel for classically allowed motion when the kinetic energy is greater than the barrier height. Only classical motion was considered in these simulations. Naturally, the quantum effect could be significant at excess energies smaller than $2V^{\#}$; this would have to be included in any consideration of diffusion of the thermalized atoms.

While hot fluorine atoms provide an interesting example of long-range migration, cage recombination is much more effective in other systems. For example, the photodissociation quantum yield of Cl_2 in an argon matrix is less than 10^{-6} [Fajardo et al., 1988]. Geminate recombination of chlorine atoms in rare gas matrices involves a number of radiative and radiationless processes studied by Bondybey and Fletcher [1976]. The process begins with UV photoexcitation of the chlorine molecule from its ground $X^1\Sigma_g^+$ state to the repulsive $C^1\Pi_{1u}$ surface, which correlates with the final state in which there are two separated chlorine atoms in their

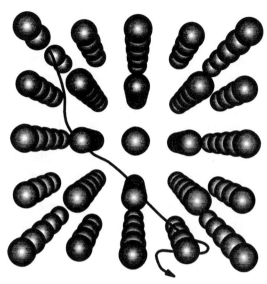

Figure 9.8. The motion of a hot F atom in argon crystal. $E_k = 2.8$ eV, $T = 12$ K. (From Alimi et al. [1990].)

ground $^2P_{3/2}$ electronic states (Figure 9.9). This level also correlates with three bound excited valence states of Cl_2, namely the $A'^3\Pi_{2u}$, $A^3\Pi_{1u}$ and $^3\Pi_{0^-u}$ surfaces. A fourth bound excited state, $B^3\Pi_{0^+u}$, correlates with the $Cl(^2P_{3/2}) + Cl(^2P_{3/2})$ asymptote (see, for example, Chang [1967]). In rare gas matrices, recombination of a Cl atom pair populates the $A'^3\Pi_{2u}$ state via fast radiationless transitions. Phosphorescence from the $v = 0$ level of this state to various vibrational levels of the ground $X^1\Sigma_g^+$ state has been observed with a radiative lifetime of 76 ms [Bondybey and Fletcher,

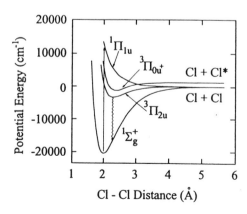

Figure 9.9. Approximate potential energy curves of low-lying electronic states of a Cl_2 molecule. (From Bondybey and Fletcher [1976].)

1976]. Interestingly, the argon matrix experiments show no evidence of $B^3\Pi_{0^+u} \leftrightarrow X^1\Sigma_g^+$ transitions, which are known from gas-phase studies [Clyne and McDermid, 1979].

When molecular chlorine is subjected to photodissociation in hydrocarbon matrices, the chlorine atom air can either recombine or react with the hydrocarbon molecules that form the cage walls, i.e., reaction (9.16). Tague and Wight [1992] reported that the branching ratio for these two competing processes can be influenced by application of an external magnetic field during the photolysis period. The magnetic field effect (MFE) can be observed not only as a variation in the quantum yield of the reaction product, but also as a complementary change in the intensity of the $A'^3\Pi_{2u} \rightarrow X^1\Sigma_g^+$ phosphorescence from Cl_2 molecules that recombine on the excited surface. The results are illustrated in Figure 9.10.

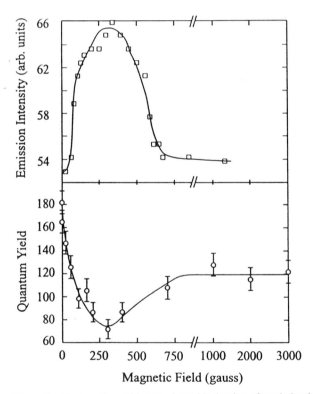

Figure 9.10. Magnetic effect on the solid-state photochlorination of methylcyclopropane at 77 K. The lower panel shows the variation of quantum yield of chain reaction initiated by Cl_2 photodissociation. Data in the upper panel represent the intensity of recombinational phosphorescence. (From Tague and Wight [1992].)

The origin of the MFE can be understood by examining the states of the separated Cl atom pair near the asymptotic region. Here, the spin-orbit states, which are characterized by the quantum number $\Omega = 0, 1, 2, 3$ are split by interaction of the nuclear spin ($I = \frac{3}{2}$ for both ^{35}Cl and ^{37}Cl) with the spin-orbit vector of each atom. The magnitude of this splitting is only about $0.01\ \text{cm}^{-1}$ (the hyperfine coupling constant of gas-phase Cl), which is much smaller than the characteristic thermal energy (kT at 77 K is $54\ \text{cm}^{-1}$). If motion through curve crossing regions (e.g., between the $C^1\Pi_{1u}$ and $A'^3\Pi_{2u}$ surface in the asymptotic region) is fast compared with the hyperfine correlation time ($\sim 5 \times 10^{-10}$ s), then the system will tend to follow the diabatic curves (i.e., the ones for which Ω is the good quantum number). In this scheme, a Cl atom pair initially prepared on the $C^1\Pi_{1u}$ surface by photodissociation will tend to remain there.

However, when the system is subjected to an external field, the magnetic sublevels in the asymptotic region approach each other and begin to mix. In fact, at 0.1 T, four of the sublevels (one each from $\Omega = 0, 1, 2, 3$) become degenerate. Since the system spends a relatively long time in this mixing region, the nuclear and electron spins become correlated, and transitions can occur from the initial repulsive surface ($\Omega = 1$) to either of the attractive surfaces ($\Omega = 0, 2$, or 3). Analogous MFEs were studied thoroughly in liquid-phase radical pair recombination (see, the monograph by Salikhov et al. [1984]) and result from level crossings that occur when the Zeeman energy becomes equal to the zero field splittings. There are two outstanding features of the study by Tague and Wight [1992]. The first is that the low-temperature solid environment leads to an unusually large MFE as a result of dissipative motion along the potential energy surfaces. The second is that the detailed mechanism for MFEs in systems with large spin-orbit coupling constants has been elucidated for the first time.

9.3. CHAIN REACTIONS

The chain polymerization of formaldehyde CH_2O was the first reaction for which the low-temperature limiting of the rate constant was observed (see reviews by Goldanskii [1976, 1979]). Each growth step involves the formation of a C–O bond between neighboring monomer molecules in the lattice, and the charge is transferred to the growing end of the polymer, viz.

$$H(OCH_2)_n^+ + CH_2O \rightarrow H(OCH_2)_n OCH_2^+ \qquad (9.17)$$

Initiation of the reaction sequence occurs by proton transfer from a photogenerated acid such as HCl, or from an excited-state acid such as nitrophenol [Mansueto et al., 1989; Mansueto and Wight, 1989]. In the original work of Goldanskii [1979] the CH_2O^+ initiator was formed by γ radiolysis.

The rate constant for each growth step is constant over the range 4.2–12 K, where $k_c = 1.6 \times 10^2 \, s^{-1}$. At higher temperatures, the apparent activation energy increases to 2.3 kcal/mol at 140 K, where $k \sim 10^5 \, s^{-1}$.

Polymerization reactions of formaldehyde and acetaldehyde have now been studied by several research groups, and the results are generally in agreement despite the variety of different experimental approaches to the problem (see, for example, Mansueto and Wight [1992]). Similar effects have been found in γ-irradiated acrylonitrile and acrolein [Gerasimov et al., 1980].

Basilevsky et al. [1982] proposed a mechanism of ionic polymerization in crystalline formaldehyde that was based on Semenov's assumption [Semenov, 1960] that solid-state chain reactions are possible only when the products of each chain step "prepare" a configuration of reactants that is suitable for the next step. Monomer crystals for which low-temperature polymerization has been observed fulfill this condition. In the initial equilibrium state the monomer molecules are located in lattice sites and the creation of a chemical bond requires surmounting a high barrier. However, upon creation of the primary cation (protonated formaldehyde), the active center shifts toward another monomer, and the barrier for addition of the next link diminishes. Likewise, subsequent polymerization steps involve motion of the cationic end of the polymer toward a neighboring monomer, which results in a low barrier to formation of the next C–O bond. Since the covalent bond lengths in the polymer are much shorter than the van der Waals distances of the monomer crystal, this polymerization process cannot take place in a strictly linear fashion. It is believed that this difference is made up at least in part by rotation of each CH_2O link as it is incorporated into the chain.

Chain growth at $T < T_c$ has been associated with tunneling of the CH_2 group at the growing end of the polymer through a distance of 0.6 Å toward the next monomer molecule. However, this mechanism was based on a possible crystalline structure of formaldehyde that is different from the experimentally determined structure, which was obtained only recently [Weng et al., 1989]. In thin-film samples of formaldehyde formed by vapor deposition, the average chain lengths (about 7–10 monomer units) are much shorter than those obtained by Goldanskii [1976] in bulk samples. However, there is disagreement about whether this effect is due to formation of disordered structures or other crystalline structures in the vapor-deposited samples (see Mansueto et al. [1989], Mansueto and Wight [1991], and Misochko et al. [1991]).

Growth of long chains ($n \geq 10^2$) in mixed $1:1$ crystals of ethylene with chlorine or bromine at 20–70 K has been studied in detail by Wight et al. [1992a, 1993]. Active radicals were generated by pulsed laser photolysis of Cl_2 or Br_2. The rate constant has been found to be $k_c = 8$–12 s^{-1} below $T_c = 45$ K. The chain grows by the radical chain mechanism

$$Cl \cdot + H_2C = CH_2 \rightarrow H_2ClC{-}CH_2 \cdot \qquad (9.18)$$

$$H_2ClC{-}CH_2 \cdot + Cl_2 \rightarrow H_2ClC{-}CH_2Cl + Cl \cdot \qquad (9.19)$$

in which a small concentration of radicals can convert relatively large amounts of Cl_2 and C_2H_4 to 1,2-dichloroethane. Although the gas-phase reaction produces both *gauche* and *anti* conformers of the product (due to the low barrier for rotation about its C–C bond), the solid-state photochemical chain reaction forms exclusively the *anti* conformer. The high stereospecificity of this reaction is attributed to the structure of the mixed crystal, which is thought to consist of an alternating quasi-one-dimensional arrangement of reactants, due to the donor–acceptor interaction [Hassel and Romming, 1962].

Several other examples of stereospecificity in solid-state low-temperature reactions studied by Wight et al. [1990], Tague et al. [1992], and Tague and Wight [1991, 1992] are given in Table 9.4. These reactions are often characterized by high specificity for formation of a single product, in contrast to analogous reactions in liquid solutions at higher temperatures here distributions of isomeric products and even polychlorinated hydrocarbons are usually observed.

We have seen that the structure of the reactant lattice is a critical factor in determining the local structure of the products (i.e., which conformational or configurational isomers of the products are formed). In

TABLE 9.4
Stereoselectivity of Solid-State Photochlorination

Reactant	Quantum Yield in $1:1$ Mixtures	Product
Ethylene	120 (50 K)	*anti*-1,2-Dichloroethane
Propene	740 (50 K)	*anti*-1,2-Dichloropropane
Acetylene	24 (60 K)	*anti*-1,2-Dichloroethylene
Cyclopropane	52 (77 K)	*anti, anti*-1,3-Dichloropropane
Methylcyclopropane	500 (77 K)	Chloromethylcyclopropane
Methylcyclohexane	24 (77 K)	*axial*-1-*chloro*-1-*methylcyclohexane*

chain-reaction systems operating at low temperature, there is an addition-al constraint that the reactant and product crystals must be commensurate with one another. If this were not the case, then chemical conversion would lead to the accumulation of strains, which would stop the chain growth. At higher temperatures, this strain typically causes fracture of the crystals. In the particular case of the ethylene/chlorine chain reaction, computer simulations have shown that the reactant and product lattices may be commensurate [Benderskii et al., 1991c; Wight et al., 1993], and that propagation of the chains occurs along an axis of the crystal in which the interreactant distances are shortest (i.e., the shortest diagonal of the *ac* plane). The arrangement of reactants, free radical intermediates, and products in this plane is shown in Figure 9.11. Because the reactant and product lattices are compatible, formation of the line of products does not preclude the growth of chains along the neighboring diagonals. The chain lengths are limited only by randomly distributed defects in the bulk crystal and can be as high as 260 ± 70 at 60–70% conversion [Wight et al., 1993].

The temperature dependence of the rate constant for reactions of C_2H_4 with Cl_2, Br_2, and HBr [Barkalov et al., 1980] are all nearly the same, despite the differences in the reactant masses. At 50–80 K the apparent activation energy is 1 kcal/mol, which is the same as for the gas-phase reaction of C_2H_4Cl with Cl_2.

A multidimensional PES for the solid-state reaction (9.19) has been calculated by Wight et al. [1993] with the aid of atom–atom potentials combined with the semiempirical LEPS method (see, e.g., Eyring et al. [1983]). Because of the substantial exoergicity of this elementary step, the

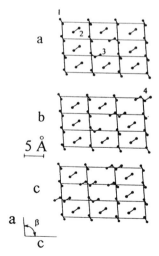

Figure 9.11. Molecular arrangement in the *ac* plane of a mixed 1:1 ethylene–chlorine crystal illustrating (a) radical pair formation, (b) single chain growth, and (c) chain growth in the vicinity of product line. Molecules labeled 1–4 are ethylene, chlorine, chloroethyl radical C_2H_4Cl, and *trans*-dichloroethane $C_2H_4Cl_2$, respectively. (From Wight et al. [1993].)

PES is characterized by an early transition state; that is, the saddle point is strongly shifted toward the reactant valley so that the reaction barrier is overcome without appreciable lengthening of the Cl–Cl bond. Since the angle between the valleys is less than 90° and the intramolecular vibration frequencies are much greater than ω_0, criterion (2.86) indicates that this reaction takes place in the vibrationally adiabatic regime.

Analysis of the MEP has shown that the relative displacement of the reactants is much smaller than the tunneling length, and the reorganization energy of the environment is just 0.12–0.17 kcal/mol (i.e., considerably smaller than $V^{\#} \sim 2$ kcal/mol). This situation permits a reduction in the number of degrees of freedom to $N = 7$, and the reaction is modeled by considering a Cl–Cl $\cdots$ C–C–C complex immersed in a fixed crystalline environment. Among the normal modes of the complex there are three high-frequency stretching vibrations of the chlorine molecule and radical ($q_1 - q_3$), which practically do not mix with the other modes. This allows a further reduction of the dimensionality to four modes. The q_7 mode, which has an intermediate frequency, is mainly due to bending of the radical. The low-frequency vibrations q_4–q_6 are assigned to intermolecular stretching and libration. The main contribution to the reaction path comes from the q_5 and q_6 vibrations, which correspond to simultaneous relative approach and rotation of reactants. It has been shown that there is a two-dimensional cut of the PES such that the MEP lies completely within it. The two coordinates in this cut are q_4 and a linear combination q_5–q_7. This cut is illustrated in Figure 9.12, along with the MEP. Motion along the reaction path is adiabatic with respect to the fast coordinates q_1–q_3 and nonadiabatic in the space of the slow coordinates q_4–q_7. Nevertheless, since the MEP has a small curvature, deviations of the tunneling trajectory from it are small. This small curvature approximation was used earlier for calculating tunneling splitting in $(HF)_2$ (see Section 8.4).

The rate constant calculated for reaction (9.19) using this approach is

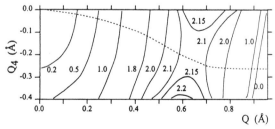

Figure 9.12. The two-dimensional cut of PES and MEP of reaction (9.18a) in coordinates q_4 and q. The coordinate q is the linear combination of q_5–q_7. (From Wight et al. [1993].)

characterized by a low-temperature limit $T_c = 20\text{--}25\,K$, $k_c = 10^{-2}\text{--}10^{-1}\,s^{-1}$. In the Arrhenius region at high temperatures the calculated activation energy, $E_a = 1.4\,kcal/mol$, is in good agreement with the experimental results. The results for the chlorine–ethylene system show that although chemical reactions in low-temperature solids may involve many degrees of freedom, it is sometimes possible to simplify the problem using the techniques reviewed in this volume to the point where the salient features of the dynamics can be elucidated.

CONCLUSION

In this volume, we have surveyed some recent developments in chemical dynamics at low temperatures that demonstrate the evolution from macroscopic descriptions of solid-state chemical processes to microscopic models. We have seen how quantum effects can be incorporated into modern theories of chemical kinetics, and how the comparison of theory and experiment can be used to elucidate a consistent quantal description of elementary chemical reactions. The most prominent feature of low-temperature dynamics is the participation of states having energies below the barrier to reaction. Deviations from the Arrhenius law result directly from the influence of these states on the transition probability from reactants to products. The tunneling transition from the ground state of the reactants becomes predominant below the crossover temperature T_c where the rate constant assumes its quantum limit. The strong dependence of the tunneling rate constant on both the barrier height and tunneling distance creates a fascinating situation in which experimentally determined tunneling rates vary over many orders of magnitude. As a result of this exponential dependence, the essential tunneling dynamics are more closely associated with intrawell motion of the reactants than with motion over the classical barrier. It is in this sense that we have come to recognize the close relationship that exists between cryo-chemistry and molecular spectroscopy,

Comparison of experimental data with tunneling models unequivocally demonstrates the multidimensional character of tunneling transitions. We have seen how motion of the tunneling particle transverse to the reaction coordinate can lead to "corner-cutting" trajectories that lie far from the minimum energy path, and how symmetric and antisymmetric coupling of nonreactive modes to the reaction coordinate can promote or suppress the tunneling rate. Even the low-frequency bath modes that are far removed from the reaction coordinate play an essential role in determining the overall dynamics of the system. We have come to recognize limiting cases based on ratios of characteristic frequencies and features of the potential energy surfaces that can be used to simplify the analysis of the dynamics (i.e., the sudden and adiabatic approximations).

Perhaps the most dramatic manifestation of quantum chemical dynamics is coherence. If the tunneling coordinate is not strongly coupled to the remaining degrees of freedom, the barrier crossing is periodic. The signature of this condition is tunneling splitting, which can be measured using a variety of high-resolution spectroscopic techniques. The concept of a rate constant can be introduced only when the coupling to the bath modes is strong enough to destroy the coherence. In this case, the rate constant is proportional to the square of the tunneling splitting. Because the spectral density of the bath is very different for isolated and condensed phase systems, it is sometimes possible for a particular tunneling reaction to be observed in both the coherent and incoherent regimes, depending on the reactant environment. Ordinarily, the tunneling matrix elements are approximately the same in both cases, but destruction of the coherence drastically lengthens the characteristic time scale for the transition.

As in every emerging field of study, there are gaps between the abstract level of understanding and the ability to provide quantitative descriptions of specific systems. In this volume, we have tried to identify a number of systems that are amenable to analysis with the existing level of theory and computational power. There are nevertheless a large number of general problems still waiting to be solved, and we shall highlight some of the most prominent of them here.

Scattering theory provides a framework for introducing a formally exact concept of the rate constant for unbound reactant and product states (i.e., gas-phase reactions) via the flux–flux correlation functions. However, the applicability of this formulation to bound potential energy surfaces (i.e., condensed phase reactions) remains an open question. This problem is closely related to the problem of predicting whether or not a tunneling transition lies in the coherent or incoherent regime, for a "true" rate constant (in the sense of an exponential relaxation of populations from a perturbed state towards equilibrium) can be defined only in the incoherent regime. Because the typical time scale of the flux–flux correlator ($\hbar\beta$) is shorter than the characteristic tunneling time scale, this quantity cannot be used to determine whether the transition is coherent or not. In this respect, rate theories based on the flux–flux correlator go no farther than the imaginary time instanton theories: They infer the long-time dynamics from propagators calculated at much shorter times (real or imaginary), which are of the order $\hbar\beta$. Therefore, one may expect a correct answer only if it is known in advance that the system possesses a rate constant. Such is the case for reactions on unbound potential surfaces.

So far, one can be much more successful in calculating a rate constant when one knows in advance that it exists, than in answering the question of whether it exists. A considerable breakthrough in this area was the solution of the spin-boson problem, which, however, has only limited relevance to any problem in chemistry because it neglects the effects of intrawell dynamics (vibrational relaxation) and does not describe thermally activated transitions. A number of attempts have been made to go beyond the two-level system approximation, but the basic question of how vibrational relaxation affects the transition from coherent oscillations to exponential decay awaits a quantitative solution. Such a solution might be obtained by numerical computation of real-time path integrals for the density matrix using the influence functional technique.

Two-dimensional semiclassical studies show that when no additional assumptions are made (such as moving along a certain predetermined path) and when the fluctuations around optimum tunneling path are taken into account, in most cases the results of two-dimensional instanton theory are very close to the exact numerical solution. Once the main difficulty in going from one dimension to two is circumvented, there appears to be no serious difficulty in extending the algorithm to higher dimensions, where the usual basis set methods fail due to the exponentially increasing number of basis functions required. Another attractive feature of the instanton method, as distinct from other quasiclassical approaches, is that it shows a direct connection between tunneling trajectories and the temperature dependence of the rate constant: Certain types of trajectories correspond to certain behaviors of $k(T)$. For example, the presence of two-dimensional trajectories always extends the Arrhenius dependence to low temperatures, thereby making the transition from activated to tunneling dynamics less sharp.

In general, for potentials with several saddle points, there exist critical temperatures for which the behavior of tunneling trajectories changes qualitatively. These temperatures correspond to the appearance of new tunneling pathways. An example is the potential with two saddle points for which two 2D trajectories eventually merge into a single 1D trajectory at a certain critical temperature. There may be several families of trajectories, and they cannot be obtained from one another by a continuous transformation. The number of these families should be determined by the topology of the surface, i.e., by the number and position of saddle points, minima, and maxima. The probem of determining these families is still unchallenged, and the present status of instanton theory forces one to rely on physical intuition for choosing relevant families of trajectories, rather than on exact topological considerations. Although one might imagine a numerical scheme that would not miss any solution, the computational effort expended by such a scheme would probably be very large for a complicated surface. For $T = 0$, the requirement of finite action restricts the choice. For example, consider the potential for two-proton transfer, which is characterized by two global *cis* minima, two local *trans* minima, and four saddle points between them. At $T = 0$ there is a trajectory that starts out at a local minimum and corresponds to the family of solutions that continuously collapse to the saddle point when temperature goes up. However, this trajectory cannot describe a tunneling event between the global minima because its energy is always no less than the energy of the local minimum, so that the action diverges when $\hbar\beta$ goes to infinity, which implies zero transition probability. Nevertheless, it is this family that describes thermally activated transitions at higher temperatures. That is, the finite action requirement suggests looking for another family of trajectories that plays a dominant role at sufficiently low temperatures.

Another problem that awaits solution is photochemical quantum dynamics, in particular, tunneling dynamics in excited electronic states. From a theorist's point of view, this implies solving the curve crossing tunneling problem for a multidimensional case.

Since the tunneling trajectories, in the general case, do not coincide with the minimum energy path, tunneling dynamics requires knowledge of large regions of multidimensional space, sometimes located far from the classical saddle that connects reactant and product valleys. In principle, the basis set methods for coupled anharmonic motions give the values for tunneling splitting for different energy levels directly. However, in practice this is extremely difficult to do in many dimensions due to growth of the number of required basis functions (see, for example, Carney et al. [1978]). The vibrationally adiabatic and sudden approximations allow one to avoid this difficulty by reduction of the problem to fewer degrees of freedom. Moreover, the optimum tunneling trajectories define the region of the potential surface needed to be calculated. They can also be useful in defining good action-angle variables for subsequent quasiclassical quantization [Ezra et al., 1987]. We note that the time-dependent self-consistent field approach is an alternative to exact variational quantum methods.

In the present volume we discuss only two limits of coherent and incoherent transitions, and have not considered the intermediate region in which the coherence is destroyed. In the classical dynamics of polyatomics, the periodicity of motion is destroyed due to increasing coupling with the bath modes, which leads to quasiperiodic motion and ultimately to chaos (see reviews by Berne et al. [1982]; Brumer and Shapiro [1988]; Noid et al. [1981]). Model problems of quantum chaos and the transition from coherent tunneling to localization are subjects of many papers (see, for example, the review by Eckhardt [1988]). At present there is no evidence of irregular behavior in low-temperature systems, most likely because the low-temperature regime is where well-behaved low-energy levels are dominant. The application of quantum chaos theory will be required in the future for studying more subtle effects associated with the shape and width of tunneling lines, especially near the crossover temperature where the tunneling competes with thermally activated hopping and transitions from excited states with irregular behavior.

Finally, let us point out that low-temperature chemical dynamics is not concerned solely with tunneling phenomena. Low-temperature radiation chemistry and photochemical reactions are predominantly determined by transformations of initially excited species, including translational, rotational, and vibrational relaxation and dephasing, as well as nonadiabatic electronic transitions in the intersystem crossing region. Molecular-dynamical descriptions of cage reactions and escape, along with Landau-Zener dynamics, in combination with tunneling dynamics, form the foundation for the general theory of chemical reactions. We hope that the experimental examples and theoretical ideas presented in this volume may foster new efforts toward a more complete understanding of chemical dynamics in the quantum regime.

REFERENCES

Abed, K. J., and Clough, S. 1987. *Chem. Phys. Lett.* **142**, 209.

Abed, K. J., Clough, S., Horsewill, A. J., and Mohammed, M. A. 1988. *Chem. Phys. Lett.* **147**, 624.

Abouaf-Marguin, L., Jacox, M. E., and Milligan, D. E. 1977. *J. Mol. Spectrosc.* **67**, 34.

Abraham, A. 1961. *The Principles of Nuclear Magnetism.* London: Oxford University Press.

Affleck, I. 1981. *Phys. Rev. Lett.* **46**, 388.

Alberding, N., Austin, R. H., Chan, S. S., Eisenstein, L., Fraunfelder, H., Gunsalus, I. C., and Nordlung, T. M. 1976. *J. Chem. Phys.* **65**, 4701.

Alexandrou, C., and Negele, J. W., 1988, *Phys. Rev.* C **37**, 1513.

Alberding, N., Austin, R. H., Beeson, K. W., Chan, S. S., Eisenstein, L., Fruanfelder, H., Good, D., Gunsalus, I. C., Nordlung, T. M., Perutz, M. F., Reynolds, A. H., and Yue, K. T. 1980. *Phys. Rev. Lett.* **44**, 1156.

Alefeld, B., Anderson, I. S., Heidemann, A., Magerl, A., and Trevino, S. F. 1982. *J. Chem. Phys.* **76**, 2758.

Alimi, R., Gerber, R. B., and Apkarian, V. A. 1990. *J. Chem. Phys.* **92**, 3551.

Allen, P. S. 1974. *J. Phys. Chem.* C **7**, L22.

Al-Soufi, W., Grellmann, K. H., and Nickel, B. 1991. *J. Phys. Chem.* **95**, 10503.

Altkorn, R. I., and Schatz, G. C. 1980. *J. Chem. Phys.* **72**, 3379.

Ambegaokar, V. 1987. *Phys. Rev.* B **37**, 1624.

Amirav, A., Even, U., and Jortner, J. 1980, *Chem. Phys.* **51**, 31.

Anderson, M., Bosio, L., Bruneaux-Poulle, J., and Fourme, R. 1977. *J. Chim. Phys. Chim. Phys. Biol.* **74**, 68.

Anderson, P. W., Halperin, B. I., and Varma, C. M. 1972. *Philos. Mag.* **25**, 1.

Andzelm, J., Huzinaga, S., Klobukowski, M., and Rodzio, E. 1985. *Chem. Phys.* **100**, 1.

Arnold, C., Gettys, N. S., Thompson, D. L., and Raff, L. M. 1986, *J. Chem. Phys.* **84**, 3803.

Arnold, V. I. 1978. *Mathematical Methods of Classical Mechanics.* Berlin: Springer.

Aslangul, C., Pottier, N., and Saint-James, D. 1985. *Phys. Lett.* A **110**, 249.

Aslanoosi, A. M., Horsewill, A. J., and Clough, S. 1989. *J. Phys. Condens. Matter* **1**, 643.

Auerbach, A., and Kivelson, S. 1985. *Nucl. Phys.* B **257**, 799.

Auerbach, A., Freed, K. F., and Gomer, R. 1987. *J. Chem. Phys.* **86**, 2356.

Austin, R. H., Beeson, K. W., Eisenstein, L., Fraunfelder, H., and Gunsalus, I. C. 1975. *Biochemistry* **14**, 5355.

Babamov, V. K., and Marcus, R. A. 1981. *J. Chem. Phys.* **74**, 1790.

Babamov, V. K., Lopez, V., and Marcus, R. A. 1983. *J. Chem. Phys.* **78**, 5621.

Baer, M. 1982. *Adv. Chem. Phys.* **49**, 191.

Balian, R., and Bloch, C. 1974. *Ann. Phys.* **85**, 514.

Barbara, P. F., Walsh, P. K., and Brus, L. E. 1989. *J. Phys. Chem.* **93**, 29.

Barkalov, I. M., Goldanskii, V. I., Kiryukhin, D. P., and Zanin, A. M. 1980. *Chem. Phys. Lett.* **73**, 273.

Barker, A. S., and Sievers, A. J. 1975. *Rev. Mod. Phys.* **47** (Suppl. 2), 1.

Baron, P. A., and Harris, D. O. 1974, *J. Mol. Spectrosc.* **49**, 70.

Bartelt, G., Eychmuller, A., and Grellmann, K. H. 1985. *Chem. Phys. Lett.* **118**, 568.

Barton, A. E., and Howard, B. J. 1982. *Farad. Discuss. Chem. Soc.* **73**, 45.

Bas', A. I., Zeldovich, Ya. B., and Perelomov, A. M. 1971. *Scattering, Reactions and Decays in Nonrelativistic Quantum Mechanics.* Moscow: Nauka (in Russian).

Basilevsky, M. V., Gerasimov, G. N., and Petrochenko, S. I. 1982. *Chem. Phys.* **72**, 349.

Baughcum, S. L., Duerst, R. W., Rowe, W. F., Smith, Z., and Wilson, E. B. 1981. *J. Am. Chem. Soc.* **103**, 6296.

Baughcum, S. L., Smith, Z., Wilson, E. B., and Duerst, R. W. 1984. *J. Am. Chem. Soc.* **106**, 2260.

Bauschlicher, C. W., Bender, C. F., and Schaefer, H. F. 1976. *J. Am. Chem. Soc.* **98**, 3072.

Bell, R. P. 1933. *Proc. R. Soc. London Ser. A* **139**, 466.

Bell, R. P. 1935. *Proc. R. Soc. London Ser. A* **148**, 241.

Bell, R. P. 1937. *Proc. R. Soc. London Ser. A* **158**, 128.

Bell, R. P. 1973. *The Proton in Chemistry*, 2nd ed. Ithaca, NY: Cornell University Press.

Bell, R. P. 1980. *The Tunnel Effect in Chemistry*. London: Chapman & Hall.

Benderskii, V. A., and Goldanskii, V. I. 1992. *Int. Rev. Phys. Chem.* **11**, 1.

Benderskii, V. A., and Makarov, D. E. 1992. *Phys. Lett. A* **161**, 535.

Benderskii, V. A., Goldanskii, V. I., and Ovchinnikov, A. A. 1980. *Chem. Phys. Lett.* **73**, 492.

Benderskii, V. A., Misochko, E. Y., Ovchinnikov, A. A., and Filippov, P. G. 1983. *Zh. Fiz. Khim.* **59**, 1079 (in Russian).

Benderskii, V. A., Goldanskii, V. I., and Makarov, D. E. 1990. *Chem. Phys. Lett.* **171**, 94.

Benderskii, V. A., Goldanskii, V. I., and Makarov, D. E. 1991a. *Chem. Phys. Lett.* **154**, 407.

Benderskii, V. A., Goldanskii, V. I., and Makarov, D. E. 1991b. *Chem. Phys. Lett.* **186**, 517.

Benderskii, V. A., Goldanskii, V. I., Makarov, D. E., and Misochko, E. Ya. 1991c. *Chem. Phys. Lett.* **179**, 334.

Benderskii, V. A., Grinevich, P. G., Makarov, D. E., and Pastur, D. L. 1992a. *Chem. Phys.* **161**, 51.

Benderskii, V. A., Goldanskii, V. I., and Makarov, D. E. 1992b. *Chem. Phys.* **159**, 29.

Benderskii, V. A., Grinevich, P. G., and Makarov, D. E. 1993a. *Chem. Phys.* **170**, 275.

Benderskii, V. A., Goldanskii, V. I., and Makarov, D. E. 1993b, Phys. Reports, in press.

Berne, B. J., Di-Leon, N., and Rosenberg, R. O., 1982, *J. Phys. Chem.* **86**, 2166.

Besnard, M., Fouassier, M., Lassegues, J. C., and Dianoux, A. J. 1991. *Mol. Phys.* **73**, 1059.

Bicerano, J., Schaefer, H. F., and Miller, W. H. 1983. *J. Am. Chem. Soc.* **105**, 2550.

Bixon, M., and Jortner, J. 1968. *J. Chem. Phys.* **48**, 715.

Blake, G. A., Busarow, K. L., Cohen, R. C., Laughlin, K. B., Lee, Y. T., and Saykally, R. J. 1988. *J. Chem. Phys.* **89**, 6577.

Blanchet, G. B., Dinavclo, N. J., and Plummer, E. W. 1982, *Surf. Sci.*, **118**, 496.

Block, P. A., Mark, D., Marshall, L. G., Pedersen, L. D., and Miller, R. E. 1992. *J. Chem. Phys.* **96**, 7321.

Blum, K. 1981. *Density Matrix Theory and Applications.* New York: Plenum.

Bolshakov, B. V., and Tolkachev, V. A. 1976. *Chem. Phys. Lett.* **40**, 468.

Bolshakov, B. V., Stepanov, A. A., and Tolkachev, V. A. 1980. *Int. J. Chem. Kinet.* **12**, 271.

Bondybey, V. E., and Fletcher, C. 1976. *J. Chem. Phys.* **64**, 3615.

Bondybey, V. E., Haddon, R. C., and English, J. H. 1984. *J. Chem. Phys.* **80**, 5432.

Borgis, D. C., and Hynes, J. T. 1991. *J. Chem. Phys.* **94**, 3622.

Borgis, D. C., Lee, S. and Hynes, J. T. 1989. *Chem. Phys. Lett.* **162**, 19.

Bosch, E., Moreno, M., Lluch, J. M., and Bertran, J. 1990. *J. Chem. Phys.* **93**, 5685.

Bratan, S., and Strohbusch, F. 1980. *J. Mol. Struct.* **61**, 409.

Brickman, J. 1981, Ber. Bunsenges., *Phys. Chem.*, *Bd.* **85**, S. 106.

Brown, R. S., Tse, A., Nakashima, T., and Haddon, R. C. 1979. *J. Am. Chem. Soc.* **101**, 3157.

Brucker, G. A., and Kelley, D. F. 1989. *Chem. Phys.* **136**, 213.

Brumer, P., and Shapiro, M. 1988, *Adv. Chem. Phys.* **70**, 365.

Brunton, G., Griller, D., Barklay, L. R. C., and Ingold, K. U. 1976. *J. Am. Chem. Soc.* **98**, 6803.

Buchenau, U., Prager, M., Nucker, N., Dianoux, A. J., Ahmad, N., and Phillips, W. A. 1986. *Phys. Rev. B* **34**, 5665.

Buchwalter, S. L., and Closs, G. H. 1979. *J. Am. Chem. Soc.* **101**, 4688.

Buckingham, A. D. 1967. *Adv. Chem. Phys.* **12**, 107.

Bunker, P. R. 1979. *Molecular Symmetry and Spectroscopy.* New York: Academic.

Bunker, P. R. 1983. *Annu. Rev. Phys. Chem.* **34**, 59.

Bunker, P. R., and Landsberg, B. M. 1977. *J. Mol. Spectrosc.* **67**, 374.

Bunker, P. R., Landsberg, B. M., and Winnewisser, B. P. 1979. *J. Mol. Spectrosc.* **74**, 9.

Bunker, P. R., Jensen, P., Hofranek, M., Lischka, P., and Karpfen, A. 1990. *J. Chem. Phys.* **93**, 6266.

Burton, G., Gray, J. A., Griller, D., Barclay, L. R. C., and Ingold, K. U. 1978. *J. Am. Chem. Soc.* **100**, 4197.

Busch, J. H., Fluder, E. M., and de la Vega, J. R. 1980. *J. Am. Chem. Soc.* **102**, 4000.

Butenhoff, T. J., and Moore, C. B. 1988. *J. Am. Chem. Soc.* **110**, 8336.

Cailet, J., and Claveric, P. 1975. *Acta Crystallogr. A* **31**, 448.

Caldeira, A. O., and Leggett, A. J. 1981. *Phys. Rev. Lett.* **46**, 211.

Caldeira, A. O., and Leggett, A. J. 1983. *Ann. Phys.* **149**, 374.

Callan, C. G., and Coleman, S. 1977. *Phys. Rev. D* **16** 1762.

Calvani, P., and Lupi, S. 1987, *J. Chem. Phys.* **87**, 1913; 1989. *J. Chem. Phys.* **90**, 5924.

Campion, A., and Williams, F. 1972. *J. Am. Chem. Soc.* **94**, 7633.

Carney, G. D., Sprandel, L. L., and Kern, C. W. 1978, *Adv. Chem. Phys.* **37**, 305.

Carreira, L. A., Jiang, G. J., Person, W. B., and Willis, J. N. 1972. *J. Chem. Phys.* **56**, 1440.

Carrington, T., and Miller, W. H. 1986. *J. Chem. Phys.* **84**, 4364.

Carrington, T., Hubbard, L. M., Shaefer, H. F., and Miller, W. H. 1984. *J. Chem. Phys.* **80**, 4347.

Carsky, P., Bartlett, R. J., Fitzgerald, G., Noga, J., and Spirko, V. 1988. *J. Chem. Phys.* **89**, 3008.

Casati, G., and Molinari, L. 1989. *Progr. of Theor. Phys. (Suppl.)* **98**, 286.

Cavagnat, D., and Pesquer, M. 1986. *J. Phys. Chem.* **90**, 3289.

Cavagnat, D., Magerl, A., Vettier, C., and Clough, S. 1986. *J. Phys. C* **19**, 6665.

Cavagnat, D., Trevino, S. F., and Magerl, A. 1989. *J. Phys. Condens. Matter* **1**, 10047.

Cesi, F., Rossi, G. C., and Testa, M. 1991. *Ann. Phys. (NY)* **206**, 318.

Chakraborthy, T., and Choudhury, M. 1990. *Chem. Phys. Lett.* **171**, 25.

Chakraborty, Z., Hedegard, P., and Nylens, M. 1988. *J. Phys. C* **21**, 3437.

Chalasinski, G., Szczesniak, M. M., and Scheiner, S. 1991. *J. Chem. Phys.* **94**, 2807.

Chance, B., Fischetti, R., and Rowers, L. 1983. *Biochemistry* **22**, 3820.

Chandler, D. 1987. *Introduction to Modern Statistical Mechanics*. New York: Oxford University Press.

Chang, T. Y. 1967. *Rev. Mod. Phys.* **39**, 911.

Chang, Y. T., Yamagushi, Y., Miller, W. H., and Schaefer, H. F. 1987. *J. Am. Chem. Soc.* **109**, 7245.

Chapman, S., Garrett, B. C., and Miller, W. H. 1975. *J. Chem. Phys.* **63**, 2710.

Chiang, C.-M., and Jackson, B. 1987. *J. Chem. Phys.* **87**, 5497.

Child, M. S. 1974. *Molecular Collision Theory*. London: Academic.

Choe, J., and Harmony, M. D. 1980. *J. Mol. Spectrosc.* **81**, 480.

Clough, S., Heidemann, A., Horsewill, A. J., Lewis, J. D., and Paley, M. N. J. 1981a. *J. Phys. C* **14**, L525.

Clough, S., Horsewill, A. J., and Heidemann, A., 1981b. *Chem. Phys. Lett.* **82**, 264.

Clough, S., Heidemann, A., Horsewill, A. J., Lewis, J. D., and Paley, M. N. J. 1982. *J. Phys. C* **15**, 2495.

Clough, S., Hediemann, A., Horsewill, A. J., and Palley, M. N. J. 1984a. *Z. Phys. B* **55**, 1.

Clough, S., Horsewill, A. J., and McDonald, P. J. 1984b. *J. Phys. C* **17**, 1115.

Clough, S., Horsewill, A. J., McDonald, R. J., and Zelaya, F. O. 1985. *Phys. Rev. Lett.* **55**, 1794.

Clyne, M. A. A., and McDermid, I. S. 1979. *J. Chem. Soc. Farad. Trans. 2* **75**, 1313, 1677.

Cohen, R. C., and Saykally, R. J. 1991. *J. Chem. Phys.* **95**, 7891.

Cohen, R. C., and Saykally, R. J. 1992. *J. Phys. Chem.* **96**, 1024.

Coker, D. F., and Watts, R. O. 1987. *J. Phys. Chem.* **91**, 2513.

Coudert, L. H. and Hougen, J. T. 1990, *J. Mol. Spectrosc.* **139**, 259.

Coudert, L. H., and Hougen, J. Y. 1988. *J. Mol. Spectrosc.* **130**, 86.

Coudert, L. H., Lovas, F. J., Suenram, R. D., and Hougen, J. T. 1987. *J. Chem. Phys.* **87**, 6290.

Coveney, P. V., Child, M. S., and Barany, A. 1985. *J. Phys. B* **18**, 4557.

Cremer, D. 1990. *J. Phys. Chem.* **94**, 5509.

Cremer, D., and Pople, J. A. 1975a. *J. Am. Chem. Soc.* **97**, 1354; 1975b. *J. Am. Chem. Soc.* **97**, 1358.

Crofton, M. W., Jagod, M.-F., Rehfuss, B. D., and Oka, T. 1989. *J. Chem. Phys.* **91**, 5139.

Crossley, M. J., Field, L. D., and Harding, M. M. 1987. *J. Am. Chem. Soc.* **109**, 2335.

Currie, J. F., Krumhansl, J. A., Bishop, A. R., and Trullinger, S. E. 1980. *Phys. Rev. B* **22**, 477.

Curtiss, L. A., and Pople, J. A. 1976. *J. Mol. Spectrosc.* **61**, 1.

Dakhnovskii, Yu. I., and Nefedova, V. V. 1991. *Phys. Lett. A* **157**, 301.

Dakhnovskii, Yu., and Ovchinnikov, A. A. 1985. *Phys. Lett. A* **113**, 147.

Dakhnovskii, Yu. I., and Semenov, M. B. 1989. *J. Chem. Phys.* **92**, 7606.

Dashen, R. F., Hasslacker, B., and Neven, A. 1975. *Phys. Rev. D* **11**, 3424.

Dashevskii, V. G., and Kitaigorodskii, A. I. 1970. *Sov. J. Struct. Chem.* **11**, 913.

Dattaggupta, S., Grabert, H., and Jung, R. 1989. *J. Phys. Condens. Mater.* **1**, 1405.

Dayton, D. C., Jucks, K. W., and Miller, R. E. 1989. *J. Chem. Phys.* **90**, 2631.

Debruner, P., and Fraunfelder, H. 1982. *Experience in Biochemical Perception.* New York: Academic, 326.

Dekker, H. 1986. *Phys. Lett. A* **114**, 295.

Dekker, H. 1987a. *J. Phys. C* **20**, 3643.

Dekker, H. 1987b. *Phys. Rev. A* **35**, 1436.

Dekker, H. 1991. *Physica A* **175**, 485.

de la Vega, J. R. 1982. *Acc. Chem. Res.* **15**, 185.

de la Vega, J. R., Bush, J. H., Schauble, J. H., Kunze, K. L., and Haggert, B. E. 1982. *J. Am. Chem. Soc.* **104**, 3295.

DeLeon, R. L., and Muenter, J. S. 1980. *J. Chem. Phys.* **72**, 6020.

DeVault, D., and Chance, B. 1966. *Biophys. J.* **6**, 825.

Dewar, M. J. S. 1984. *J. Am. Chem. Soc.* **106**, 209.

Dewar, M. J. S., Merz, K. M., and Stewart, J. J. P. 1984. *J. Am. Chem. Soc.* **106**, 4040.

Deycard, S., Hughes, L., Lusztyk, J., and Ingold, K. U. 1987. *J. Am. Chem. Soc.* **109**, 4954.

Deycard, S., Lusztyk, J., Ingold, K. U., Zebretto, F., Zgierskii, M. Z., and Siebrand, W. 1988. *J. Am. Chem. Soc.* **110**, 6721.

Deycard, S., Lusztyk, J., and Ingold, K. U. 1990. *J. Am. Chem. Soc.* **112**, 4284.

Dhawan, L. L., and Prakash, S. 1984. *J. Phys. F* **14**, 2329.

Dick, B. G. 1977. *Phys. Rev. B* **16**, 3359.

Difoggio, R., and Gomer, R. 1982. *Phys. Rev. B* **25**, 3490.

Dixon, D. A., and Komornicki, A. 1990. *J. Phys. Chem.* **94**, 5630.

Doba, T., Ingold, K. U., Siebrand, W., and Wildman, T. A. 1984. *J. Phys. Chem.* **88**, 3165.

Dodd, R. K., Eilbeck, J. C., Gibbon, J. D., and Morris, H. C. 1982. *Solitons and Nonlinear Wave Equations.* New York: Academic.

Dogonadze, R. R., and Kuznetsov, A. M. 1975. *Prog. Surf. Sci.* **6**, 1.

Doll, J. D. and Freeman, D. L. 1984. *J. Chem. Phys.* **80**, 2239.

Doll, J. D., and Freeman, D. L. 1988. *Adv. Chem. Phys.* **73**, 120.

Doll, J. D., George, T. F., and Miller, W. H. 1972. *J. Chem. Phys.* **58**, 1972.

Doll, J. D., Coalson, R., and Freeman, D. L. 1985. *Phys. Rev. Lett.* **55**, 1.

Doll, J. D., Freeman, D. L., and Gillan, M. J. 1988. *Chem. Phys. Lett.* **143**, 277.

Douketis, C., Hutson, J. M., Orr, B. J., and Scoles, G. 1984. *Mol. Phys.* **52**, 763.

Dubinskaya, A. M. 1990. *Sov. Sci. Rev. B Chem.* **14**, 37.

Dyke, T. R. 1977. *J. Chem. Phys.* **66**, 492.

Dyke, T. R., Howard, B. J., and Klemperer, W. 1972. *J. Chem. Phys.* **56**, 2442.

Dyke, T. R., Mock, K. M., and Muenter, J. S. 1977. *J. Chem. Phys.* **66**, 498.

Eckhardt, B. 1988, *Phys. Rep.* **163**, 205.

Eisenberger, H., Nickel, B., Ruth, A., Al-Soufi, W., and Grellmann, K. H. 1991. *J. Phys. Chem.* **95**, 10509.

Emsley, J. 1984. *Struct. Bonding* **57**, 147.

Engelholm, G. G., Luntz, A. C., Gwinn, W. D., and Harris, D. O. 1969. *J. Chem. Phys.* **50**, 2446.

Escribano, R., Bunker, P. R., and Gomez, P. C. 1988. *Chem. Phys. Lett.* **150**, 60.

Eyring, H. 1935. *J. Chem. Phys.* **3**, 107.

Eyring, H., Lin, S. H., and Lin, S. M. 1983. *Basic Chemical Kinetics.* New York: Wiley.

Ezra, G. S., Graig, C. M., and Fried, L. E. 1987, *J. Phys. Chem.* **91**, 3721.

Fajardo, M. E., Withnall, R., Feld, J., Okada, F., Lawrence, W., Wiedemann, L., and Apkarian, V. A. 1988. *Laser Chem.* **9**, 1.

Feld, J., Kunttu, H., and Apkarian, V. A. 1990, *J. Chem. Phys.* **93**, 1009.

Felker, P. M., and Zewail, A. H. 1985. *J. Chem. Phys.* **82**, 2961, 2975, 2995.

Felker, P. M., and Zewail, A. H. 1988. *Adv. Chem. Phys.* **70**, 265.

Feynman, R. P. 1972. *Statistical Mechanics.* Reading, MA: Addison-Wesley.

Feyman, R. P., and Hibbs, A. R. 1965. *Quantum Mechanics and Path Integrals.* New York: McGraw-Hill.

Feyman, R. P., and Kleinert, H. H. 1986. *Phys. Rev. A* **34**, 5080.

Feynman, R. P., and Vernon, F. I. 1963. *Ann. Phys. (NY)* **24**, 118.

Fillaux, F., and Carlile, C. J. 1989. *Chem. Phys. Lett.* **162**, 188.

Fillaux, F., and Carlile, C. J. 1990. *Phys. Rev. B* **42**, 5990.

Firth, D. W., Barbara, P. F., and Trommsdorf, H. P. 1989. *Chem. Phys.* **136**, 349.

Flory, P. J. 1969. *Statistical Mechanics of Chain Molecules.* New York: Interscience.

Flynn, C. P., and Stoneham, A. M. 1970. *Phys. Rev. B* **1**, 3967.

Fong, F. K. 1974. *J. Am. Chem. Soc.* **96**, 7638.

Ford, G. W., Lewis, J. T., and O'Connel, R. F. 1988. *Phys. Lett. A* **128**, 29; *Phys. Rev. A* **37**, 4419.

Forel, M. T., and Fokassier, M. 1967. *Spectrochim. Acta* **23A**, 1977.

Fraser, G. T. 1989. *J. Chem. Phys.* **90**, 2097.

Fraser, G. T., Nelson, D. D., Jr., Charo, A., and Klemperer, W. 1985, *J. Chem. Phys.* **82**, 2535.

Fraser, G. T., and Pine, A. S. 1989. *J. Chem. Phys.* **91**, 3319.

Fraser, G. T., and Suenram, R. D. 1992. *J. Chem. Phys.* **96**, 7287.

Frei, H., and Pimental, G. C. 1983. *J. Chem. Phys.* **78**, 3698.

Frisch, M. J., Del Bene, J. E., Binkley, J. S., and Schaefer, H. F. 1986. *J. Chem. Phys.* **84**, 2279.

Frydman, L., Olivieri, F. C., and Diaz, L. E. 1988. *J. Am. Chem. Soc.* **110**, 336.

Fukai, Y. and Sugimoto, H. 1985. *Adv. Phys.* **34**, 263.

Fuke, K., and Kaya, K. 1989. *J. Phys. Chem.* **93**, 614.

Fukui, K. 1970. *J. Phys. Chem.* **74**, 4161.

Furue, H., and Pacey, P. D. 1986. *J. Phys. Chem.* **90**, 397.

Gajdos, J., and Bleha, T. 1983. *Collection Czehoslov. Chem. Commun.* **48**, 71.

Gallo, M. M., Hamolton, T. P., and Schaefer, H. F. 1990. *J. Am. Chem. Soc.* **112**, 8714.

Garg, A., Onuchic, J. N., and Ambegaokar, V. 1985. *J. Chem. Phys.* **83**, 4491.

Garrett, B. C., and Truhlar, D. G. 1983. *J. Chem. Phys.* **79**, 4931.

Garrett, B. C., and Truhlar, D. G. 1991. *J. Phys. Chem.* **95**, 10374.

Garrett, B. C., Redman, M. J., Steckler, R., Truhlar, D. G., Baldridge, K. K., Bartol, D., Schmidt, M. W., and Gordon, M.S. 1988. *J. Phys. Chem.* **92**, 1476.

Geoffroy, M., Kispert, L. D., and Hwang, J. S. 1979. *J. Chem. Phys.* **70**, 4238.

Gerasimov, G. N., Dolotov, S. N., and Abkin, D. A. 1980. *Int. J. Radiat. Phys. Chem.* **15**, 405.

Gerber, R. B., and Alimi, R. 1990. *Chem. Phys. Lett.* **173**, 393.

Gerber, R. B., Buch, V., and Ratner, M. A. 1982. *J. Chem. Phys.* **77**, 3022.

Gilchrist, J. 1991. *J. Mol. Liquid* **49**, 67.

Gillan, M. J. 1987. *J. Phys. C* **20**, 3621.

Gillispie, G. D., Van Benthem, M. H., and Vangness, M. 1986. *J. Phys. Chem.* **90**, 2569.

Girardet, C., and Lakhlifi, A. 1985. *J. Chem. Phys.* **83**, 5506.

Girardet, C., and Lakhlifi, A. 1989. *J. Chem. Phys.* **91**, 1423.

Girardet, C., Abouaf-Marguin, L., Gautier-Roy, B., and Maillard, D. 1984. *Chem. Phys.* **83**, 415, 431.

Gisin, N. 1981. *J. Phys. A* **14**, 2259.

Gisin, N. 1982. *Physica A* 111, 364.

Glasstone, S., Leidler, K. J., and Eyring, H. 1941. *The Theory of Rate Processes*. New York: McGraw Hill.

Goldanskii, V. I. 1959. *Dokl. Akad. Nauk. USSR* **124**, 1261; **127**, 1037 (in Russian).

Goldanskii, V. I. 1976. *Annu. Rev. Phys. Chem.* **27**, 85.

Goldanskii, V. I. 1979. *Nature* **279**, 109.

Goldanskii, V. I., Trakhtenberg, L. I., and Flerov, V. N. 1989a. *Tunneling Phenomena in Chemical Physics*. New York: Gordon & Breach.

Goldanskii, V. I., Benderskii, V. A., and Trakhtenberg, L. I. 1989b. *Adv. Chem. Phys.* **75**, 349.

Goldman, M. 1970. *Spin Temperature and Nuclear Magnetic Resonance in Solids*. London: Oxford University Press.

Gomez, M., Bowen, S. P., and Krumhausl, J. A. 1967. *Phys. Rev.* **153**, 1009.

Gomez, P. C., and Bunker, P. R. 1990. *Chem. Phys. Lett.* **165**, 351.

Grabert, H. and Weiss, U. 1984. *Phys. Rev. Lett.* **53**, 1787.

Grabert, H., Weiss, U., and Hanggi, P. 1984a. *Phys. Rev. Lett.* **52**, 2193.

Grabert, H., Weiss, U., Hanggi, P., and Riseborough, P. 1984b. *Phys. Lett. A* **104**, 10.

Grabert, H., Olschowski, P., and Weiss, U. 1987. *Phys. Rev. B* **36**, 1931.

Grabert, H., Schramm, D. N., and Ingold, K. U. 1988. *Phys. Rep.* **168**, 115.

Graf, F., Meyer, R., Ha, T.-K., and Ernst, R. R. 1981. *J. Chem. Phys.* **75**, 2914.

Graser, G. T., Nelson, D. D., Charo, A., and Klemperer, W. 1985. *J. Chem. Phys.* **82**, 2535.

Grellmann, K. H., Schmitt, O., and Weller, H. 1982. *Chem. Phys. Lett.* **88**, 40.

Grellmann, K. H., Weller, H., and Tauer, E. 1983. *Chem. Phys. Lett.* **95**, 195.

Grellmann, K. H., Mordzinski, A., and Heinrich, A. 1989. *Chem. Phys.* **136**, 201.

Grondey, S., Prager, M., Press, W., and Heidemann, A. 1986. *J. Chem. Phys.* **85**, 2204.

Grote, R. F., and Hynes, J. T. 1980. *J. Chem. Phys.* **73**, 2715.

Gutzwiller, M. C. 1967. *J. Math. Phys.* **8**, 1979.

Gwo, D. H. 1990. *Mol. Phys.* **71**, 453.

Haase, G., Asscher, M., and Kosloff, R. 1989. *J. Chem. Phys.* **90**, 3346.

Hancock, G. C., and Truhlar, D. G. 1989. *J. Chem. Phys.* **90**, 3498.

Hancock, G. C., Rejto, P., Steckler, R., Brown, F. B., Schwenke, D. W., and Truhlar, D. G. 1986. *J. Chem. Phys.* **85**, 4997.

Hancock, G. C., Truhlar, D. G., and Dykstra, C. E. 1988. *J. Chem. Phys.* **88**, 1786.

Hancock, G. C., Mead, C. A., Truhlar, D. G., and Varandas, A. J. S. 1989. *J. Chem. Phys.* **91**, 3492.

Hanggi, P. 1986. *J. Stat. Phys.* **42**, 105.

Hanggi, P., and Hontscha, W. 1988. *J. Chem. Phys.* **88**, 4094.

Hanggi, P., and Hontscha, W. 1991. *Ber. Bunsenges. Phys. Chem.* **95**, 379.

Hanggi, P., Talkner, P., and Borkovec, M. 1990. *Rev. Mod. Phys.* **62**, 251.

Harris, D. O., Engelholm, G. G., Tolman, C. A., Luntz, A. C., Keller, A., Kim, H., and Gwinn, W. D. 1969. *J. Chem. Phys.* **50**, 2438.

Harris, J., and Anderson, S. 1985. *Phys. Rev. Lett.* **55**, 1583.

Hassel, O., and Romming, C. 1962. *Quart. Rev.* **14**, 1.

Haussler, W., and Huller, A. 1985. *Z. Phys. B* **59**, 177.

Havenith, M., Cohen, R. C., Busarow, K. L., Gwo, D.-H., Lee, Y. T., and Saykally, R. J. 1991. *J. Chem. Phys.* **94**, 4776.

Hayashi, S., Umemura, J., Kato, S., and Morokuma, K. 1984. *J. Phys. Chem.* **88**, 1330.

Heidemann, A., Press, W., Lushington, K. J., and Morrison, J. A. 1981. *J. Chem. Phys.* **75**, 4003.

Heidemann, A., Lushington, K. J., Morrison, J. A., Neumaer, K., and Press, W. 1984. *J. Chem. Phys.* **81**, 5799.

Heidemann, A., Magerl, A., Prager, M., Richter, D., and Springer, T. (Eds). 1987. *Springer Proceedings in Physics*, Vol. 19, *Quantum Aspects of Molecular Motions in Solids*. Berlin: Springer.

Heidemann, A., Prager, M., and Monkenbusch, M. 1989. *Z. Phys. B* **76**, 77.

Heller, E. J. 1990. *J. Chem. Phys.* **92**, 1718.

Hennig, J., and Limbach, H. H. 1979. *J. Chem. Soc. Farad. Trans. 2*, **75**, 752.

Hewson, A. C. 1984. *J. Phys. C* **15**, 3841, 3855.

Hildebrandt, R. L., and Shen, Q. 1982. *J. Mol. Spectrosc.* **86**, 587.

Hill, J. R., and Dlott, D. D. 1988. *J. Chem. Phys.* **89**, 830.

Hipes, P. G., and Kupperman, A. 1986. *J. Phys. Chem.* **90**, 3630.

Hofranek, N., Lischka, P., and Karpfen, A. 1988. *Chem. Phys.* **121**, 137.

Holloway, M. K., Reynolds, C. H., and Merz, K. M. 1989. *J. Am. Chem. Soc.* **111**, 3466.

Holmgren, S. L., Waldman, M., and Klemperer, W. 1978, *J. Chem. Phys.* **69**, 1661.

Holstein, T. 1959. *Ann. Phys.* **8**, 325.

Holstein, T. 1978. *Philos. Mag. B* **37**, 49.

Hontscha, W., Hanggi, P., and Pollak, E. 1990. *Phys. Rev. B* **41**, 2210.

Horsewill, A. J., and Aibout, A. 1989a. *J. Phys. Condens. Matter* **1**, 9609.

Horsewill, A. J., and Aibout, A. 1989b. *J. Phys. Condens. Matter* **1**, 10533.

Horsewill, A. J., Aslanoosi, A. M., and Carlie, C. J. 1987. *J. Phys. C* **20**, L869.

Horsewill, A. J., Green, R. M., and Aslanoosi, A. M. 1989. *Chem. Phys.* **138**, 179.

Hoshi, N., Yamauchi, S., and Hirota, N. 1990. *J. Phys. Chem.* **94**, 7523.

Hougen, J. T. 1985. *J. Mol. Spectrosc.* **114**, 395.

Hougen, J. T. 1987. *J. Mol. Spectrosc.* **123**, 197.

Hougen, J. T., and Ohashi, J. T. 1985. *J. Mol. Spectrosc.* **109**, 134.

Howard, B. J., and Pine, A. S. 1985. *Chem. Phys. Lett.* **146**, 582.

Huang, Z. S. and Miller, R. E. 1988. *J. Chem. Phys.* **88**, 8008.

Huller, A. 1980. *Z. Phys. B* **36**, 215.

Huller, A., and Baetz, L. 1988. *Z. Phys. B* **72**, 47.

Huller, A., and Raich, J. 1979. *J. Chem. Phys.* **71**, 3851.

Hund, F. 1927. *Z. Phys.* **43**, 805.

Hutson, J. M. 1988. *J. Chem. Phys.* **89**, 4550.

Hutson, J. M. 1990. *J. Chem. Phys.* **92**, 157.

Idziak, S., and Pislewski, N. 1987. *Chem. Phys.* **111**, 439.

Ikeda, T., and Lord, R. C. 1972. *J. Chem. Phys.* **56**, 4450.

Ikeda, T., Lord, R. C., Malloy, T. B., and Ueda, T. 1972. *J. Chem. Phys.* **56**, 1434.

Im, H., and Bernstein, E. R., 1988, *J. Chem. Phys.* **88**, 7337.

Isaakson, A. D., and Truhlar, D. G. 1982. *J. Chem. Phys.* **76**, 1380.

Ischtwan, I. and Collins, M. A. 1988. *J. Chem. Phys.* **89**, 2881.

Itzkovskii, A. S., Katunin, A. Ya., and Lukashevich, I. I. 1986. *Sov. Phys. JETP* **91**, 1832.

Ivlev, B. I., and Ovchinnikov, Yu. N. 1987. *Sov. Phys. JETP* **93**, 668.

Jaquet, R., and Miller, W. H. 1985. *J. Phys. Chem.* **489**, 2139.

James, H. M., and Keenan, T. A. 1959. *J. Chem. Phys.* **31**, 12.

Jensen, P., and Bunker, P. R. 1982. *Mol. Spectrosc.* **94**, 114.

Johnson, C. S. 1967. *Mol. Phys.* **12**, 25.

Johnston, H. S. 1960. *Adv. Chem. Phys.* **3**, 131.

Johnston, H. S., and Parr, C. 1963. *J. Am. Chem. Soc.* **85**, 2544.

Johnston, H. S., and Rapp, D. 1961. *J. Am. Chem. Soc.* **83**, 1.

Jortner, J., and Pullman, B. (Eds.). 1986. *Tunneling.* Dordrecht: Reidel.

Juanos i Timoneda, J., and Hynes, J. T. 1991. *J. Phys. Chem.* **95**, 10431.

Kagan, Yu. M., and Klinger, M. I. 1974. *J. Phys. C* **7**, 2791.

Kanamori, H., Endo, Y., and Hirota, E. 1990. *J. Chem. Phys.* **92**, 197.

Kapphan, S. 1974. *J. Phys. Chem. Solids* **35**, 621.

Kapphan, S., and Luty, F. 1972. *Phys. Rev. B* **6**, 1537.

Kappula, H., and Gläser, W. 1972. In *Inelastic Scattering of Neutrons in Solids and Liquids.* Vienna: IAEA, p. 841.

Kato, S., Kato, H., and Fukui, K. 1977. *J. Am. Chem. Soc.* **99**, 684.

Kensy, V., Gonzales, M. M., and Grellmann, K. H. 1993. *Chem. Phys.* **170**, 381.

Khatri, S. S., and Verma, A. L. 1983. *J. Phys. C* **16**, 2157.

Kilpatrick, J. E., Ptizer, K. S., and Spitzer, K. 1947. *J. Am. Chem. Soc.* **69**, 2485.

Kim, H., and Gwinn, W. D. 1969. *J. Chem. Phys.* **51**, 1815.

King, H. F., and Hornig, D. F. 1969. *J. Chem. Phys.* **44**, 4520.

Kitaigorodskii, A. I. 1961. *Organic Chemical Crystallography*. New York: Consultants Bureau.

Kitaigorodskii, A. I., Mirskaya, K., and Nauchitel, V. 1970. *Sov. Phys. Crystallogr.* **14**, 769.

Klaffer, J., and Shlesinger, M. E. 1986. *Proc. Natl. Acad. Sci. USA* **83**, 848.

Klinger, M. I. 1985. *Soviet Phys. Usp.* **28**, 391.

Kobashi, K., Etters, R. D., and Yamamoto, Y. 1984. *J. Phys. C* **17**, 13.

Kooner, Z. S., and van Hook, W. A. 1988. *J. Chem. Phys.* **92**, 6414.

Kramers, H. 1940. *Physica* **7**, 284.

Krueger, H. and Weitz, E. 1992. *J. Chem. Phys.* **96**, 2846.

Kubo, R. 1957. *J. Phys. Soc. Jpn.* **12**, 570.

Kubo, R. 1965. *Statistical Mechanics*. Amsterdam: North Holland.

Kubo, R., and Nakajima, S. 1957. *J. Phys. Soc. Jpn.* **12**, 1203.

Kubo, R., and Toyazawa, Y. 1955. *Progr. Theor. Phys.* **13**, 160.

Kubo, R., Toda, M., and Nashitsume, N. 1985. *Statistical Physics: Nonequilibrium Statistical Mechanics*. Berlin: Springer.

Kunze, K. L., and de la Vega, J. R. 1984. *J. Am. Chem. Soc.* **106**, 6528.

Laane, J. 1987. *Pure Appl. Chem.* **59**, 1307.

Laidler, K. J. 1969. *Theories of Chemical Kinetics*. New York: McGraw Hill.

Laing, J. R., Yuan, J. M., Zimmerman, H. DeVries, P. L., and George, T. F. 1977. *J. Chem. Phys.* **66**, 2801.

Landau, L. D. 1932. *Phys. Z. Sowjetunion* **2**, 46.

Landau, L. D., and Ligfshitz, E. M. 1981. *Quantum Mechanics*. Elmsford, NY: Pergamon.

Langer, G. S. 1969. *Ann. Phys.* **54**, 258.

Larkin, A. I., and Ovchinnikov, Yu. N. 1984. *Sov. Phys. JETP* **59**, 420.

Lascola, R., and Nesbitt, D. J. 1991. *J. Chem. Phys.* **95**, 7917.

Lauderdale, J. G., and Truhlar, D. G. 1985. *Surf. Sci.* **164**, 558.

Lautie, A., and Novak, A. 1980. *Chem. Phys. Lett.* **71**, 290.

Lavrushko, A. G., and Benderskii, V. A. 1978. Unpublished.

Lee, K.-P., Miyazaki, T., Fueki, K., and Gotoh, K. 1987. *J. Phys. Chem.* **91**, 180.

Lee, T. J., and Schaefer, H. F. 1986. *J. Chem. Phys.* **85**, 3437.

Legen, A. C. 1980. *Chem. Rev.* **80**, 231.

Leggett, A. J., Chakravarty, S., Dorsey, A. T., Fisher, M. P. A., Garg, A., and Zwerger, M.1987. *Rev. Mod. Phys.* **59**, 1.

Leopold, K. R., Fraser, G. T., Lin, F. J., Nelson, D. D., and Klemperer, W. 1984. *J. Chem. Phys.* **81**, 4922.

Le Roy, R. J., Sprague, E. D., and Williams, F. 1972. *J. Phys. Chem.* **76**, 545.

Le Roy, R. J., Murai, H., and Williams, F. 1980. *J. Am. Chem. Soc.* **102**, 2325.

Levine, A. M., Hontscha, W., and Pollak, E. 1989. *Phys. Rev. B* **40**, 2138.

Levit, S., Negele, J. W., and Paltiel, Z. 1980a. *Phys. Rev. C* **21** 1603.

Levit, S., Negele, J. W., and Paltiel, Z. 1980b. *Phys. Rev. C* **22,** 1979.

Levy, D. H. 1980. *Annu. Rev. Phys. Chem.* **31**, 197.

Li, L., Lipert, R. J., Lobuc, J., Chupka, W. A., and Colson, S. D. 1988. *Chem. Phys. Lett.* **151**, 335.

Lide, D. R. 1957. *J. Chem. Phys.* **27**, 343.

Limbach, H-H., Hennig, J., Gerritzer, G., and Rumpel, H. 1982. *Faraday Discuss. Chem. Soc.* **74**, 229.

Lindh, R., Roos, B. O., and Kraemer, W. P. 1987. *Chem. Phys. Lett.* **139**, 407.

Lippincott, E. R., and Schroeder, R. 1955. *J. Chem. Phys.* **23**, 1097.

Lippincott, E. R., and Schroeder, R. 1957. *J. Phys. Chem.* **61**, 921.

Longuet-Higgins, H. C. 1963. *Mol. Phys.* **6**, 445.

Look, D. C., and Lowe, L. J. 1966. *J. Chem. Phys.* **44**, 3437.

Loth, K., Graf, F., and Gunthard, H. 1976. *Chem. Phys.* **13**, 95.

Lovejoy, C. M., and Nesbitt, D. J. 1989. *J. Chem. Phys.* **91**, 2790.

Loyd, R., Mathur, S. N., and Harmony, M. D. *J. Mol. Spectrosc.* **72**, 359.

Lucas, D., and Pimentel, G. C. 1979. *J. Phys. Chem.* **83**, 2311.

Lynden-Bell, R. N. 1964. *Mol. Phys.* **8**, 71.

MacPhail, R. A. and Variyar, J. E. 1989, *Chem. Phys. Lett.* **161**, 239.

Makarov, D. E., and Makri, N. 1993. *Phys. Rev. A*, in press.

Makri, N. 1989. *Chem. Phys. Lett.* **159**, 409.

Makri, N. 1991a. *J. Chem. Phys.* **94**, 4949.

Makri, N. 1991b. *Comput. Phys. Commun.* **63**, 331.

Makri, N. 1992. *Chem. Phys. Lett.* **193**, 389.

Makri, N. 1993. *J. Phys. Chem.* **97**, 2417.

Makri, N., and Miller, W. H. 1987a. *J. Chem. Phys.* **87**, 5781.

Makri, N., and Miller, W. H. 1987b. *Chem. Phys. Lett.* **139**, 10.

Makri, N., and Miller, W. H. 1989a. *J. Chem. Phys.* **90**, 904.

Makri, N., and Miller, W. H. 1989b. *J. Chem. Phys.* **91**, 7026.

Mansueto, E. S., and Wight, C. A. 1989. *J. Am. Chem. Soc.* **111**, 1900.

Mansueto, E. S. and Wight, C. A. 1991. *J. Photochem. Photobiol. A: Chem.* **60**, 251.

Mansueto, E. S., and Wight, C. A. 1992. *J. Phys. Chem.* **96**, 1502.

Marcus, R. A. 1964. *Annu. Rev. Phys. Chem.* **15**, 155.

Marcus, R. A. 1985. *NATO ASI Ser.*, *Ser. B* **120**, 293.

Marshall, M. D., Charo, A., Leung, H. O., and Klemperer, W. 1985. *J. Chem. Phys.* **83**, 4924.

Mathiensen, H., Norman, N., and Pedersen, B. F. 1967. *Acta Chem. Scand.* **21**, 127.

McDonald, P. J., Horsewill, A. J., Dunstan, D. J., and Hall, N. 1989. *J. Phys. Condens. Matter.* **1**, 2441.

McGreery, J. H., and Wolken, G. 1975. *J. Chem. Phys.* **63**, 2340.

McLaughlin, D. W. 1972. *J. Math. Phys.* **13**, 1091.

Meier, B. H., Graf, F., and Ernst, R. R. 1982. *J. Chem. Phys.* **76**, 767.

Melnikov, V. I., and Meshkov, S. V. 1983. *JETP Lett.* **38**, 130.

Menand, A., and Kingham, D. R. 1984. *J. Phys. D* **17**, 203.

Menand, A., and Kingham, D. R. 1985. *J. Phys. C* **18**, 4539.

Mermin, N. D. 1991. *Physica A* **177**, 561.

Merz, K. M., and Reynolds, C. H. 1988. *J. Chem. Soc. Chem. Commun.* **90**.

Meschede, L., and Limbach, H-H. 1991. *J. Phys. Chem.* **95**, 10267.

Meschede, L., Gerritzen, D., and Limbach, H-H. 1988. *Ber. Bunsen Ges. Phys. Chem.* **92**, 469.

Metropolis, N., Rosenbluth, A. W., Rosenbluth, M. N., Teller, H., and Teller, E. 1953. *J. Chem. Phys.* **21**, 1087.

Meyer, R., and Bauder, A. 1982. *J. Mol. Spectrosc.* **94**, 136.

Meyer, R., and Ernst, R. R., 1987. *J. Chem. Phys.* **86**, 784.

Miller, W. H. 1974. *J. Chem. Phys.* **61**, 1823.

Miller, W. H. 1975a. *J. Chem. Phys.* **62**, 1899.

Miller, W. H. 1975b. *J. Chem. Phys.* **63**, 996.

Miller, W. H. 1976. *Acc. Chem. Res.* **9**, 36.

Miller, W. H. 1979. *J. Phys. Chem.* **83**, 960.

Miller, W. H. 1983. *J. Phys. Chem.* **87**, 3811.

Miller, W. H., and George, T. F. 1972. *J. Chem. Phys.* **56**, 5668.

Miller, W. H., Schwartz, S. D., and Tromp, J. W. 1983. *J. Chem. Phys.* **79**, 4889.

Mills, I. M. 1984. *J. Phys. Chem.* **88**, 532.

Misochko, E. Y., Filippov, P. G., Benderskii, V. A., Ovchinnikov, A. A., Barkalov, I. M., and Kirukhin, D. P., 1980, *Dokl. Akad. Nuak USSR* **253**, 163 (in Russian).

Misochko, E. Y., Benderskii, V. A., Goldanskii, V. I., and Kononikhina, V. V. 1991. *Dokl.Akad. Nauk USSR* **316**, 403 (in Russian).

Miyazaki, T. 1991. *Chem. Phys. Lett.* **176**, 99.

Miyazaki, T., and Lee, K.-P. 1986. *J. Phys. Chem.* **90**, 400.

Miyazaki, T., Lee, K.-P., Fueki, K., and Takeuchi, A. 1984. *J. Phys. Chem.* **88**, 4959.

Miyazaki, T., Iwata, N., Lee, K.-P., and Fueki, K. 1989. *J. Phys. Chem.* **93**, 3352.

Miyazaki, T., Morikita, H., Fueki, K., and Hiraku, T. 1991a. *Chem. Phys. Lett.* **182**, 35.

Miyazaki, T., Hiraku, T., Fueki, K., and Tsuchihashi, Y. 1991b. *J. Phys. Chem.* **95**, 26.

Mnyukh, Yu. V. 1963. *J. Phys. Chem. Solids* **24**, 631.

Mordzinski, A., and Grellmann, K. H. 1986. *J. Phys. Chem.* **90**, 5503.

Mordzinski, A., and Kühnle, W. 1986. *J. Phys. Chem.* **90**, 1455.

Morillo, M., and Cukier, R. I. 1990. *J. Chem. Phys.* **92**, 4833.

Morillo, M., Yang, D. Y., and Cukier, R. I. 1989. *J. Chem. Phys.* **90**, 5711.

Muttalib, K. A., and Sethna, J. 1985. *Phys. Rev. B* **32**, 3462.

Myhre, P. C., McLaren, K. L., and Yannoni, C. S. 1985. *J. Am. Chem. Soc.* **107**, 59.

Nagaoka, S., Terao, T., Imashior, F., Saika, A., Hirota, N., and Hayashi, S. 1983. *J. Chem. Phys.* **79**, 4694.

Nakamura, H. 1991. *Int. Rev. Phys. Chem.* **10**, 123.

Narayanamurti, V., and Pohl, R. O. 1970. *Rev. Mod. Phys.* **42**, 201.

Narayanamurti, V., Seward, W. D., and Pohl, R. O. 1966. *Phys. Rev.* **148**, 481.

Nayashi, S. 1983. *J. Chem. Phys.* **79**, 4694.

Nelson, D. D. and Klemperer, W. 1987. *J. Chem. Phys.* **87**, 139.

Nelson, D. D., Fraser, G. T., and Klemperer, W. 1985a. *J. Chem. Phys.* **83**, 945.

Nelson, D. D., Fraser, G. T., and Klemperer, W. 1985b. *J. Chem. Phys.* **83**, 6201.

Nelson, D. D., Klemperer, W., Fraser, G. T., Lovas, F. J., and Suenram, R. D. 1987a. *J. Chem. Phys.* **87**, 6364.

Nelson, D. D., Fraser, G. T., and Klemperer, W. 1987b. *Science* **238**, 1670.

Nelson, R., and Pierce, L. 1965. *J. Mol. Spectrosc.* **18**, 344.

Nesbitt, D. J., and Naaman, R. 1989. *J. Chem. Phys.* **91**, 3801.

Nikitin, E. E., and Korst, N. N. 1965. *Theor. Exp. Chem.* **1**, 5 (in Russian).

Noid, D. W., Koszykowsky, M. L., and Marcus, R. A. 1981. *Ann. Rev. Phys. Chem.* **32**, 267.

Novak, A. 1974. *Structur. Bonding* **18**, 177.

Novick, S. E., Davies, P., Harris, S. J., and Klemperer, W. 1973. *J. Chem. Phys.* **59**, 2273.

Odutola, J. A., Hu, T. A., Prinslow, D., O'dell, S. E., and Dyke, T. R. 1988. *J. Chem. Phys.* **88**, 5352.

Ohshima, Y., Matsumoto, Y., Takami, M., and Kuchitsu, M. 1988. *Chem. Phys. Lett.* **147**, 1.

Oppenlander, A., Ramband, C., Trommsdorf, H. P., and Vial, J. C. 1989. *Phys. Rev. Lett.* **63**, 1432.

Ovchinnikov, A. A., and Ovchinnikova, M. Ya. 1982. *Adv. Quant. Chem.* **16**, 161.

Ovchinnikova, M. Ya. 1965. *Dokl. Phys. Chem.* **161**, 259.

Ovchinnikova, M. Ya. 1979. *Chem. Phys.* **36**, 15.

Pacey, P. D. 1979. *J. Chem. Phys.* **71**, 2966.

Paddon-Row, M. N., and Pople, J. A. 1985. *J. Phys. Chem.* **89**, 2768.

Parris, P. E., and Silbey, R. 1985. *J. Chem. Phys.* **83**, 5619.

Pekar, S. I. 1954. *Untersuchung uber die Electronen theorie der Kristalle.* Berlin: Academ.-Verlag.

Pertsin, A. J., and Kitaigorodskii, A. I. 1987. *The Atom–Atom Potential Method: Application to Organic Molecular Solids.* Berlin: Springer.

Perutz, R. N. 1985. *Chem. Rev.* **85**, 77, 97.

Peternelj, J., and Jencic, I. 1989. *J. Phys. A* **22**, 1941.

Peternelj, J., Jencic, I., Cvikl, B., and Pinter, M. M. 1987. *Phys. Rev. B* **36**, 25.

Pfafferott. G., Oberhammer, H., Boggs, J. E., and Caminati, W. 1985. *J. Am. Chem. Soc.* **107**, 2305.

Phillips, W. A. 1972. *J. Low-Temp. Phys.* **7**, 381.

Phillips, W. A. (Ed.). 1981. *Amorphous Solids.* Berlin: Springer.

Pine, A. S., and Howard, B. J. 1986. *J. Chem. Phys.* **84**, 590.

Pine, A. S., and Lafferty, W. J. 1983. *J. Chem. Phys.* **78**, 2154.

Pine, A. S., Lafferty, W. J., and Howard, B. J. 1984. *J. Chem. Phys.* **80**, 2939.

Platz, M. S., Senthilnathan, V. P., Wright, B. B., and McCurdy, C. W. 1982. *J. Am. Chem. Soc.* **104**, 6494.

Pollak, E. 1986a. *Phys. Rev. A* **33**, 4244.

Pollak, E. 1986b. *J. Chem. Phys.* **85**, 865.

Polyakov, A. M. 1977. *Nucl. Phys. B* **121**, 429.

Pople, J. A. 1987. *Chem. Phys. Lett.* **137**, 10.

Poupko, R., Luz, Z., and Zimmermann, H. 1982. *J. Am. Chem. Soc.* **104**, 5307.

Prager, M., and Heidemann, A. 1987. *Neutron Tunneling Spectroscopy: A Compilation of Available Data*. ILL, International Report, 87 PR 15T.

Prager, M., and Langel, W. 1986. *J. Chem. Phys.* **85**, 5279.

Prager, M., and Langel, N. 1989. *J. Chem. Phys.* **90**, 5889.

Prager, M., Heidemann, A., and Hausler, W. 1986. *Z. Phys. B* **64**, 447.

Prager, M., Stanislawski, J., and Hausler, W. 1987. *J. Chem. Phys.* **86**, 2563.

Prager, M., Hempelmann, R., Langen, H., and Muller-Warmuth, W. 1990. *J. Phys. Condens. Matter* **2**, 8625.

Prass, B., Colpa, J. P., and Stehlik, D. 1988. *J. Chem. Phys.* **88**, 191.

Prass, B., Colpa, J. P., and Stehlik, D. 1989. *Chem. Phys.* **136**, 187.

Press, W. 1972. *J. Chem. Phys.* **56**, 2597.

Press, W. 1981. *Single-Particle Rotations in Molecular Crystals. Springer Trends in Modern Physics*, Vol. 92, Berlin: Springer.

Press, W., and Kollmar, A. 1975. *Solid State Commun.* **17**, 405.

Punkkinen, M. 1980. *Phys. Rev. B* **21**, 54.

Puska, M. J., and Nielmien, R. M. 1984. *Phys. Rev.* **B 29**, 5382.

Quack, M., and Suhm, M. A. 1990. *Chem. Phys. Lett.* **171**, 517.

Quack, M., and Suhm, M. A. 1991. *J. Chem. Phys.* **95**, 28.

Rajaraman,R. 1975. *Phys. Rep.* **21**, 227.

Rajaraman, R. 1982. *Solitons and Instantons*. Amsterdam: North Holland.

Rambaud, C., Oppenlander, A., Pierre, M., Trommsdorf, H. P., and Vial, J. C. 1989. *Chem. Phys.* **136**, 335.

Rambaud, C., Oppenlander, A., Trommsdorf, H. P., and Vial, J. C. 1990. *J. Luminesc.* **45**, 310.

Ranfagni, A., Magnai, D., and Englman, R. 1984. *Phys. Rep.* **108**, 165.

Ray, D., Robinson, R. L., Gwo, D.-H., and Saykally, R. J. 1986. *J. Chem. Phys.* **84**, 1171.

Razavy, M. 1978. *Can. J. Phys.* **56**, 311, 1372.

Razavy, M., and Pimpale, A. 1988. *Phys. Rep.* **168**, 305.

Redington, R. L. 1990. *J. Chem. Phys.* **92**, 6447.

Redington, R. I., and Bock, C. W. 1991. *J. Phys. Chem.* **95**, 10284.

Redington, R. I., and Redington, T. E. 1979. *J. Mol. Spectrosc.* **78**, 229.

Redington, R. L., Chen, Y., Scherer, G. I., and Field, R. W. 1988. *J. Chem. Phys.* **88**, 627.

Rettner, C. T., and Stein, H. 1987. *J. Chem. Phys.* **87**, 770.

Richter, D., 1986. *Hyperfine Inter.* **31**, 168.

Rijks, W., and Wormer, P. E. S. 1989. *J. Chem. Phys.* **90**, 6507; 1990. *J. Chem. Phys.* **92**, 5774.

Rios, M. A., and Rodriguez, J. 1990. *Theochemistry* **204**, 137.

Robinson, G. W., and Frosh, R. P. 1963. *J. Chem. Phys.* **37**, 1962; **38**, 1187.

Robinson, R. L., Gwo, D-H., and Saykally, R. J. 1987. *J. Chem. Phys.* **87**, 5156.

Roginsky, S. Z., and Rozenkevitsch, L. V. 1930. *Z. Phys. Chem. B* **10**, 47.

Rom, N., Moiseyev, N., and Lefebvre, R. 1991. *J. Chem. Phys.* **95**, 3562.

Rossetti, R., and Brus, L. E. 1980. *J. Chem. Phys.* **73**, 1546.

Rossetti, R., Haddon, R. C., and Brus, L. E. 1980. *J. Am. Chem. Soc.* **102**, 6913.

Rossetti, R., Rayford, R. Haddon, R. C., and Brus, L. E. 1981. *J. Am. Chem. Soc.* **107**, 4303.

Rowe, J. M., Rush, J. J., Shapiro, S. M., Hinks, D. G., and Susman, S. 1980. *Phys. Rev. B* **21**, 4863.

Rumpel, H., Limbach, H-H., and Zachmann, G. 1989. *J. Phys. Chem.* **93**, 1812.

Saitoh, T., Mori, K., and Itoh, R. 1981. *Chem. Phys.* **60**, 161.

Salikhov, K. M., Molin, Yu. N., Sagdeev, R. Z., and Buchachenko, A. L. 1984. *Spin Polarization and Magnetic Effects in Radical Reactions.* Amsterdam: Elsevier.

Sana, M., Larry, G., and Villaveces, J. L. 1984. *Theor. Chim. Acta* **65**, 109.

Sarai, A. 1982. *J. Chem. Phys.* **76**, 5554.

Sasetti, M., and Weiss, U. 1990. *Phys. Rev. A* **41**, 5383.

Sato, N., and Iwata, S. 1988. *J. Chem. Phys.* **89**, 2832.

Savel'ev, V. A., and Sokolov, N. D. 1975. *Chem. Phys. Lett.* **34**, 281.

Schlabach, M., Wehrle, B., and Limbach, H. H. 1986. *J. Am. Chem. Soc.* **108**, 3856.

Schmid, A. 1983. *J. Low. Temp. Phys.* **49**, 609.

Schmid, A. 1986. *Ann. Phys.* **170**, 333.

Schmidt, M. W., Gordon, M. S., and Dupuis, M. 1985. *J. Am. Chem. Soc.* **107**, 2585.

Schmuttenmaer, C. A., Cohen, R. C., Loeser, J. G., and Saykally, R. I. 1991. *J. Chem. Phys.* **95**, 9.

Schroeder, R., and Lippincott, E. R. 1957. *J. Phys. Chem.* **61**, 021.

Schwentner, N., and Apkarian, V. A. 1989. *Chem. Phys. Lett.* **154**, 413.

Scold, K. 1968. *J. Chem. Phys.* **49**, 2443.

Sekiya, H., Takesue, H., Nishimura, Y., Li, Z-H. Mori, A., and Takeshita, H. 1990a. *J. Chem. Phys.* **92**, 2790.

Sekiya, H., Nagashima, Y., and Nishima, Y. 1990b. *J. Chem. Phys.* **92**, 5761.

Sekiya, H., Saskai, K., Nishimura, Y., Mori, A., and Takeshita, H. 1990c. *Chem. Phys. Lett.* **174**, 133.

Sekiya, H., Nagashima, Y., Tsuji, T., Nishimura, Y., Mori, A., and Takeshita, H. 1991. *J. Phys. Chem.* **95**, 10311.

Semenov, N. N. 1960. *Khim. Tekhnol. Polimerov* **7–8**, 196 (in Russian).

Senthilnathan, V. P., and Platz, M. S. 1982. *J. Am. Chem. Soc.* **102**, 7637.

Sethna, J. 1981. *Phys. Rev. B* **24**, 692.

Shian, W-I., Duestler, E. N., Paul, I. C., and Curtin, D. Y. 1980. *J. Am. Chem. Soc.* **102**, 4546.

Shida, N., Barbara, P. F., and Almlof, J. E. 1989. *J. Chem. Phys.* **91**, 4061.

Shida, N., Barbara, P. F., and Almlof, J. 1991a. *J. Chem. Phys.* **94**, 3633.

Shida, N., Almlof, J., and Barbara, P. F. 1991b. *J. Phys. Chem.* **95**, 10457.

Shimanouchi, H., and Sasada, Y. 1973. *Acta Crystallogr. B* **29**, 81.

Shimoda, K., Nishikawa, T., and Itoh, T. 1954. *J. Phys. Soc. Jpn.* **9**, 974.

Shimshoni, E., and Gefen, Y. 1991. *Ann. Phys. (N.Y.)* **210**, 16.

Shirley, J. H. 1965. *Phys. Rev. B* **138**, 979.

Shore, L. M., and Sander, H. B. 1972. *Phys. Rev. B* **6**, 1551.

Siebrand, W. 1967. *J. Chem. Phys.* **47**, 2411.

Siebrand, W., Wildman, T. A., and Zgierski, M. Z. 1984. *J. Am. Chem. Soc.* **106**, 4083, 4089.

Silbey, R., and Harris, R. H. 1983. *J. Chem. Phys.* **78**, 7330.

Silbey, R., and Harris, R. 1984. *J. Chem. Phys.* **80**, 2615.

Silbey, R., and Harris, R. H. 1989. *J. Phys. Chem.* **93**, 7062.

Silbey, R., and Trommsdorf, H. P. 1990. *Chem. Phys. Lett.* **165**, 540.

Silvera, I. F. 1980. *Rev. Mod. Phys.* **52**, 393.

Simonius, M. 1978. *Phys. Rev. Lett.* **26**, 980.

Singer, S. J., Freed, K. F., and Beind, Y. B. 1985. *Adv. Chem. Phys.* **61**, 1.

Skinner, J. L., and Trommsdorf, H. P. 1988. *J. Chem. Phys.* **89**, 897.

Skodje, R. T., Truhlar, D. G., and Garrett, B. C. 1981a. *J. Chem. Phys.* **77**, 5955.

Skodje, R. T., Truhlar, D. G., and Garrett, B. C. 1981b. *J. Phys. Chem.* **85**, 3019.

Smedarchina, Z., Siebrand, W., and Zebretto, F. 1989. *Chem. Phys.* **136**, 289.

Smith, B. J., Swanton, D. J., Pople, J. A., Schaefer, H. F., and Radom, L. 1990. *J. Chem. Phys.* **92**, 1240.

Smith, D. 1973. *J. Chem. Phys.* **58**, 3833.

Smith, D. 1975. *J. Chem. Phys.* **63**, 5003.

Smith, D. 1985. *J. Chem. Phys.* **82**, 5131.

Smith, D. 1990. *J. Chem. Phys.* **93**, 10.

Smith, G. R., and Guillory, W. A. 1977. *J. Mol. Spectrosc.* **68**, 223.

Smith, G. R., and Guillory, W. A. 1977. *Intern. J. Chem. Kin.* **9**, 953.

Sokolov, N. D., and Savel'ev, V. A. 1977. *Chem. Phys.* **22**, 383.

Someda, K., and Nakamura, H. 1991. *J. Chem. Phys.* **94**, 4258.

Somorjai, R. L., and Hornig, D. F. 1962. *J. Chem. Phys.* **36**, 1980.

Spirko, V., Stone, J. M. R., and Papousek, D. 1976. *J. Mol. Spectrosc.* **60**, 159.

Sprague, E. D., and Williams, F. 1971. *J. Am. Chem. Soc.* **93**, 787.

Sridharan, K. R., Sobol, W. T., and Pinter, M. M. 1985. *J. Chem. Phys.* **82**, 4886.

Steckler, R. and Truhlar, D. G. 1990. *J. Chem. Phys.* **93**, 6570.

Steidl, P., Von Borczyskowski, C., Fujara, F., Prass, B., and Stehlik, D. 1988. *J. Chem. Phys.* **88**, 792.

Strauss, H. L. 1983. *Annu. Rev. Phys. Chem.* **34**, 301.

Stuchebrukhov, A. A. 1991. *J. Chem. Phys.* **95**, 4258.

Stueckelberg, E. C. G. 1932. *Helv. Phys. Acta* **5**, 369.

Stunzas, P. A., and Benderskii, V. A. 1971. *Sov. Opt. Spectrosc.* **30**, 1041.

Suarez, A., and Silbey, R. 1991a. *J. Chem. Phys.* **94**, 4809.

Suarez, A., and Silbey, R. 1991b. *J. Chem. Phys.* **95**, 9115.

Tachibana, A., and Fukui, K. 1979. *Theor. Chim. Acta* **51**, 189.

Taddej, G., Bonadeo, H., Marzocchi, M. P., and Califano, S. 1973. *J. Chem. Phys.* **58**, 966.

Tague, Jr., T. J., Kligmann, P. M., Collier, C. P., Ovchinnikov, M. A., and Wight, C. A. 1992, *J. Phys. Chem.* **96**, 1288.

Tague, Jr., T. J. and Wight, C. A. 1991. *Chem. Phys.* **156**, 141.

Tague, Jr., T. J. and Wight, C. A. 1992. *J. Photochem. Photobiol. A: Chem.* **66**, 193.

Tague, T. J., and Wight, C. A. 1992. *J. Am. Chem. Soc.* **114**, 9235.

Takayanagi, T., Masaki, N., Nakamura, K., Okamoto, M., Sato, S. and Schatz, G. C. 1987. *J. Chem. Phys.* **86**, 6133.

Takeda, S., and Chihara, H. 1983. *J. Magn. Reson.* **54**, 285.

Takeda, S., Soda, G., and Chihara, H. 1980. *Solid State Commun.* **36**, 445.

Takeuchi, H., Allen, G., Suzuki, S., and Dianoux, A. J. 1980. *Chem. Phys.* **51**, 197.

Teichler, H., and Klamt, A. 1985. *Phys. Lett.* **108A**, 281.

Tokumura, K., Watanabe, Y., and Ito, M. 1986. *J. Phys. Chem.* **90**, 2362.

Topper, R. Q., and Truhlar, D. G. 1992. *J. Chem. Phys.* **97**, 3647.

Topper, R. Q., Tawa, G. J., and Truhlar, D. G. 1992. *J. Chem. Phys.* **97**, 3668.

Toriyama, K., and Iwasaki, M. 1979. *J. Am. Chem. Soc.* **101**, 2516.

Toriyama, K., Nunome, K., and Iwasaki, M. 1977. *J. Am. Chem. Soc.* **99**, 5823.

Townes, C. H. and Shawlow, A. L. 1955. *Microwave Spectroscopy*. New York: McGraw Hill.

Trakhtenberg, L. I., Klochikhin, V. L., and Pshezhetskii, S. Ya. 1982. *Chem. Phys.* **69**, 121.

Trevino, S. F., and Rymes, W. H. 1980. *J. Chem. Phys.* **73**, 3001.

Trevino, S. F., Prince, E., and Hubbard, C. R. 1980. *J. Chem. Phys.* **73**, 2996.

Tringides, M., and Gomer, R. 1986. *Surf. Sci.* **166**, 440.

Tromp, J. W., and Miller, W. H. 1986. *J. Phys. Chem.* **90**, 3482.

Truhlar, D. G. 1990. In *Proceed NATO Workshop on the Dynamics of Polyatomic van der Waals Complexes*. NATO Ser. B227, 159.

Truhlar, D. G. and Garrett, B. C. 1984. *Annu. Rev. Phys. Chem.* **27**, 1; 1987. *J. Chem. Phys.* **84**, 365.

Truhlar, D. G., Isaakson, A. D., Scodje, R. T., and Garrett, B. C. 1982. *J. Phys. Chem.* **86**, 2252.

Truhlar, D. G. and Kuppermann, A. 1971. *J. Am. Chem. Soc.* **93**, 1.

Truong, T., and Truhlar, D. G. 1987. *J. Phys. Chem.* **91**, 6229.

Tsuboi, M., Hirakawa, A. Y., Ino, T., Sasaki, T., and Tamagane, K. 1964. *J. Chem. Phys.* **41**, 2721.

Tsuji, T., Sekiya, H., Nishimura, Y., Mori, A., and Takeshita, H. 1991. *J. Chem. Phys.* **95**, 4802.

Tully, J. C. 1981, in Potential Energy Surf. Dyn. Calc. Chem. React. Mol. Energy Transfer, Truhlar, D. G., ed., New York: Plenum, pp. 805–816.

Tully, J. C., and Preston, R. K. 1971. *J. Chem. Phys.* **55**, 562.

Turner, P., Baughcum, S. L., Coy, S. L., and Smith, Z. 1984. *J. Am. Chem. Soc.* **106**, 2265.

Ulstrup, L. 1979. *Charge Transfer Process in Condensed Media*. Berlin: Springer.

Vacatello, M., Aritabile, G., Corrodini, P., and Tuzi, A. 1980. *J. Chem. Phys.* **73**, 548.

Vainshtein, A. I., Zakharov, V. I., Novikov, V. A., and Shifman, M. A. 1982. *Sov. Phys. Usp.* **25**, 195.

Viswanathan, R., and Raff, L. M. 1983. *J. Phys. Chem.* **87**, 3251.

Voth, G. A., Chandler, D., and Miller, W. H. 1989a. *J. Phys. Chem.* **93**, 7009.

Voth, G. A., Chandler, D., and Miller, W. H. 1989b. *J. Chem. Phys.* **91**, 7749.

Wang, S. C., and Gomer, R. 1985. *J. Chem. Phys.* **83**, 4193.

Wardlaw, D. M., and Marcus, R. A. 1988. *Adv. Chem. Phys.* **70**, 231.

Warshel, A., and Lifson, S. 1970. *J. Chem. Phys.* **53**, 582.

Wartak, M. S., and Krzeminski, S. 1989. *J. Phys. A* **22** L1005.

Waxman, D., and Leggett, A. 1985. *Phys. Rev. B* **32**, 4450.

Waxman, D., 1985, *J. Phys. C* **18**, L421.

Weinhaus, F., and Meyer, H. 1973. *Phys. Rev. B* **7**, 2974.

Weitekamp, D. P., Bielecki, A., Zax, D., Zilm, K., and Pines, A. 1983. *Phys. Rev. Lett.* **50**, 1807.

Whitman, D. W., and Carpenter, B. K. 1982. *J. Am. Chem. Soc.* **104**, 6473.

Whittal, M. W. G., and Gehring, G. A. 1987. *J. Phys. C* **20**, 1619.

Wight, C. A., Kligmann, P. M., Botcher, T. R., and Sedlacek, A. J. 1990, *J. Phys. Chem.* **94**, 2487.

Wight, C. A., Misochko, E. Y., Vetoshkin, E. V., and Goldanskii, V. I. 1993. *Chem. Phys.* **170**, 393.

Wigner, E. 1932. *Z. Phys. Chem. B* **19**, 1903.

Wigner, E., 1938. *Trans. Farad. Soc.* **34**, 29.

Wigner, E. P., 1959. *Group Theory and Its Application to the Quantum Mechanics of Atomic Spectra.* New York: Academic.

Wolynes, P. G. 1981. *Phys. Rev. Lett.* **47**, 968.

Wolynes, P. G. 1987. *J. Chem. Phys.* **86**, 1957.

Wurger, A. 1989. *Z. Phys. B* **76**, 65.

Wurger, A. 1990. *J. Phys. Condens. Matter* **2**, 2411.

Wurger, A., and Heidemann, A. 1990. *Z. Phys. B* **80**, 113.

Yakimchenko, O. E., and Lebedev, Ya. S. 1971. *Int. J. Radiat. Phys. Chem.* **3**, 17.

Yamamoto, T. 1960. *J. Chem. Phys.* **33**, 281.

Yamamoto, T., Kataoka, Y., and Okada, K. 1977. *J. Chem. Phys.* **66**, 2701.

Yannoni, C. S., Macho, V., and Myhre, P. C. 1982. *J. Am. Chem. Soc.* **104**, 7380.

Zaskulnikov, V. M., Vyazovkin, V. L., Bolshakov, B. V., and Tolkachev, V. A. 1981. *Int. J. Chem. Kinet.* **13**, 707.

Zerbetto, F., and Zgierski, M. Z. 1989. *Chem. Phys.* **130**, 45.

Zerbetto, F., Zgierski, M., and Siebrand, W. 1989. *J. Am. Chem. Soc.* **111**, 2799.

Zener, C. 1932. *Proc. R. Soc. London Ser. A* **137**, 696.

Zhang, D. A., Prager, M., and Weiss, A. 1991. *J. Chem. Phys.* **94**, 1765.

Zhu, C., and Nakamura, H. 1992. *J. Chem. Phys.* **97**, 8497.

Zwanzig, R. 1964. *Physica* **30**, 1109.

Zwart, E., Ter Muelen, J. J./, and Meerts, W. L. 1990. *Chem. Phys. Lett.* **166**, 500.

Zweers, A. E., and Brom, H. B., 1977. *Physica B* **85**, 223.

AUTHOR INDEX

Numbers in *italics* show the pages on which the complete references are listed.

SUBJECT INDEX